Game Choices

THE INTERSECTION OF REAL OPTIONS AND GAME THEORY

Game Choices

THE INTERSECTION OF REAL OPTIONS AND GAME THEORY

Edited by Steven Grenadier

RISK BOOKS

Published by Risk Books, a division of Risk Publications.

Haymarket House
28–29 Haymarket
London SW1Y 4RX
Tel: +44 (0)20 7484 9700
Fax: +44 (0)20 7484 9758
E-mail: books@risk.co.uk
Home Page: http://www.riskpublications.com

ISBN 1 899 332 94 4

British Library Cataloguing in Publication Data
A catalogue record for this book is available from the British Library

Risk Books Commissioning Editor: Bridie Selley
Desk Editor: Martin Llewellyn

Typeset by Marie Doherty

Printed and bound in Great Britain by Bookcraft (Bath) Ltd, Somerset.

To Rhonda and Brian

Contents

PART IV: CORPORATE SECURITIES AS STRATEGIC OPTIONS

PART V: FOUNDATIONS

Authors

Ronald Anderson is professor of financial economics at Université Catholique de Louvain where he is also director of the Institute for Economic and Social Research (IRES). He is also a fellow of the Centre for Economic Policy Research in the CEPR's Financial Markets Programme. His other publications include: "Numerical Analysis of Strategic Contingent Claims Models", in the *Journal of Computational Economics*, with Cheng Tu (1998); "A Comparative Study of Structural Models of Corporate Bond Yields", in the *Journal of Banking and Finance*, with Suresh Sundaresan (2000) and *Transition Banking: The Financial Development of Central and Eastern Europe* with Chantal Kegels, (Oxford University Press, 1998).

Professor Ankum was professor in Finance at the Faculty of Economics of the University of Amsterdam from 1970 to 1994. During some of that time he was also Chairman of the Chamber of Commerce in the Zaan region. He is still Chairman of the board of the Bureau of (Consumer) Credit Registration and Chairman of the Board of the Dutch Property Market Index. His publications include *Inflation, Cost Accounting and Income Determination: An Analysis of their Mutual Interdependence* (1969) and with A.G.Z. Kemna, *Introduction to Option Theory* (1990).

Andrew Caplin is professor of economics at New York University. He has held positions previously at Harvard, Princeton and Columbia. Much of his research has investigated the aggregate impact of microeconomic indivisibilities. Other topics of long-standing interest include the economics of residential real estate markets, and economics and psychology. He enjoys straddling the boundaries that traditionally separate branches of social science. He gained his BA at Cambridge University and his PhD from Yale University.

Prajit K. Dutta is a professor of economics at Columbia University where he has taught since 1987. His research interests are in game theory and its applications and is currently researching the economics of global warming and the economics of the Internet. He is the author of *Strategies and Games: Theory and Practice* (MIT Press, 1999) and numerous research articles. He holds a PhD from Cornell University and an undergraduate degree from the University of Delhi.

David Emanuel is the manager of quantitative modelling for Williams Energy Marketing and Trading in Tulsa. He has been on the faculties of the University of British Columbia, New York University and the University of Texas at Dallas. He has also worked in the futures industry and developed the SPAN margining system for the Chicago Mercantile Exchange. For several years he was a financial consultant and co-founded the firm EMA, based in Dallas. David has degrees from Cambridge, London and Stanford.

Drew Fudenberg is a professor at Harvard University. Prior to that he was at the MIT and has been a visiting professor at the University of California. Professor Fudenberg is principal investigator for the National Science Foundation Grants. He is editor of *Econometrica* and has been associate editor of a number of academic journals including the *Journal of Economic Theory*, the *Quarterly Journal of Economics* and *Games and Economic Behaviour*. He received his PhD in Economics from the MIT.

Steven Grenadier is an associate professor of finance, with tenure, at the Graduate School of Business at Stanford University. His research focuses on applying option pricing theory to real investment analysis. Subjects have included the impact of competition and private information on option exercise decisions, the underpinnings of real estate cycles and the analysis of lease contracts. His articles have been published in the *Journal of Finance*, the *Journal of Financial Economics*, the *Review of Financial Studies* and the *Journal of Real Estate Economics and Finance*. He teaches courses in investments and finance theory. He is a senior financial economist with Financial Engines, Inc., a director of E*Trade Funds, a director of the Real Options Group and a consultant to several Fortune 500 companies. He received his Bachelor's degree from the University of California, Berkeley and his PhD from Harvard University.

Nalin Kulatilaka is a professor of finance at the School of Management at Boston University. Professor Kulatilaka's current research examines the strategic use of real options and risk management tools. He is the co-author of *Real Options: Managing Strategic Investments in an Uncertain World* (1999). He received his degrees from Imperial College, Harvard University and the MIT.

Pierre Mella-Barral is currently lecturer in finance at the London School of Economics. He has been published in the *Review of Financial Studies* and *Journal of Banking and Finance*. He studied engineering in Paris and took his PhD in economics at the University of Cambridge.

Alberto Moel is a lecturer in finance at Harvard Business School (HBS). Previously he has worked at JP Morgan in New York and Mexico City, the

Toshiba Corporation in Japan and IBM in New York, Cambridge, and Zurich. His current research interests focus on real option valuation of natural resource investments, the exercise of real options in the mining industry, the determinants of cross-border stock exchange listing and models of credit risk pricing. He has written cases on the valuation of a copper mine, the effects of a currency shock on an emerging market firm's capital structure, the new product development process in financial services, cross-border capital raising, and on the application of modern financial techniques to savings and loans and the mortgage markets. Before joining the HBS faculty, he was the Charles M. Williams Fellow in Finance at HBS. Alberto holds SB, SM and ScD degrees from MIT and an MBA from HBS, where he graduated with distinction as a Baker Scholar.

Enrico Perotti holds the chair in international finance at the University of Amsterdam. He has been a visiting professor at the IMF, the University of Michigan, the London Business School and the Central European University. He has worked as a consultant for the IMF, the World Bank, the European Commission, the New York Stock Exchange and the EBRD as well as various private financial institutions and eastern European governments. He is currently advising the EC and the Russian Central Bank on bank reform in Russia. His research interests are in the area of corporate finance, banking, and international finance. His work, both theoretical and empirical, has been published in journals such as the *American Economic Review*, the *Journal of Financial Economics*, the *Review of Financial Studies* and the *Journal of International Money and Finance*. He received his PhD in finance from the MIT in 1990.

William Perraudin is professor of financial economics at Birkbeck College, University of London and Special Advisor on Regulatory Policy to the Bank of England. His research interests include continuous time pricing, credit risk modelling, strategic contingent claims models and risk management. Prior to Birkbeck, he taught at Cambridge University and worked in the City and for the International Monetary Fund. He has also acted as consultant for investment banks, central banks and multilateral organisations.

Aldo Rustichini has been in the faculty of the department of economics at Northwestern University, New York University, CORE (Louvain-La-Neuve), CentER (Tilburg University) and Boston University. He is associate editor of *Journal of Economic Theory*, *Journal of Mathematical Economics* and *Review of Economic Dynamics*. His research interests are game theory, decision theory, political economy and economic dynamics. He has a PhD in mathematics from the University of Minnesota.

Han T.J. Smit is associate professor of finance at Erasmus University. He has been a visiting scholar at Boston University and Columbia University. Dr Smit has experience in researching corporate finance, strategy and uses a combination of real options and game theory to model competitive strategies. He has published several articles on financial management. He received an MBA from Erasmus University and his PhD from the University of Amsterdam.

Suresh M. Sundaresan is the Chase Manhattan Bank Professor of Finance and Economics at the Graduate School of Business at Columbia University. He has published papers in the areas of options, swaps, futures, forwards, auctions, corporate debt, term structure of interest rates and asset pricing. His current research focuses on liquidity, default risk, bankruptcy code and asset pricing issues under default risk. He teaches the fixed-income securities course and the Advanced Derivatives course at Columbia Business School. He has conducted training programs and consulted in the areas of fixed-income securities, credit risk, derivatives and risk management. He earned his engineering degree from Madras University and a post graduate diploma in management from the Indian Institute of Management, Calcutta. In 1980 he graduated with a PhD from GSIA at Carnegie Mellon University.

Lenos Trigeorgis is professor of finance at the University of Cyprus and until recently was a visiting professor of finance at the University of Chicago, Columbia University and Bocconi University. He previously taught at Boston University and the University of Massachusetts. He has been published in numerous journals on corporate finance, competition and strategy and capital budgeting; he serves on the editorial boards of several journals. He has edited several books and is the author of *Real Options* (1996). Dr Trigeorgis is the president of the Real Options Group (ROG) and he holds a PhD (DBA) from Harvard University.

Jean Tirole is scientific director of the Institut d'Economie Industrielle, University of Social Sciences in Toulouse. He is also affiliated with CERAS (Paris) and the MIT where he holds a visiting position. Before moving to Toulouse in 1991, he was professor of economics at MIT. In 1998, he was president of the *Econometric Society*, whose executive committee he has served on since 1993. Jean has published over a hundred professional articles in economics and finance, as well as six books including *The Theory of Industrial Organization*, *Game Theory*, with Drew Fudenberg, *A Theory of Incentives in Procurement and Regulation*, with Jean-Jacques Laffont, *The Prudential Regulation of Banks*, with Mathias Dewatripont and *Competition in Telecommunications* again with Jean-Jacques Laffont. His research covers industrial organisation, regulation, game theory, banking and finance and macroeconomics. He received his PhD in economics from MIT in 1981, engineering degrees from Ecole Polytechnique (Paris) and from Ecole Nationale des Ponts et Chaussées and a "Doctorat de 3ème cycle" in decision mathematics from the University Paris IX.

Peter Tufano is a professor of business administration at the Harvard Business School. His research focuses on financial innovation and the use of financial engineering techniques by corporations. Peter studies how firms can use strategic risk management practices and creative security design to achieve competitive advantage. His paper on risk management practices in the gold mining industry shared the Smith-Breeden prize for the best paper published in the *Journal of Finance* in 1996. Peter has also studied the industrial organisation of financial markets by examining competition among financial intermediaries, including investment banks and mutual funds. His article, "Financial Innovation and First Mover Advantages", published in the *Journal of Financial Economics*, examines the benefits of new product development to innovative investment banks. He is a Faculty Research Fellow at the National Bureau of Economic Research, and was the Berol Faculty Fellow at the Harvard Business School. He gained his PhD in business economics from Harvard University and his MBA from Harvard Business School with distinction as a Baker Scholar. He graduated summa cum laude in economics from Harvard College in 1979.

Joseph T. Williams is a principal of Professors Capital, a collection of private entities with investments in real estate throughout Southern California. Previously, he was professor of finance at the University of British Columbia. For the past decade his research has been focused on real estate finance, including applications of costly search, dynamic agency, and real options. His publications also include theory and empirical work on investments and corporate finance. He was a founder of the *Review of Financial Studies* and the first president of its supporting society. He has a PhD in economics from Harvard University.

An Introduction to Option Exercise Games

Steven R. Grenadier

Stanford University

The real options approach to analysing investment under uncertainty has become part of the mainstream literature of financial economics. The subject of real options now typically comprises an entire chapter in corporate finance textbooks.[1] Essentially, the real options approach posits that the opportunity to invest in a project is analogous to an American call option on the investment opportunity. Once that analogy is made, the vast and rigorous machinery of financial options theory is at the disposal of real investment analysis. The real options approach is well summarised in Dixit and Pindyck (1994) and Trigeorgis (1996).[2]

While the real options approach has proven useful in providing an analytical framework for analysing the timing of investment decisions, a notable failure of the approach has been an almost universal lack of strategic considerations. In standard real options models investment (exercise) strategies are formulated in isolation, without regard to the potential impact of other firms' exercise strategies. Agents are assumed to be perfectly informed, and the exercise of one agent's option is assumed to have no impact on the value of other agents' options. Such stark assumptions are at odds with the realities of many real-world investment contexts. Market realities do not always allow one to formulate an optimal exercise strategy in isolation.

In order to highlight the importance of including strategic considerations into real options models, consider the following two examples of well-known applications.

Example 1: Real estate development in an oligopoly

A classic case of a real option is real estate investment. As in Titman (1985) and Williams (1991), the development of real estate is analogous to an American call option on a building, where the exercise price is equal to the construction cost. Taken literally, the standard, myopic real options approach implies that developers should ignore the construction behaviour of their competitors. However, in real estate markets, developers are likely to face considerable competition from competitors, and the development activities of competitors will have a fundamental impact on one's development options. By extending the options framework to account for strategic interaction, a much richer set of investment implications is obtained. While the standard real options models dictate that a developer should wait until the development option is considerably in-the-money (ie, the value of a building is much larger than the cost of construction), competition and the fear of pre-emption will likely force developers to build much earlier. In addition, while standard real options models imply that development will be simultaneous, game-theoretic models allow for the possibility of sequential development. Competitive models of real estate development can also help explain boom-and-bust behaviour in commercial construction, as well as why rational developers may construct new buildings in the face of declining demand and market values.

Example 2: Oil exploration and asymmetric information

Another well-known case of a real option is oil exploration. Paddock, Siegel and Smith (1988) model the option-like characteristics of offshore oil leases. In addition to possessing the standard features of options (ongoing oil price uncertainty, large lump-sum exercise costs), it also possesses important private information externalities. Due to the leasing practices of the US government, it is frequently the case that two or more firms lease adjacent tracts of land that may contain an oil deposit, (Hendricks and Kovenock (1989)). Prior to obtaining the lease, each firm has private information about the probability of finding a deposit. Each firm obtains a noisy signal of the likelihood of striking oil by combining seismic surveys with in-house expertise. However, the existence and magnitude of the deposit can only be known through drilling. Firms then face a trade-off between the benefits of drilling and potentially obtaining oil earlier, and the benefits of waiting for other firms to drill first and reveal information about the size of the deposit. Therefore, in equilibrium, firms must determine their optimal drilling exercise times in a context of private information and ongoing uncertainty. Such an equilibrium is constructed in Grenadier (1999).

Until recently, there have been relatively few real options applications concerned with strategic equilibrium in exercise policies. There are two possible reasons for this omission. First, this is partially a result of the fact that such issues are typically ignored in the financial options literature, as most financial options are widely held side-bets between agents external to the underlying firm, and therefore their exercise does not influence the characteristics of the underlying security or the options themselves.[3] Second, the application of game theory to continuous-time models is not well developed, and often quite tricky.

By augmenting the real options framework with strategic considerations, many of the most well-known predictions of standard real option analysis are overturned. For example, perhaps the most cited result of the real options literature is the invalidation of the standard net present value rule of investing in any project with a non-negative net present value. Because the future value of the asset is uncertain, there is an opportunity cost to investing today. This is often referred to as the "option to wait". Thus, the optimal investment rule, as described in the real options literature, is to invest when the asset value exceeds the investment cost by a potentially large option premium. As stated in Dixit and Pindyck (1994), "Hence the simple NPV rule is not just wrong; it is often very wrong". However, clearly this result depends on the lack of competitive access to the project. If firms fear pre-emption, then the option to wait becomes less valuable. In the limit of perfect competition, the "option to wait" becomes valueless, and the traditional net present value rule obtains. To understand investment in industries with competitive pressure, a game-theoretic analysis of equilibrium exercise strategies is essential.

A related implication of real options models that do not consider the impact of competitive forces is that they imply virtually no downside risk to investment. That is, since a monopolist invests at a substantial option premium, the likelihood for large asset value reversals is remote. Thus, in the traditional real options framework, such *ex post* losses are highly infrequent. However, we know that in many real world asset markets, investments that initially appeared promising turned out poorly as the market quickly turned downward. It is therefore very difficult for the standard real options framework to explain boom-and-bust markets such as real estate, where periodic bouts of overbuilding result in waves of high vacancy and foreclosure rates. It is found, however, that given the much smaller option premiums in competitive environments, such reversals become much more likely.

This book provides a collection of some of the most important work in the area of combining real options models with game theory. The selections include papers that provide a theoretical underpinning for strategic real options models, state-of-the-art applications of the theory and practical applications. Much of the work is quite recent and it is undoubtedly true that this is an area of rapid innovation. It will be exciting to see the future trajectory of research in this area in the coming years.

ILLUSTRATION: A SIMPLE OPTION EXERCISE GAME

It should prove useful to provide a simple example of a model of a real options game. The model will rely heavily on the work of Dixit and Pindyck (1994, Chapter 9), Smets (1993) and Huisman and Kort (2000). The model will describe the equilibrium exercise strategies for two firms possessing options to invest in a production opportunity. Depending on the initial state of the industry, the Nash equilibria can take a variety of forms, including pure and mixed strategies as well as simultaneous and sequential exercise.

Assumptions and set-up of model

Consider an industry comprised of two identical firms. Each firm possesses an option to invest in a technology that will produce a unit of output. The cost of investment is K. The cashflow stream from the investment is uncertain, and the investment is irreversible. Most importantly, the payoff from any firm's investment depends on the strategy of its competitor. If only one firm invests, it achieves the (perhaps temporary) payoffs of a monopolist. If both firms invest the firms receive the payoffs of a duopoly and each firm must then determine its optimal exercise strategy fully cognisant of its competitor's similar calculation. In a Nash equilibrium, each firm's exercise strategy is optimal, conditional on its competitor's exercise strategy.

The price of a unit of output, P(t), fluctuates stochastically over time so as to clear the market:

$$P(t) = X(t) \cdot D[Q(t)] \tag{1}$$

where D is the inverse demand function, X(t) is an exogenous shock process to demand, and Q(t) is the industry supply process. The inverse demand function will be downward sloping, $D' < 0$, ensuring a "first-mover advantage" to the investment option. In this particular duopoly example, Q(t) can either be 0, 1, or 2, depending on whether or not the firms have exercised their options.

The market is characterised by evolving uncertainty in the state of demand. Even demand at the next instant is uncertain as of today. Let the shock to industry demand, X(t), evolve as a geometric Brownian motion process:

$$dX = \alpha X dt + \sigma X dz \tag{2}$$

where α is the instantaneous conditional expected percentage change in X per unit time, σ is the instantaneous conditional standard deviation per unit time, and dz is the increment of a standard Wiener process.

In the model, each firm must contemplate its value if it is the first to exercise (Leader), the second to exercise (Follower), or if it exercises at the same time as its competitor (Simultaneous). I will begin by deriving the value of a firm in each of these positions. For the moment, I simply take the positions as given. Later, I derive the equilibrium set of exercise strategies where firms choose their roles.

To begin, I value the payoff of choosing to be the Follower. As is standard in solving dynamic games, I work backwards in a dynamic-programming fashion. Thus, assume that one firm has already exercised. Since the Leader has already exercised his option, the Follower can construct his exercise strategy without fear of pre-emption. Therefore, this is a standard option pricing problem in which the Follower's strategy is to maximise his option's value.

Let F(X) denote the value of the Follower. For simplicity, I assume risk neutrality, so that all assets are priced so as to yield an expected rate of return equal to the risk-free rate, r. This seemingly restrictive assumption can easily be relaxed by adjusting the drift rate to account for a risk premium in the manner of Cox and Ross (1976).

Using standard arguments (ie, Dixit and Pindyck (1994), Grenadier (1995)), F(X) must solve the following equilibrium differential equation:

$$0 = \tfrac{1}{2}\sigma^2 X^2 F_{xx} + \alpha X F_x - rF \tag{3}$$

Differential equation (3) must be solved subject to appropriate boundary conditions. These boundary conditions serve to ensure that the Follower chooses the optimal exercise strategy:

$$F(X_F) = \frac{X_F D(2)}{r - \alpha} - K \tag{4}$$

$$F'(X_F) = \frac{D(2)}{r - \alpha}$$

Here, X_F is the value of X(t) that triggers entry. That is, the optimal exercise strategy of the Follower is to exercise the first moment that $X(t) \geq X_F$. The first boundary condition is the value-matching condition. It simply states that at the moment the Follower's option is exercised, the payoff is $[X_F D(2)/(r - \alpha)] - K$. Note that the term $[X_F D(2)/(r - \alpha)]$ is simply the discounted expected present value of the duopoly cashflow in perpetuity. The second boundary condition is the smooth-pasting or high-contact condition.[4] This condition ensures that the exercise trigger is chosen so as to maximise the value of the option.

Closed-form solutions for F(X) and the Follower's exercise trigger X_F are easily obtained. The solutions can be written as:

$$F(X) = \begin{cases} \left(\dfrac{K}{\beta - 1}\right) \cdot \left(\dfrac{X}{X_F}\right)^{\beta} & \text{for } X < X_F \\[2ex] \dfrac{XD(2)}{r - \alpha} - K & \text{for } X \geq X_F \end{cases} \tag{5}$$

where

$$X_F = \left(\frac{\beta}{\beta - 1}\right)\left(\frac{r - \alpha}{D(2)}\right) K$$

$$\beta = \frac{-(\alpha - \sigma^2/2) + \sqrt{(\alpha - \sigma^2/2)^2 + 2r\sigma^2}}{\sigma^2} > 1 \qquad \textbf{(6)}$$

and where $\alpha < r$ to ensure convergence.

Now, consider the value of the Leader, after the Leader has exercised and conditional upon the Follower pursuing the optimal Follower strategy. Until the Follower exercises, the Leader receives the monopoly cashflow of $X \cdot D(1)$. However, once the trigger X_F is reached, the Leader will face competition, and will only receive the duopoly payout flow $X \cdot D(2)$ from then onward.

Let $L(X)$ denote the value of the Leader, after it has exercised. Prior to the exercise of the Follower, $L(X)$ must solve the following equilibrium differential equation:

$$0 = \tfrac{1}{2}\sigma^2 X^2 L_{XX} + \alpha X L_X - rL + XD(1) \qquad \textbf{(7)}$$

Differential equation (7) must be solved subject to the boundary condition:

$$L(X_F) = \frac{X_F D(2)}{r - \alpha} \qquad \textbf{(8)}$$

This boundary condition reflects the fact that when the Follower exercises at the trigger X_F, the value of the Leader equals the present value of the duopoly cashflow in perpetuity.

The solution for $L(X)$ can be expressed as:

$$L(X) = \begin{cases} \dfrac{XD(1)}{r - \alpha} + \dfrac{\beta}{\beta - 1}\,\dfrac{D(2) - D(1)}{D(2)}\,K\left(\dfrac{X}{X_F}\right)^{\beta} & \text{if } X < X_F \\[2ex] \dfrac{XD(2)}{r - \alpha} & \text{if } X \geq X_F \end{cases} \qquad \textbf{(9)}$$

At the moment the Leader exercises and pays the exercise price K, the Leader receives the payoff $L(X) - K$.

Finally, consider the value of the firms if they both exercise immediately and simultaneously. Denote this value by $S(X)$. $S(X)$ simply equals the previously mentioned present value of the duopoly cashflow in perpetuity:

$$S(X) = \frac{XD(2)}{r - \alpha} \qquad \textbf{(10)}$$

At the moment both firms exercise, each firm receives the payoff $S(X) - K$.

We can now consider the relative values of the firms at the moment of first exercise. If there is sequential exercise, the Leader receives a payoff of L(X) – K and the Follower receives F(X). Depending on the initial entry time (which is determined by equilibrium considerations and derived below), the Leader's payoff may be greater or less than that of the Follower. If there is simultaneous exercise, each receives a payoff of S(X) – K. The following proposition describes the relative valuations.

Proposition 1. *There exists a unique point,* $X_L \in (0, X_F)$, *with the following properties:*

$$
\begin{aligned}
L(X) - K &< F(X) && \textit{for } X < X_L \\
L(X) - K &= F(X) && \textit{for } X = X_L \\
L(X) - K &> F(X) && \textit{for } X_L < X < X_F \\
L(X) - K &= F(X) && \textit{for } X \geq X_F
\end{aligned}
$$

In addition,

$$
\begin{aligned}
S(X) - K &< \min[L(X) - K, F(X)] && \textit{for } X < X_F \\
S(X) - K &= L(X) - K = F(X) && \textit{for } X \geq X_F
\end{aligned}
$$

Proof: See the appendix to Grenadier (1996) for a similar result.

Proposition 1 demonstrates that there is a unique value of $X \in (0, X_F)$, defined as X_L, at which the payoffs to both Leader and Follower are equal. At any point below X_L each firm would prefer being the Follower; at any point above X_L each would prefer (or be indifferent to) being the Leader. At X_L, the benefits of a potentially temporary monopoly just equal the costs of paying the exercise price earlier. Proposition 1 also demonstrates that for any $X < X_F$, the value of simultaneous exercise is strictly less than the value of being a Leader or Follower. For any $X \geq X_F$, however, the value of Leading, Following, or Simultaneous exercise are all equal.

Equilibrium exercise strategies

Given the relative values of the Leader and Follower for a given current value of X, I can now construct the equilibrium set of exercise strategies. In particular, I focus on symmetric, Markov, sub-game perfect equilibrium exercise strategies in which each firm's exercise strategy, conditional upon the other's exercise strategy, is value-maximising. Such continuous-time games can be quite complex. Thus, the presentation below will be mostly heuristic.

At each point in time, let firms 1 and 2 play a mixed-strategy exercise game. Firms 1 and 2 choose probabilities $p_1(X)$ and $p_2(X)$, respectively, denoting their probability of exercise at that moment in time. For notational simplicity,

I shall ignore the exercise probabilities' dependence on the argument X and write them simply as p_1 and p_2. The rules are as follows. If neither firm wishes to exercise, $p_1 = p_2 = 0$, then no firm exercises, and the firms play a new game (and choose new probabilities) upon the next realisation of X(t). If at least one firm chooses a non-zero probability, then a multi-stage game is played. In each round (which is instantaneous), each firm exercises with their respective probability. In each round, four outcomes can occur, depending on the realised outcome of the combination of random strategies. One possible outcome is that firm 1 exercises and firm 2 does not; the second possible outcome is that firm 2 exercises and firm 1 does not; the third possible outcome is that firms 1 and 2 simultaneously exercise and the fourth possible outcome is that neither firm exercises. In the case in which neither firm exercises, the game moves to the next round and the game is repeated until at least one firm exercises.

The equilibrium strategies will depend on the initial value of X. First, suppose that $X \geq X_F$. In this case, the equilibrium is very simple. From the derivation of the optimal Follower's strategy, it is clear that exercise by one firm will result in immediate exercise by the other firm. The symmetric Nash equilibrium is for each firm to exercise immediately, with probability one. Thus, in this range, the equilibrium involves the pure strategy $p^*(X) = p_1 = p_2 = 1$, and exercise is always simultaneous. This is clearly a Nash equilibrium. Given that your competitor exercises, the optimal Follower strategy is to also exercise immediately.[5]

Second, suppose that $X \in (X_L, X_F)$. From Proposition 1, in this range the value of being a Leader is strictly greater than that of being a Follower. Thus, each firm has an incentive to pre-empt the other. Of course, each firm must trade-off this benefit of immediate exercise with the possibility that both firms exercise simultaneously, resulting in the worst possible outcome. We will find that the equilibrium exercise strategies will be mixed strategies, with $p_1, p_2 \in (0, 1)$. Consider the expected payoff to firm i from the exercise game, where $i = 1, 2$ and $j \neq i$, and where $\max(p_i, p_j) > 0$. First, consider the first round of the game (the game moves on to later rounds if no exercise occurs in a given round). With probability $p_i \cdot (1 - p_j)$, firm i emerges as the Leader and exercises with payoff $L(X) - K$. With probability $p_i \cdot p_j$, both firms exercise simultaneously and receive the payoff $S(X) - K$. With probability $(1 - p_i) \cdot p_j$, firm i emerges as the Follower with a value of $F(X)$. Finally, with probability $(1 - p_i) \cdot (1 - p_j)$, the game is repeated and the players move on to the next round with the same probabilities of outcomes. Writing out the multi-round probabilities, we can write firm i's expected payoff, $V_i(p_i, p_j)$, conditional on strategies p_i, p_j, as follows:

$$V_i(p_i, p_j) = \{p_i(1 - p_j)[L(X) - K] + p_i p_j[S(X) - K] + (1 - p_i)p_j F(X)\} \times \sum_{k=0}^{\infty} [(1 - p_i) \cdot (1 - p_j)]^k$$

$$= \frac{p_i(1 - p_j)[L(X) - K] + p_i p_j [S(X) - K] + (1 - p_i)p_j F(X)}{1 - (1 - p_i)\cdot(1 - p_j)} \qquad \textbf{(11)}$$

The equilibrium set of mixed strategies is for each firm to exercise with the following probability:

$$p^*(X) \equiv p_1 = p_2 = \frac{L(X) - F(X) - K}{L(X) - S(X)}, \text{ for } X_L \in (X_L, X_F) \qquad \textbf{(12)}$$

To demonstrate that this is an equilibrium, suppose that firm j exercises with probability $p^*(X)$. Then, $V_i(p_i, p^*(X)) = F(X)$, for any choice of p_i. Thus, the *ex ante* expectation of firm i of its value from responding with any exercise probability p_i is F(X). This indifference across strategies permits a mixed-strategy equilibrium. Thus, firm i can randomise and can do no better than by choosing $p_i = p^*(X)$. By symmetry, the same argument holds for firm j.

Note that in this range, both simultaneous and sequential exercise can emerge. The probability of a simultaneous equilibrium with both firms obtaining $S(X) - K$ is $p^*(X)/[2 - p^*(X)]$. The probability of a sequential equilibrium with firm 1 emerging as the Leader with value $L(X) - K$ and firm 2 emerging as the Follower with value F(X) is $[1 - p^*(X)]/[2 - p^*(X)]$. Similarly, the probability of a sequential equilibrium with firm 2 emerging as the Leader with value $L(X) - K$ and firm 1 emerging as the Follower with value F(X) is $[1 - p^*(X)]/[2 - p^*(X)]$. As X approaches X_L, $p^*(X)$ goes to zero and sequential exercise occurs with probability one. Each player will have a 50% chance of becoming the Leader and a 50% chance of becoming the Follower. As X approaches X_F, $p^*(X)$ goes to one, and simultaneous exercise occurs with probability one.

Finally, suppose that $X \leq X_L$. The equilibrium exercise strategies are for each firm to wait until X(t) enters the region (X_L, ∞) before exercising. Thus, $p^*(X) = 0$ for all $X \leq X_L$. Therefore, if the current value of $X \leq X_L$, both firms will wait until the first moment that X(t) exceeds the threshold X_L. At that moment, both firms exercise with probability $p^*(X_L + \varepsilon)$, for infinitesimal ε. As discussed above, one firm (determined by the outcome of the mixed-strategy game) will exercise immediately, and the other firm will wait until X_F is reached. The values of both firms will equal F(X).

To demonstrate that this is an equilibrium, consider the following simple argument. In the range where $X \leq X_L$, the value of being a Follower is at least as great as the value of being a Leader. The strategy of not exercising in this range is a dominant strategy. By pursuing the equilibrium strategy of waiting, the firm achieves the value of F(X), whereas if it exercises with any positive probability it would achieve a value which is never greater than this (ie, $L(X) - K$ and $S(X) - K$ are less than or equal to F(X) in this range).

The equilibrium exercise strategies for this game are now summarised.

Equilibrium: *A symmetric, Markov, sub-game perfect equilibrium is a function of the current value of X. The equilibrium strategies are:*

(i) *If* $X \geq X_F$, *then each firm exercises immediately. There will be simultaneous exercise.*

(ii) *If* $X \in (X_L, X_F)$, *then each firm chooses a mixed-strategy by exercising immediately with probability* $p^*(X) = [L(X) - F(X) - K]/[L(X) - S(X)]$. *There can be either simultaneous or sequential exercise.*

(iii) *If* $X \leq X_L$, *then neither firm exercises. Each firm waits until the first moment that* $X(t)$ *exceeds the trigger* X_L *before exercising with the strategy outlined above in (ii).*

A numerical example

In order to make the equilibrium concept more concrete, consider the following numerical example. I begin by setting six structural parameters of the problem equal to the following numerical values: $r = 0.05$, $\alpha = 0.02$, $\sigma = 0.20$, $K = 1$, $D(1) = 1.0$, and $D(2) = 0.75$. We can now determine the triggers X_F and X_L. Using (6), we find that $X_F = 0.10883$. To determine X_L, we need to find the level of X, between 0 and X_F, that satisfies $F(X) = L(X) - K$. Using the first lines of (5) and (9) to give us F(X) and L(X), respectively, we search for X_L.[6] We find that $X_L = 0.0643$.

Now that we have the triggers X_L and X_F, we can investigate the equilibrium exercise strategies for different initial levels of X. Here are three examples.

Equilibrium with $X = 0.12$

Since X is above X_F, we are in range (i) of the equilibrium. The equilibrium exercise strategies are for both firms to exercise immediately. Using equation (10), which gives us S(X), and subtracting the exercise price $K = 1$, the value of each firm under simultaneous exercise is $S(0.12) - K = 2.0$.

Equilibrium with $X = 0.08$

Since X is between X_L and X_F, we are in range (ii) of the equilibrium. The equilibrium exercise strategies are for both firms to play an exercise game with probability $p^*(X)$, whose formula is in equation (12). Using the first lines of (5), (9) and (10) to give us F(X), L(X) and S(X), respectively, we can solve for $p^*(0.08) = 0.471$. From the discussion near the end of the previous section, this exercise probability implies that each firm has a 34.6% chance of being the Leader, a 34.6% chance of being the Follower, and a 30.8% chance of exercising simultaneously with its competitor. The expected payoff from this strategy is equal to $0.346 \times [L(0.08) - K] + 0.346 \times F(0.08) + 0.308 \times [S(0.08) - K] = 1.058$. Note that the expected payoff from the game, in equilibrium, is equal to the value of being a Follower, as $F(0.08) = 1.058$.

Equilibrium with $X = 0.06$

Since X is below X_L, we are in range (iii) of the equilibrium. The equilibrium dictates that neither firm should exercise until $X(t)$ rises above the trigger X_L, where $X_L = 0.06432$. If $X(t)$ reaches X_L, each firm will have a 50% chance of being the Leader and a 50% chance of being the Follower. Since according to Proposition 1 the value of the Leader and Follower are equal at X_L, this is equivalent to being a Follower with certainty. The first line of equation (5) provides the value of being a Follower, $F(X)$. Thus, the value of each firm, in equilibrium is equal to $F(0.06) = 0.671$.

Extension: a pure-strategy equilibrium

The introduction of mixed-strategy equilibria into continuous-time games introduces additional complexity into the framework. However, as demonstrated in the model above, such mixed-strategy equilibria can be handled in a very tractable fashion. Nevertheless, in order to rule out mixed-strategies, the prevailing method in the literature (ie, Dixit and Pindyck (1994, Chapter 9), Smets (1993), Grenadier (1996)) is to rule out simultaneous exercise.[7] The assumption that is used to rule out simultaneous exercise is that when both firms attempt to exercise simultaneously, a fair coin is flipped and the winner becomes the Leader and the loser becomes the Follower.

To see the impact on the equilibrium from the coin flip assumption, consider the following discussion. First, in the range $X \geq X_F$, the equilibrium is essentially unchanged. The winner of the coin flip will exercise immediately, and the loser of the coin flip will exercise an instant thereafter. This is because the optimal Follower's strategy is to exercise immediately in this range. Empirically, this "bang-bang" equilibrium will be indistinguishable from the simultaneous exercise equilibrium. Second, in the range $X \leq X_L$, the equilibrium strategies remain for both firms to defer exercise. It is only in the range from $X_L < X < X_F$ that the equilibrium is impacted. In this range, the equilibrium is now a pure-strategy equilibrium where each firm attempts to exercise immediately with probability one. The winner of the coin flip will exercise immediately, while the loser of the coin flip follows the optimal Follower strategy of exercising at the trigger X_F. To demonstrate that this is an equilibrium, consider the optimal response of firm i to firm j exercising with probability one ($p_j = 1$). If firm i exercises with probability p_i, it will receive the expected payoff of $\frac{1}{2}p_i\,[L(X) - K] + \frac{1}{2}p_iF(X) + (1 - p_i)F(X)$. The first two terms represent the cases where both firms attempt to exercise immediately, and the coin flip determines the case of the tie. The third term represents the case where only the competitor attempts to exercise immediately. Rewriting the expected payoff yields

$F(X) + \frac{1}{2}p_i[L(X) - K - F(X)]$. Since from Proposition 1, $L(X) - K > F(X)$ in this range, the optimal response is to set p_i as high as possible: $p_i = 1$. Thus, this is a symmetric pure-strategy equilibrium where each attempts to exercise immediately in this intermediate range.

OUTLINE OF THE BOOK

Part I, *Simple Introductory Models of Option Games*, features two chapters that provide an intuitive introduction to models that combine game theory and real options. By focusing on discrete-time decisions, the analytical machinery required to solve such models is substantially reduced. Kulatilaka and Perotti (Chapter 1) provide a simple, and quite illuminating analysis of growth options under competition. While a discussion of growth options is prevalent in the corporate finance literature, much of the analysis ignores the key role of strategic competition. In their model, one or both firms posses an initial growth option that permits future production cost reductions. The state of the market in the future is unknown at the outset. At a future date, the state of the market is realised, and both firms produce output under a Cournot equilibrium. The chapter considers two possible strategic aspects of investments. First, the growth option may allow a firm to obtain a future competitive advantage, thus potentially preempting future competition. Second, product market competition is permitted and a Cournot outcome is obtained. It also demonstrates that, depending on the market environment, increasing uncertainty can either increase or decrease investment in growth options. While the model is simplified by considering only European options (and thus the timing of potential exercise is no longer endogenous), much of the intuition would continue to hold under a more traditional continuous-time American option framework. Smit and Ankum (Chapter 2) provide an intuitive introduction to some of the issues involved in real options under competition. Their model examines some of the issues that arise when firms hold non-exclusive options to invest in projects. Using simple numerical examples, they are able to present many of the results that were later obtained in more mathematically intensive settings.

Part II, *Real Options and Game-Theoretic Industry Equilibria* includes three models of continuous-time real option equilibria under competition. Much of the groundwork for research in this area was laid by Smets (1993). Smets looked at a game in which two symmetric firms were allowed to choose the time at which to switch from exporting a good to pursuing foreign direct investment in a stochastically growing foreign market. Building on the real options framework of Brennan and Schwartz (1985), McDonald and Seigel (1986), and Dixit (1989a, 1989b), Smets adds the concept of a stopping equilibria based on the work of Dutta and Rustichini (1991).

Grenadier (Chapter 3) provides a model of a real option exercise game, and applies it to understanding real estate markets. As first made apparent by Titman (1985), the development of real estate is analogous to the exercise of a call option. However, in Grenadier's model, the development decision is made more realistic by the inclusion of competition. Specifically, the model analyses the sub-game perfect equilibrium development strategies in a market composed of two competing developers. Depending on the initial parameters, two types of equilibria can emerge. The first is a sequential equilibrium in which development occurs over time. The second is a simultaneous ("development cascade") equilibrium in which both developers build at the same moment. The model is able to help explain some important empirical properties of real estate markets such as boom-and-bust behaviour, and why building booms can ensue even when rents and building values are declining.

Williams (Chapter 4) provides one of the first rigorous treatments of equilibrium in real options markets. His model accounts for four features of real options that distinguish them from financial options.[8] First, exercising the option to develop an asset increases the supply of the asset, thus impacting the output price for all other developed assets. Second, if firms that develop the assets have increasing costs or limited capacity, then the cost of development depends on the aggregate demand for development. Third, the supply of undeveloped assets may be constrained. Fourth, in a less than perfectly competitive market, the actions of each option holder can impact the value of the options held by other agents. Williams obtains a solution for a sub-game perfect Nash equilibrium in such an environment.

A very important limiting case of strategic options is the case of perfect competition, where the number of competitors approaches infinity. Under perfect competition, each firm takes the underlying output price process as given, and chooses its optimal investment (exercise) strategy so as to maximise its value. In equilibrium, the price process must be such that at any point of investment, free entry ensures that the net present value of entry is precisely zero. Such a competitive equilibrium under dynamic uncertainty was solved by Lucas and Prescott (1971) using a central planning approach. Leahy (Chapter 5) provides a very general and instructive real option approach to the determination of competitive equilibrium under uncertainty. In particular, it is shown that a very useful simplifying result holds where the equilibrium investment strategies are identical to those that a myopic firm would choose without consideration of its competitors' reactions.

Part III, *Models of Asymmetric Information and Search* augments the real option approach with considerations of information, bargaining, and matching. Grenadier (Chapter 6) extends the real options literature by including private, asymmetric information into the framework. Traditionally, both real and financial option pricing models assume that all agents are perfectly informed about all of the option's parameters. However, in many realistic applications, agents do not possess perfect information and in such markets, it is likely to be the case that observed option exercise behaviour will provide an information revelation mechanism by which all agents can infer private information. For example, firms contemplating drilling for oil on adjacent tracts of land may wish to learn from the drilling (exercise) decisions of its competitors. Similarly, real estate developers may wish to learn about other developers' opinions of the viability of the market by watching their development activities. Grenadier provides a Bayesian Nash equilibrium for real options markets in which agents learn by observing competitors' exercise strategies. In equilibrium, it is typically the case that information is revealed by the most informed agents to the least informed agents. Thus, less informed agents can piggyback off the revealed private information of others. It is also possible for herds (or "informational cascades") to form, in which agents ignore their private information and all exercise simultaneously, each finding their private information overwhelmed by what is learned from others. Such herding behaviour is economically wasteful, and can lead to excess investment. Lambrecht and Perraudin (1999), in a different setting, develop a model which combines incomplete information and the threat of preemption. Firms compete for an investment project under an information structure in which firms do not know the investment costs of their competitors.

Williams (Chapter 7) provides a model of real asset pricing, in a market in which buyers and sellers engage in sequential search in an intertemporal setting. In the model, once a buyer and seller meet, they engage in bilateral bargaining over the purchase price. If there exists a price that both buyer and seller prefer to no transaction, then they agree on the Nash bargaining solution. The equilibrium average transaction price and liquidity premium are derived. The model is also extended to deal with the optimal development of new assets. Williams finds that in the equilibrium with search, development can occur sooner and undeveloped assets have lower market values than in the standard solution without search.

Moel and Tufano (Chapter 8) and Caplin and Leahy (Chapter 9) illustrate practical examples of how strategic analysis can play a role in a real options context. Moel and Tufano provide a case study of an auction for rights to explore and develop a mining property in Peru. In addition to valuing the rights, participants had to contemplate their bidding strategy in the face of competition. Interestingly, it is shown that the bidding rules can induce the successful bidder to make uneconomic investments. Caplin and Leahy focus on the informational spill-over aspects of competitive investment by providing a model used to explain the dramatic economic turn-around of Lower Sixth Avenue in New York. This area went from a long period of under-use, to a rapid rise in prospects following the arrival of the first superstore to locate in the area. Using a search model with information externalities, they are able to explain both the long period of under-use and the rapid buildup. With information externalities, firms do not take into account the value to others of the information revealed by their activities, and are too selective in equilibrium. However, when a firm's actions reveal a potentially profitable market, the information revealed to the market can lead to a cascade of investment. While the model is not solved in a traditional real options framework, it is clearly closely related and could be solved in such a framework.

Part IV, *Corporate Securities as Strategic Options*, deals with models of financial contracts whose payoffs are influenced by strategic considerations. Emanuel (Chapter 10), provides one of the first attempts at modelling the strategic aspects of a corporate security: warrants issued by firms. As previously discussed, one of the reasons that, until recently, little work had been done in considering the strategic aspect of real options is that there has been little need for strategic considerations in the financial options literature. Standard financial options such as calls and puts are written and purchased by agents external to the company whose stocks underlie the options. For example, a standard call option on a company's stock is not issued by the company and the call's exercise has no impact on the company's prospects or value. Thus, strategic analysis of option exercise is not needed for most of the standard cases of financial options. However, for the case of warrants and convertible securities, strategic issues do come into play. For example, a warrant is a call option that is issued by a company on its own stock. When warrants are exercised, the company issues new shares and receives cash from the exercise. Thus the exercise of warrants held by other investors can potentially impact the value of one's own warrant. One needs to consider the potential dilutive impact of the new share issuance, as well as what the firm will do with the exercise proceeds. Other notable papers that also deal with warrant exercise and competitive issues are Constantinides (1984), Constantinides and Rosenthal (1984) and Spatt and Sterbenz (1988).

A more recent application of strategic option analysis is seen in work on option-based games involving debt and equityholders of firms. Since almost its inception, the option pricing literature has been used to value corporate debt contracts. Black and Scholes (1973), Merton (1973, 1974) and Black and Cox (1976) applied option models to value the capital structure of a levered firm. In its most basic form, risky debt can be interpreted as a portfolio comprising a riskless bond plus a short position in a put option on the value of the firm's underlying assets. A common feature of most option pricing models used for the pricing of debt contracts is the assumption that the threshold at which bankruptcy occurs is exogenous. Similarly, it is typically assumed that the contracted payments in the event of financial distress are not subject to renegotiation.

Because of the costs of transferring ownership in the event of reorganisation, there exists the potential for strategic behaviour in the determination of debt payments and liquidation policy. Since liquidation is costly, equityholders may be able to act strategically and force concessions from debtholders by paying less than the contracted coupon payment. Since debtholders recognise this, they are likely to respond to such behaviour in order to maximise the value of their claims. This leads to a "debt service game" in which the equityholders and debtholders form a sub-game perfect equilibrium in which equityholders may pay less than the contracted interest payment and in which debtholders may choose to not force liquidation, even if they were contractually permitted. Each claimholder acts to maximise the value of their interests in the firm. Such models result in higher default premia than do non-strategic models of debt valuation.

Given the potential for strategic behaviour in the servicing of debt, models which combine option pricing methodologies and game-theoretic equilibrium notions seem to be potentially quite useful. Chapters 11 and 12 provide two models of such strategic debt service games. Anderson and Sundaresan (Chapter 11), determine an equilibrium to a debt service game in a discrete-time binomial setting. Debtholders and equityholders behave non-cooperatively and the point of reorganisation is determined endogenously. Mella-Barral and Perraudin (Chapter 12), provide an equilibrium to a debt service game in a continuous-time setting. In this framework, they are able to derive simple closed-form solutions for the firm's claims.

Part V, *Foundations*, provides two foundational chapters for much of the game-theoretic analysis in this book. Fudenberg and Tirole (Chapter 13) and Dutta and Rustichini (Chapter 14), provide very important methodological contributions to the continuous-time game theory literature. While continuous-time modelling is pervasive in the option pricing literature, much of the work on continuous-time game theory is quite recent. Fudenberg and Tirole develop a formal model of games of timing, in which a continuous-time version of mixed-strategy equilibria is presented. It is this notion of mixed-strategy equilibria that was used in the illustrative model presented earlier in this article. While the timing game considered in Fudenberg and Tirole is one of certainty, it is still a critical foundational piece for games under uncertainty. Dutta and Rustichini provide a characterisation of equilibria for two-player stochastic games. Specifically, they analyse the class of Markov-Perfect equilibria that holds in games in which players can make discontinuous changes in the variables that affect their payoffs. In such a setting, they find the existence of equilibria that take the form of the two-sided (s, S) rule, where actions are taken when the state variable reaches a fixed threshold. In an earlier paper, Dutta and Rustichini (1991) characterise a class of Markov-perfect equilibria called stopping equilibria, where they show how the solution to a single-agent-stopping problem can be used to describe all the equilibria in this class. Such stopping equilibria form the basis of much of the results in the strategic exercise game literature.

The chapters in this book illustrate how the intersection of real options and game theory provides powerful new insights into the behaviour of economic agents under uncertainty. While the real options paradigm has contributed greatly to our understanding of real and financial decisions, it is surely limited by its lack of focus on strategic considerations. By augmenting the real options framework with the large body of work in game theory, much richer, and arguably more empirically consistent predictions of behaviour are obtained. I hope that these chapters will stimulate interest and lead to further developments in this burgeoning area of research.

1 For example, see Brealey and Myers (2000) and Van Horne (1998).
2 The application of the real options approach to investment is quite broad. Brennan and Schwartz (1985) use an option pricing approach to analyse investment in natural resources. McDonald and Siegel (1986) provided the standard continuous-time framework for analysis of a firm's investment in a single project. Majd and Pindyck (1987) enrich the analysis with a time-to-build feature. Dixit (1989b) uses the real option approach to examine entry and exit from a productive activity. Triantis and Hodder (1990) analyse manufacturing flexibility as an option. Titman (1985) and Williams (1991) use the real options approach to analyse real estate development.
3 A notable exception is the case of warrants and convertible securities (Emanuel (1983), Constantinides (1984), Spatt and Sterbenz (1988)).
4 See Merton (1973) for a derivation of the high-contact condition.
5 Note that there is no "collusive" equilibrium that dominates this outcome; the optimal simultaneous exercise game is for both firms to exercise at the first moment X(t) equals or exceeds the trigger X_F.
6 This solution is simple to find using any software program with numerical root finding capabilities.
7 The exception is Huisman and Kort (2000), where mixed-strategies are permitted.
8 Some, but not all of these features are also shared by warrants and convertible securities issued by firms.

BIBLIOGRAPHY

Anderson, R. and S. Sundaresan, 1996, "Design and Valuation of Debt Contracts", *Review of Financial Studies* 9, pp. 37–68.

Black, F. and J.C. Cox, 1976, "Valuing Corporate Securities: Some Effects of Bond Indenture Provisions", *Journal of Finance* 31, pp. 351–67.

Black, F. and M. Scholes, 1973, "The Pricing of Options and Corporate Liabilities", *Journal of Political Economy* 81, pp. 637–59.

Brealey, R.A. and S.C. Myers, 2000, *Principles of Corporate Finance*, Sixth Edition (New York: McGraw–Hill).

Brennan, M.J. and E. Schwartz, 1985, "Evaluating Natural Resource Investments", *Journal of Business* 58, pp. 135–57.

Caplin, A. and J. Leahy, 1993, "Miracle on Sixth Avenue: Information Externalities and Search", *Economic Journal* 108, pp. 60–74; reprinted as Chapter 9 of the present volume.

Constantinides, G., 1984, "Warrant Exercise and Bond Conversion in Competitive Markets", *Journal of Financial Economics* 13, pp. 371–97.

Constantinides, G. and R. Rosenthal, 1984, "Strategic Analysis of the Competitive Exercise of Certain Financial Options", *Journal of Economic Theory* 32, pp. 128–38.

Cox, J.C. and S.A. Ross, 1976, "The valuation of options for alternative stochastic processes", *Journal of Financial Economics* 3, pp. 145–66.

Dixit, A., 1989a, "Hysteresis, Import Penetration, and Exchange Rate Pass-through", *Quarterly Journal of Economics* 104, pp. 205–28.

Dixit, A., 1989b, "Entry and Exit Decisions Under Uncertainty", *Journal of Political Economy* 97, pp. 620–38.

Dixit, A. and R. Pindyck, 1994, *Investment under Uncertainty* (Princeton University Press).

Dutta, P. and A. Rustichini, 1991, "A Theory of Stopping Time Games with Applications to Product Innovation and Asset Sales", Discussion Paper 523 (Department of Economics, Columbia University).

Dutta, P. and A. Rustichini, 1995, "(s,S) Equilibria in Stochastic Games", *Journal of Economic Theory* 67, pp. 1–39; reprinted as Chapter 14 of the present volume.

Emanuel, D., 1983, "Warrant Valuation and Exercise Strategy", *Journal of Financial Economics* 12, pp. 211–35; reprinted as Chapter 10 of the present volume.

Fudenberg, D. and J. Tirole, 1985, "Preemption and Rent Equalization in the Adoption of New Technology", *Review of Economic Studies* 52, pp. 383–401; reprinted as Chapter 13 of the present volume.

Grenadier, S.R., 1995, "Valuing Lease Contracts: A Real-Options Approach", *Journal of Financial Economics* 38, pp. 297–331.

Grenadier, S.R., 1996, "The Strategic Exercise of Options: Development Cascades and Overbuilding in Real Estate Markets", *Journal of Finance* 51, pp. 1653–79; reprinted as Chapter 3 of the present volume.

Grenadier, S.R., 1999, "Information Revelation Through Option Exercise", *Review of Financial Studies* 12, pp. 95–130; reprinted as Chapter 6 of the present volume.

Hendricks, K. and D. Kovenock, 1989, "Asymmetric Information, Information Externalities, and Efficiency: The Case of Oil Exploration", *RAND Journal of Economics* 20, pp. 164–82.

Huisman, K. and P. Kort, 2000, "Effects of Strategic Interactions on the Option Value of Waiting", Working Paper (Tilburg University, The Netherlands).

Kulatilaka, N. and E. Perotti, 1998, "Strategic Growth Options", *Management Science* 44, pp. 1021–31; reprinted as Chapter 1 of the present volume.

Lambrecht, B. and W. Perraudin, 1999, "Real Options and Preemption Under Incomplete Information", Working Paper (University of Cambridge).

Leahy, J., 1993, "Investment in Competitive Equilibrium: The Optimality of Myopic Behavior", *Quarterly Journal of Economics* 108, pp. 1105–33.

Lucas, R.E., Jr. and E.C. Prescott, 1971, "Investment Under Uncertainty", *Econometrica* 39, pp. 659–81.

Majd, S. and R. Pindyck, 1987, "Time to Build, Option Value, and Investment Decisions", *Journal of Financial Economics* 18, pp. 7–28.

McDonald, R. and D. Siegel, 1986, "The Value of Waiting to Invest", *Quarterly Journal of Economics* 101, pp. 707–27.

Mella-Barral, P. and W. Perraudin, 1997, "Strategic Debt Service", *Journal of Finance* 52, pp. 531–56.

Merton, R.C., 1973, "Theory of Rational Option Pricing", *Bell Journal of Economics and Management Science* 4, pp. 141–83.

Merton, R.C., 1974, "On the Pricing of Corporate Debt: The Risk Structure of Interest Rates", *Journal of Finance* 29, pp. 449–70.

Moel A. and P. Tufano, 2000, "Bidding for the Antamina Mine", in M. Brennan and L. Trigeorgis, (eds.), *Project Flexibility, Agency, and Competition* (Oxford University Press).

Paddock, J.L., D.R. Siegel and J.L. Smith, 1988, "Option Valuation of Claims on Real Assets: The Case of Offshore Petroleum Leases", *Quarterly Journal of Economics* 103, pp. 479–508.

Smets, F., 1993, "Essays on Foreign Direct Investment", Unpublished dissertation (Yale University).

Smit, H. and L.A. Ankum, 1993, "A Real Options and Game-Theoretic Approach to Corporate Investment Strategy under Competition", *Financial Management* 22, pp. 241–50; reprinted as Chapter 2 of the present volume.

Spatt, C. and F. Sterbenz, 1988, "Warrant Exercise, Dividends, and Reinvestment Policy", *Journal of Finance* 43, pp. 493–506.

Titman, S., 1985, "Urban Land Prices under Uncertainty", *American Economic Review* 75, pp. 505–14.

Trigeorgis, L., 1996, *Real Options* (Cambridge: MIT Press).

Triantis, A. and J. Hodder, 1990, "Valuing Flexibility as a Complex Option", *Journal of Finance* 45, pp. 549–65.

Van Horne, J.C., 1998, *Financial Management and Policy*, Eleventh Edition (Prentice Hall).

Williams, J.T., 1991, "Real Estate Development as an Option", *Journal of Real Estate Finance and Economics* 4, pp. 191–208.

Williams, J.T., 1993, "Equilibrium and Options on Real Assets", *Review of Financial Studies* 6, pp. 825–50; reprinted as Chapter 4 of the present volume.

Williams, J.T., 1995, "Pricing Real Assets with Costly Search", *Review of Financial Studies* 6, pp. 825–50; reprinted as Chapter 7 of the present volume.

Part I

Simple Introductory Models of Option Games

1

*Strategic Growth Options**

Nalin Kulatilaka; Enrico C. Perotti

Boston University; University of Amsterdam and CEPR

We provide a strategic rationale for growth options under uncertainty and imperfect competition. In a market with strategic competition, investment confers a greater capability to take advantage of future growth opportunities. This strategic advantage leads to the capture of a greater share of the market, either by dissuading entry or by inducing competitors to "make room" for the stronger competitor. As a result of this strategic effect, payouts are in a rough sense more convex than in the case of no investment in a growth option. When the strategic advantage is strong, increased uncertainty encourages investment in growth options: higher uncertainty means more opportunity rather than simply larger risk. If the strategic effect is weak the reverse is true. On the other hand an increase in systematic risk discourages the acquisition of growth options. Our results contradict the view that volatility is a strong disincentive for investment.

The goal of this chapter is to investigate the decision to make an irreversible investment under imperfect competition and uncertainty. The real option literature has led to a significant advance in understanding the valuation of investment relative to the static NPV approach. (See Dixit and Pindyck (1994) for an excellent survey.) However, real option analysis has often been based on two very specific assumptions: (a) the firm has a monopoly over an investment opportunity and (b) the product market is perfectly competitive. As a result, investment does not affect either prices or market structure.

*This chapter was first published in *Management Science* 44(8), 1998, pp. 1021–31 and is reprinted with the permission of Institute for Operations Research and the Management Sciences. The authors wish to thank Marvin Freedman, Michael Manove, Alan Marcus and Stewart Myers for helpful discussions and Avinash Dixit and two anonymous referees for their careful comments. They also wish to thank the Editor and Associate Editor for the considerable patience and inspiration. All remaining errors are the authors' own.

On the other hand it is often recognised that early investment is associated with a greater ability to expand in the future; however, usually these "growth options" have been introduced as exogenous technical advantages and modelled in an *ad hoc* fashion. This chapter seeks to contribute to the theoretical foundations of growth options by explicitly modelling the strategic advantage gained by investment vis-à-vis competitors.

The classic real option framework is an accurate description of investment decisions such as the development of a fully owned natural resource, for which monopolistic access to the investment opportunity is secure and the impact on market structure is minimal. However, when there are other potential competitors, not investing may lead some other producer to seize the opportunity. Moreover, under imperfect competition the commitment of an irreversible investment typically has strategic pre-emptive effects; immediate action may discourage entrants and enhance market share and profits (Gilbert, 1989). In the early contribution by Dixit (1979), investment confers a future cost advantage vis-à-vis potential entrants, creating a strong pre-emptive effect.[1] Interestingly, most of the subsequent literature has not considered the impact of exogenous uncertainty on this strategic decision.[2]

We interpret the strategic value of initial investment as the acquisition of growth opportunities relative to competitors, which will be exercised if market conditions are favourable. We then contrast the value of acquiring such strategic growth options against the alternative of no investment, and offer some novel insights on the impact of uncertainty on the attractiveness of strategic investment.

In this chapter an initial investment results in the acquisition of a "capability" which allows the firm take better advantage of future growth opportunities. Specifically, in our model an initial investment in a growth option reduces future production costs, so that expansion can take place at a lower cost than for competitors without growth options. Examples of strategic investment leading to future comparative advantages may be research into building a technological advantage, an advertising campaign leading to identification and name recognition by consumers, and organisational and logistic planning leading to lower costs in building production capacity. The acquisition of this strategic advantage endogenously leads to the capture of a greater share of the market, either by dissuading entry or by inducing competitors to "make room" for the stronger competitor. This is particularly valuable in states of high demand when profits per unit of output are higher. Thus strategic investment in conditions of uncertainty can be viewed as a commitment to a more aggressive future strategy.

Our main contribution is to show that, contrary to the result found in the real options literature, the effect of uncertainty on the relative value of the strategic growth options is ambiguous under imperfect competition. An important difference under imperfect competition is that profits are convex in demand, since oligopolistic firms respond to better market conditions by increasing both output and prices. Thus expected cashflows increase with volatility, as high marginal revenues at higher levels of demand more than compensate for low revenues at low levels of demand.

When strategic investment has a significant pre-emptive effect, it leads to higher market share, and thus a greater (relative) convexity of *ex post* profits relative to the case of no investment. As a result, even though the value of not investing increases with rising uncertainty, the value of the growth option increases even more. When instead the investment confers only a modest strategic advantage, the potential profit gain is less significant relative to the cost of the investment; an increase in volatility will increase the value of not investing and thus raise the threshold for investment in the growth option. For intermediate levels of strategic advantage, increased uncertainty favours strategic investment unless the probability shift results in a much higher likelihood of entry. Finally, since maximum losses are bounded above by the initial investment, at a very high level of uncertainty a further increase favours strategic investment.

These results are confirmed in the case when more firms have access to the investment opportunity and are robust to various specifications of strategic advantages gained by the investment (Kulatilaka and Perotti, 1991). However, when a systematic risk component to the volatility is introduced, we find that an increase in systematic risk discourages strategic investment, as it leads to higher risk exposure.

The rest of the chapter is organised as follows. It first examines a simple benchmark model of monopoly investment under uncertainty. This allows us to introduce several key features arising from market power. Second it studies the more general case of imperfect competition and the strategic effect of investment in capability. We explore the impact of uncertainty on the value of a strategic growth option and obtain a closed-form solution for the case of a lognormally distributed demand. Third it considers the possibility of simultaneous strategic investment and the effect of systematic risk. Finally it offers some conclusive remarks.

A BENCHMARK MODEL: MONOPOLY INVESTMENT IN A GROWTH OPTION

We consider as a benchmark the extreme but simple case of a firm with monopoly in both the investment opportunity and in the product market. In this context the growth option has no strategic effect. We progressively relax these assumptions later.

At time 0, a single firm (denoted by M) has the opportunity to make an initial irreversible investment of amount I, which confers a capability for more efficient (specifically, lower cost) production.[3] This will turn out to be equivalent to the purchase of a growth option. Until time 1, when the market opens, there is uncertainty over the scale of future demand. We assume that the demand for the good is linear in prices and increasing in the random variable θ. Let P(Q) be the inverse demand function expressing the market price as a function of total supply Q:

$$P(Q, \theta) = \theta - Q$$

where θ is distributed on $(0,\infty)$, with expected value $E_0[\theta] \equiv \theta_0 > 0$. Uncertainty is fully resolved at time 1 prior to production.

If no initial investment is made the firm will produce only when the market is profitable, producing at a unit cost of K. The firm will choose an output level $Q_M^N = \frac{1}{2}(\theta - K)$ with associated profits $\pi_M^N = \frac{1}{4}(\theta - K)^2$; it will not produce if $\theta < \theta_M^* \equiv K$.

In contrast, an initial investment reduces the future unit cost to κ (where $\kappa < K$), due to learning, logistic and product development improvements. If $\theta < \kappa$ the firm will not produce; else, it will choose an output level $Q_M^I = \frac{1}{2}(\theta - \kappa)$ with associated profits $\pi_M^I = \frac{1}{4}(\theta - \kappa)^2$. Thus Q_M^N is less than Q_M^I because investing in the growth option enhances the incentive to expand production.

Notice that these payout functions are continuous, monotonically increasing, and convex in θ. An increase in θ has a more than proportional effect on payouts because a firm with market power responds to higher demand by increasing both output and prices.

We initially assume no systematic risk (or risk neutrality) and a zero interest rate. Then the net present value of the capability investment is:

$$V_M^I \equiv E_0\left[\pi_M^I\right] - I = E_0\left[\tfrac{1}{4}(\theta - \kappa)^2 \middle| \theta \geq \kappa\right] \text{prob}(\theta \geq \kappa) - I$$

The correct investment criterion calls for a comparison between the net present values of making the investment with the NPV of not making the investment. The latter value is

$$V_M^N \equiv E_0\left[\pi_M^N\right] = E_0\left[\tfrac{1}{4}(\theta - K)^2 \middle| \theta \geq K\right] \text{prob}(\theta \geq K)$$

To solve for the optimal investment decision we define the *ex post net* gain to investment (the relative value of investment) as $\Delta^M(\theta) \equiv \pi_M^I - I - \pi_M^N$. Then the net present value of the decision to acquire the growth option (relative to not investing) is the expectation of Δ^M:$G(\theta_0) \equiv E[\Delta^M(\theta)]$. The level of expected demand $\theta_0 = \Theta^M$ such that $G(\Theta^M) = 0$ is then a point of indifference.

It is easy to show that a unique value for Θ^M exists under some simple regularity conditions; namely for the set of distributions with a strictly positive support on θ where higher mean implies first order stochastic dominance (ie, for which given two random variables x_1, x_2, $E(x_1) > E(x_2) \Leftrightarrow x_1$ first order stochastically dominates x_2).

Proposition 1: *Strategic investment is optimal when θ_0 exceeds the unique expected demand threshold Θ^M.*

Proof: From the definition of *ex post* profit functions we know that $\lim_{\theta_0 \to 0} G(\theta_0) = -\frac{1}{4}(K^2 - \kappa^2) - I < 0$ and $\lim_{\theta_0 \to 0} G(\theta_0) = \infty$. It is sufficient to show that $d\{G(\theta_0)\}/d\theta_0 > 0$; then uniqueness is established by the intermediate value theorem. Since Δ^M is an increasing and differentiable function of θ, the condition $d/d\theta_0 \{E\, \Delta[\theta(\theta_0)]\} > 0$ is satisfied for all distributions of θ under consideration. Since $G(\theta_0) > 0$ for all $\theta_0 > \Theta^M$, in this range investment in the growth option has a higher NPV than the alternative of not investing. Figure 1 plots $\pi^I_M - I$, π^N_M, and Δ^M against θ for the case where $\kappa = 0$.

Figure 1 *Ex post* net profit functions: benchmark monopoly case

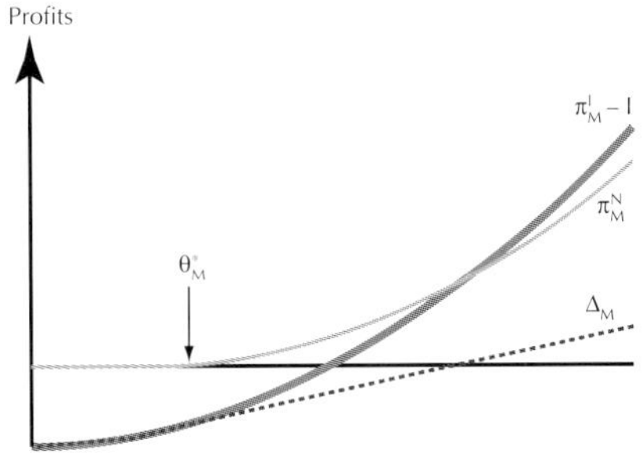

We now examine the impact of greater uncertainty on the relative attractiveness of investment. We define an increase in uncertainty over θ as one which does not affect its mean, thus adopting the concept of a mean-preserving spread as in Rothschild and Stiglitz (1970). Note that both π^I_M and π^N_M are convex and differentiable in θ; so by Jensen' s inequality the impact of a mean-preserving spread increases their expected value. Similarly, if the net gain function Δ^M were convex or concave, the net effect of increasing uncertainty would be unambiguous. However, the net gain function has two kinks; at the points κ and K. The first kink occurs at the level of demand when a monopolist which invested in growth option starts production; the second is the level at which a monopolist which did not invest starts production.

Therefore, the impact of greater uncertainty on the relative value of investment is uncertain because these entry points create discontinuities in the rate of change in marginal profits. As the next section confirms, this ambiguity is always present under imperfect competition since entry has always an impact on price and thus marginal profitability.

STRATEGIC GROWTH OPTIONS UNDER IMPERFECT COMPETITION

We now consider, within the framework of the model in the previous section, the possibility of a competitor entering at time 1. Now investment will have a strategic effect.

Firm 1 has a monopoly over the investment opportunity. (We relax this assumption later.) Firm 1 chooses whether to make a strategic investment at time 0, anticipating its impact on future market structure. A second firm (firm 2) may choose to enter the market at time 1, with a unit production cost of K. We assume that if both firms produce, the market outcome is Cournot competition.[4]

Consider first the case where firm 1 makes no initial investment, so that *ex post* it has no strategic advantage vis-à-vis the competitor. If both firms choose to produce, they face the same production cost K. As long as $\theta \geq K$, the outcome is a symmetric Cournot equilibrium in which they share the market equally; each firm produces an amount $Q_1^N = (\theta - K)/3$, which yield profits equal to $\pi_1^N = \frac{1}{9}(\theta - K)^2$. If $\theta < K$ neither will produce, as the marginal revenue falls below cost. Hence, $\theta^* \equiv K$ can be interpreted as the symmetric Cournot entry point, below which no production takes place.

If instead firm 1 invests I at time 0, market interaction is affected by its strategic advantage, which is acknowledged by firm 2 when making its output decision. It is easy to see that if both firms produce, firm 1 will choose an output level $Q_1^I = \frac{1}{3}(\theta + K - 2\kappa)$ with associated profits $\pi_1^I = \frac{1}{9}(\theta + K - 2\kappa)^2$. It is now optimal for firm 2 to choose a lower quantity, $Q_2^I = \frac{1}{3}(\theta - 2K + \kappa)$, yielding profit, $\pi_2^I = \frac{1}{9}(\theta - 2K + \kappa)^2$. Moreover, the Cournot entry point for the competitor is now higher and is defined by $\theta^{**} \equiv 2K - \kappa$.

Note the source of the market share gained by firm 1 when it invests at 0. Q_2^I is less than Q_1^I for various related reasons: first, because firm 2 faces a higher production cost; second, because it recognises firm 1's greater incentive to expand production (the post-entry dissuasion effect of strategic investment). Finally, firm 2 does not enter unless $\theta > \theta^{**}$, which is higher than θ^*. This is the entry-dissuasion effect. As a result, firm 1 acts as a monopolist for $\theta^{**} > \theta > \kappa$, charging a price $\frac{1}{2}(\theta - \kappa)$ and earning profits $\frac{1}{4}(\theta - \kappa)^2$. In general, the cost advantage derived from strategic investment increases firm 1's market share and profits for all θ. We conclude that strategic investment can be seen as offering an enhanced market share.[5]

In summary, the *ex post* gross profits of firm 1 under not investing (π_1^N) and investment (π_1^I) are given by:

$$\pi_1^N = \begin{cases} 0 & \text{if} \quad \theta < \theta^* \\ \frac{1}{9}(\theta - K)^2 & \text{if} \quad \theta \geq \theta^* \end{cases}$$

$$\pi_1^I = \begin{cases} 0 & \text{if} \quad \theta \leq \kappa \\ \frac{1}{4}(\theta - \kappa)^2 & \text{if} \quad \kappa < \theta < \theta^{**} \\ \frac{1}{9}(\theta + K - 2\kappa)^2 & \text{if} \quad \theta \geq \theta^{**} \end{cases}$$

As before, these payout functions are continuous and monotonically increasing in θ, and profit increases more than proportionately with demand. Note that profits rise faster with demand if firm 1 has invested at time 0. As little generality is gained by a positive κ, we henceforth normalise its value to 0.

Figure 2 Net gain function

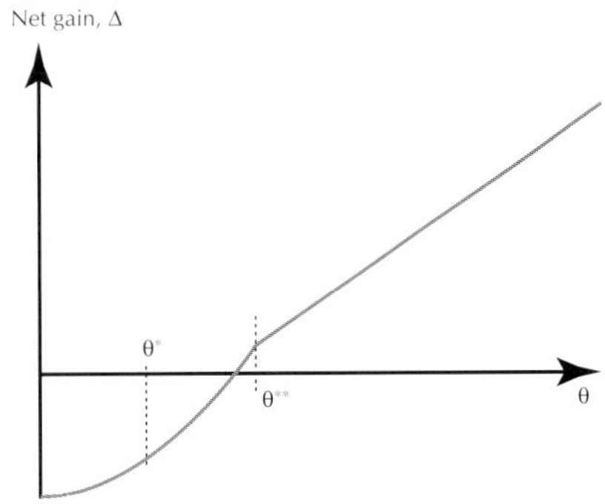

To understand the trade-off in the optimal investment problem we investigate the characteristics of the *ex post net* gain to strategic investment, defined as $\Delta \equiv \pi_1^I - I - \pi_1^N$. Figure 2 plots the function Δ.

We note that Δ exhibits two kinks. The first kink occurs at $\theta^* = K$ and corresponds to the beginning of production in the no-investment, symmetric Cournot case. This is the same as the kink in the monopoly case: at this point marginal profitability is continuous but not differentiable.[6] The relative gain to investment between θ^* and θ^{**} remains convex in θ but profits rise at a lower rate.

A more dramatic discontinuity occurs in the strategic investment case when demand is above the higher entry threshold θ^{**}. Here entry by the competitor results in a drop in the rates of increase of both price and output of the investing firm and creates a discontinuity in marginal profitability. In summary, the net gain function is piece-wise convex in the region of entry dissuasion ($\theta \le \theta^{**}$) and is linear for $\theta > \theta^{**}$.

The optimal investment decision facing firm 1 at time 0 requires comparing the relative NPV of making the strategic investment.

$$V_1^N \equiv E_0\left[\pi_1^N\right] = E_0\left[\tfrac{1}{9}(\theta - K)^2 \middle| \theta \ge \theta^*\right] \mathrm{prob}(\theta \ge \theta^*)$$

$$V_1^I \equiv E_0\left[\pi_1^I\right] - I = E_0\left[\tfrac{1}{9}(\theta + K)^2 \middle| \theta \ge \theta^{**}\right] \mathrm{prob}(\theta \ge \theta^{**})$$
$$+ E_0\left[\tfrac{1}{4}\theta^2 \middle| \theta^{**} > \theta\right] \mathrm{prob}(\theta^{**} > \theta) - I$$

Let $H(\theta_0) \equiv E(\Delta)$ denote the *expected net gain* from strategic investment, $V_1^I - V_1^N$. The threshold value of *expected demand* such that firm 1 is indifferent whether to make the strategic investment or not is denoted by Θ and defined by $H(\Theta) \equiv 0$.

Under the regularity conditions defined in the Proposition 1, it is easy to show that strategic investment is optimal when θ_0 exceeds the unique threshold Θ.

The proof follows that of Proposition 1. Although in this case Δ is non-differentiable at θ^{**}, this does not alter the result as it occurs only at a countable number of points.

We now consider the impact of a mean-preserving spread on the relative value of investment. From the valuation expression it is immediate to see that the value of acquiring the growth option is strictly increasing with the uncertainty over θ. This is a result of convexity of profits created by market power and Jensen's inequality. This suggests that as uncertainty in demand increases the incentive to invest will unambiguously increase. However, as with the real options literature, the NPV of not investing is also increasing with increasing demand uncertainty. Therefore, it is the shape of the difference in *ex post* marginal profits (ie, the curvature of Δ) which will determine the overall effect of uncertainty.

As we saw also in the benchmark monopoly case, since the net gain function Δ is neither convex nor concave the net effect of increasing uncertainty is ambiguous. The intuition is that the rate of change in marginal profits is different to the left and right of a kink. Thus when a mean-preserving spread shifts around probability mass to more extreme values, the position of the expected level of θ relative to the kink matters a great deal.

In the benchmark case this is easy to see. Figure 1 (where the first kink disappears because κ is set to zero) indicates that Δ^M (the NPV of investing over non-investing) is convex for $\theta < \theta^*$ and linear for $\theta > \theta^*$. When the threshold Θ^M is in an area where Δ^M is convex in θ, more uncertainty will increase the incentive to invest; vice versa when the threshold lies above θ^*, where the loss of profit on the downside (where Δ^M is convex in θ) is larger than the potential gain on the upside (where Δ^M is linear in θ).

In the case of imperfect competition, the effect of entry by competitors creates a second kink. Table 1 shows that at the entry point θ^{**} the marginal net gain from investment drops discontinuously. When expected demand is around this point, the downside losses from a spread in probability mass will exceed the upside gain as the profits are convex for $\theta < \theta^{**}$ and linear above this range. For values of θ below θ^{**} the payout to strategic investment is piece-wise convex; thus in this range a mean-preserving spread encourages investment.

Therefore, the direction of the impact of uncertainty on the relative value of investment is critically dependent on the magnitude of the strategic advantage, measured by the cost advantage K. When K is large, the entry

Table 1 Net gain function: $\Delta = \pi_1^I - I - \pi_1^N$

Range	Δ	$\partial\Delta/\partial\theta \equiv \Delta'$	$\partial^2\Delta/\partial\theta^2 \equiv \Delta''$
$\theta < \theta^*$	$\frac{\theta^2}{4} - I$	$\frac{\theta}{2}$	1/2
$\theta^* < \theta < \theta^{**}$	$\frac{5\theta^2}{36} + \frac{2\theta K}{9} - \frac{K^2}{9} - I$	$\frac{5\theta}{18} + \frac{2K}{9}$	5/18
$\theta > \theta^{**}$	$\frac{4\theta K}{9} - I$	$\frac{4}{9}K$	0

dissuasion range of demand ($\theta < \theta^{**}$) is larger, and thus the convex area of Δ is larger. As a result, the impact of higher uncertainty tends to favour investment. The opposite is true for a small K.

The conclusion is that in the case of strategic investment with strong pre-emptive effects, higher uncertainty will tend to decrease Θ and encourage investment; the reverse is true in the case of a weak strategic effect. In general the result is ambiguous and its exact impact depends both on the nature of the market structure *and* the nature of the distribution.

An example: lognormal distribution

We now examine the behaviour of the value functions and the strategic investment threshold under the assumption that θ is lognormally distributed with expected value $E_0(\theta) = \theta_0$ and variance σ^2, ie, $\ln(\theta/\theta_0) \sim N(-\frac{1}{2}\sigma^2, \sigma^2)$. By construction, an increase in σ will not affect the expected value of θ and would, therefore, amount to a mean-preserving spread. In all our simulations, I is normalised to 1, and κ is normalised to zero. Then the value functions have the following analytical solutions:

$$V^N(\theta_0,\sigma) = \tfrac{1}{9}\left[\theta_0^2 e^{\sigma^2} N(d_1) - 2K\theta_0 N(d_2) + K^2 N(d_3)\right]$$

and

$$V^I(\theta_0,\sigma) = \frac{\theta_0^2 e^{\sigma^2}}{4}\left[1 - \tfrac{5}{9}N(d_4)\right] + \tfrac{1}{9}\left[2K\theta_0 N(d_5) + K^2 N(d_6)\right] - I$$

where

$$d_1 = \frac{2\ln\left(\dfrac{\theta_0}{K}\right) + 3\sigma^2}{2\sigma}, \quad d_2 = d_1 - \sigma, \quad d_3 = d_2 - \sigma$$

$$d_4 = \frac{2\ln\left(\dfrac{\theta_0}{2K}\right) + 3\sigma^2}{2\sigma}, \quad d_5 = d_4 - \sigma, \quad d_6 = d_5 - \sigma$$

Table 2 summarises our simulations on the effect of a mean-preserving increases in volatility on the threshold point, Θ. This effect depends on the relative impact of higher uncertainty on V^I and V^N. The table clearly indicates that the net effect depends on K. A high value of K implies that strategic investment has a strong entry dissuasion effect and a marked market share advantage in a contested market.

At high levels of K, strategic investment is preferred even for low expected values of demand, and entry dissuasion is strong so that the competitor's

Table 2 Sensitivity of investment threshold, Θ

	Capacity cost advantage: K – κ					
Cumulative volatility	**0.3**	**0.8**	**1.06**	**1.3**	**1.8**	**3.0**
0.0	7.5000	2.8125	2.1214	2.0638	2.0047	2.0000
0.2	7.5000	2.8127	2.2046	2.0516	1.9785	1.9605
0.4	7.5000	2.8295	2.2677	2.0605	1.9139	1.8539
0.6	7.5000	2.8671	2.3117	2.0682	1.8494	1.7149
0.8	7.5002	2.9040	2.3380	2.0656	1.7887	1.5768
1.0	7.5017	2.9318	2.3495	2.0533	1.7306	1.4518
1.2	7.5058	2.9491	2.3494	2.0337	1.6749	1.3422
1.4	7.5124	2.9569	2.3408	2.0093	1.6221	1.2475
1.6	7.5201	2.9572	2.3266	1.9822	1.5729	1.1662
1.8	7.5272	2.9518	2.3090	1.9541	1.5277	1.0967
2.0	7.5328	2.9427	2.2895	1.9265	1.4868	1.0376
2.2	7.5364	2.9312	2.2697	1.9002	1.4503	0.9875
2.4	7.5381	2.9185	2.2503	1.8757	1.4181	0.9452
2.6	7.5379	2.9055	2.2319	1.8535	1.3900	0.9097
2.8	7.5363	2.8928	2.2150	1.8336	1.3657	0.8799

entry threshold θ^{**} is relatively high. Figure 3 plots the expected values of strategic investment and no investment for such a case at two levels of uncertainty. As uncertainty increases, the value of V^N unambiguously rises; but so does V^I. The overall effect in this case is that the threshold level Θ drops with higher uncertainty. In this case, higher uncertainty leads to greater upside opportunities which outweigh the higher downside risk.

At low level of strategic advantage (low K), the entry dissuasion region is small, and the market share gain is also limited. Then strategic investment is justified only under high expected demand, when entry is almost certain. In this case growth option is less valuable. Interestingly, as volatility increases without bounds, strategic investment again is favoured, as the break-even point starts falling. This is the effect of bounded losses under strategic investment.

At intermediate levels of K, the impact of uncertainty depends on the position of the break-even point relative to the *ex post* entry threshold, θ^{**}, and is in general ambiguous.

These results are consistent with our earlier general conclusions. The choice of not investing becomes more valuable under higher volatility, reflecting a lower risk exposure; however, uncertainty *may be* favourable to investment, when the extra profits from strong market share due to the deterrence or *commitment effect* outweigh its downside risks. The ambiguity of our results arises from the very essence of imperfect competition. Strategic growth options do not exhibit the continuous features of expansion options under perfect competition: now an individual firm's investment decision has a significant impact on market structure and thus on the market price.

Figure 3 Effect of increased volatility on NPV and investment threshold

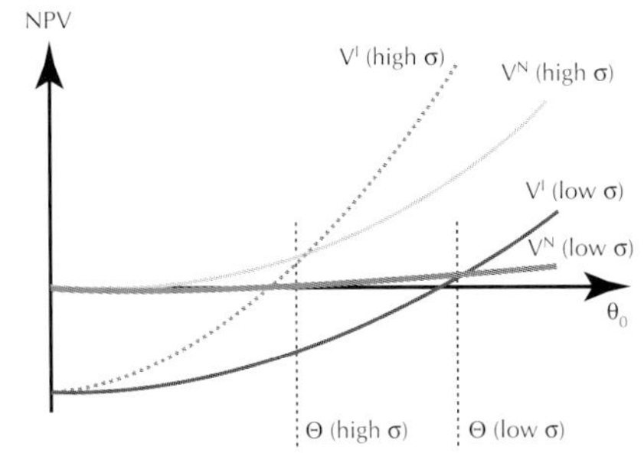

EXTENSIONS

Simultaneous strategic entry

We now extend the context of the basic model to the case when neither firm enjoys a monopoly on pre-emptive investment. In other words, both firms are able to invest at time 0 in order to reduce future capital investment costs to κ.[7]

The payout to investment by one firm depends on the competitor's investment decision and its *ex post* output decision. Since the two firms decide simultaneously, neither can condition its strategy on the other's decision. The final market outcomes are monopoly, asymmetric or symmetric cost Cournot equilibrium and no production, respectively. Each firm's belief about the other's strategy now plays a key role. In order to model *ex ante* identical firms, we ignore equilibria driven by asymmetric beliefs and focus only on symmetric equilibria.

There are three possible symmetric equilibria depending on the expected level of demand, θ_0. Define the following *ex ante* payout functions:

$$\phi_1 \equiv E_0\left[\frac{(\theta - \kappa)^2}{9}\middle|\theta > \kappa\right]\text{prob}(\theta > \kappa)$$

(Simultaneous investment)

$$\phi_2 \equiv E_0\left[\frac{(\theta - \kappa)^2}{4}\middle|2K - \kappa > \theta > \kappa\right]\text{prob}(2K - \kappa > \theta > \kappa)$$

(Strategic investment and no entry by competitor)

$$\phi_3 \equiv E_0\left[\frac{(\theta + K - 2\kappa)^2}{9}\middle|\theta \geq 2K - \kappa\right]\text{prob}(\theta > 2K - \kappa)$$

(Strategic investment and late entry by competitor)

$$\phi_4 \equiv E_0\left[\frac{(\theta - 2K - \kappa)^2}{9}\middle|\theta > 2K - \kappa\right]\text{prob}(\theta > 2K - \kappa)$$

(No investment while competitor invests)

$$\phi_5 \equiv E_0\left[\frac{(\theta - \kappa)^2}{9}\middle|\theta \geq K\right]\text{prob}(\theta > K)$$

(No investment and *ex post* entry by both)

As before, in what follows we set $\kappa = 0$.

Proposition 2: *The optimal investment policy is no investment by both firms when* $\theta_0 \le \theta^R$, *randomised investment by each firm with probability* $y(\theta_0)$ *when* $\theta^R \le \theta_0 \le \theta^S$, *and simultaneous investment by both when* $\theta_0 \ge \theta^S$, *where:*

$$\theta^R \equiv E_0(\theta) \quad \text{s.t.} \quad I + \phi_5(\theta^R) = \phi_2(\theta^R) + \phi_3(\theta^R)$$
$$\theta^S \equiv E_0(\theta) \quad \text{s.t.} \quad \phi_1(\theta^S) = \phi_4(\theta^S) + I$$

In addition, the equilibrium probability of entry, $y(\theta_0)$, for $\theta^R \le \theta_0 \le \theta^S$ is given by

$$y(\theta_0) = -\frac{I + \phi_5 - \phi_2 - \phi_3}{\phi_1 + \phi_5 - \phi_2 - \phi_3 - \phi_4}$$

Proof: See Appendix.

The optimal investment strategy is similar to the basic model. Clearly, higher expected demand favours strategic investment and reduces the value of not investing. As before, the impact of volatility depends on the degree of advantage gained by exercising the strategic investment option. Under a strong strategic advantage, higher volatility shifts downward the threshold for investment. Intuitively, when strategic gain is significant there is a stronger pre-emptive effect; thus the threshold level of expected demand at which investment is preferred, Θ, is low.

In contrast, a weak strategic advantage means that the competitive gain of pre-emptive investment is weaker on the upside, and the area of deterrence is smaller; so the reduction in profitability on the downside is greater. Since the value of not investing *increases* with volatility, the local effect of greater uncertainty in this case will be to discourage investment.

The simulation results in Table 3 confirm these effects of uncertainty on the probability of investment, around a parameter range when a randomised investment strategy may be optimal for both firms.

Table 3 Sensitivity of probability of investment (expected demand = 3.0)

Cumulative volatility	Cost advantage: K – κ				
	0.7	0.8	0.9	1.0	1.1
0.1000	0.00	0.23	0.56	0.75	0.87
0.2000	0.00	0.23	0.56	0.75	0.87
0.3000	0.00	0.23	0.55	0.75	0.87
0.4000	0.00	0.22	0.55	0.75	0.88
0.5000	0.00	0.21	0.54	0.75	0.89
0.6000	0.00	0.19	0.54	0.76	0.91
0.7000	0.00	0.17	0.54	0.78	0.94
0.8000	0.00	0.15	0.54	0.80	0.98
0.9000	0.00	0.13	0.55	0.83	1.00
1.0000	0.00	0.12	0.57	0.87	1.00
1.1000	0.00	0.11	0.59	0.92	1.00

Systematic risk

If an increase in uncertainty over θ induces an increase in non-diversifiable risk, our results need to be qualified. Assume that in equilibrium, the risk premium associated with uncertainty in θ is proportional to its volatility. That is, the equilibrium rate of growth of θ equals $r + \lambda\sigma$, where λ is the market price of risk associated with θ and r is the risk-free rate of interest. Since the payouts to the firm are non-linear functions of θ, the risk adjusted discount rate depends on the realisation of θ which is unknown at time 0. Hence, we transform the valuation problem into its risk-neutral representation, and thereby achieve risk-adjustment via adjustment to probabilities rather than the discount rate.[8]

It can easily be shown that under a risk-neutral probability measure, θ would still be lognormally distributed but its rate of return will fall short of the risk-free rate of interest by adjustment factor:

$$\ln\left(\frac{\theta}{\theta_0}\right) \sim N\left(-\frac{1}{2}\sigma^2 - 2\lambda\sigma, \sigma^2\right)$$

As a result of this risk adjustment, the valuation equations are modified as follows:

$$V^W(\theta_0,\sigma) = \tfrac{1}{9}\left[\theta_0^2 e^{\sigma^2-2\lambda\sigma}N(d_1) - 2Ke^{-\lambda\sigma}\theta_0 N(d_2) + K^2N(d_3)\right]$$

and

$$V^I(\theta_0,\sigma) = \frac{\theta_0^2 e^{\sigma^2-2\lambda\sigma}}{4}\left[1 - \tfrac{5}{9}N(d_4)\right] + \tfrac{1}{9}\left[2Ke^{-\lambda\sigma}\theta_0 N(d_5) + K^2N(d_6)\right] - I$$

where

$$d_1 = \frac{2\ln\left(\frac{\theta_0}{K}\right) + 3\sigma^2 - 2\lambda\sigma}{2\sigma}$$

$$d_4 = \frac{2\ln\left(\frac{\theta_0}{2K}\right) + 3\sigma^2 - 2\lambda\sigma}{2\sigma}$$

Note that the effect of risk-adjustment is equivalent to reducing θ_0, the expected value of θ. From the first section we know that a decrease in expected value, θ_0, will reduce $E(\pi^I)$ more than $E(\pi^W)$. Therefore, the impact of a higher risk premium will be to reduce the relative value of strategic investment. The intuition is that the strategic investment option requires the firm to bear more risk.

CONCLUDING REMARKS

This chapter has proposed a reconciliation of the real options and the strategic approach to the optimal use of strategic investment. The real options literature assumes perfect competition and analyses the effect of exogenous uncertainty. In contrast, the strategic approach endogenises market structure, however, it often ignores uncertainty, which is relevant for valuation even when the firm is risk neutral because of the option value of flexibility.

We show that the proper valuation of real investment must take into account both its strategic value (the pre-emptive effect of commitment) and the alternative value of not investing (a form of flexibility). We interpret the effect of strategic investment as lowering not just production costs but also the strike price of future expansion options. This is because choice also has a *strategic* influence on competitors' output decisions, inducing them to be less aggressive. This increases the investor's market share and, therefore, the value of its expansion option.

Our results on the effect of uncertainty on the valuation of strategic investment may be surprising in the light of current practice, which tends to view volatility as a strong disincentive for new investment.

In a richer intertemporal setting it would be possible to analyse how the *evolution* of strategic considerations should be contrasted with the opportunity cost of waiting to learn more about uncertain market conditions. We plan to do so in later research.

Empirically it may be true that real investment tends to fall when uncertainty rises, however, this is probably due to a concomitant decrease in expectations over market conditions. We intend to extend further our analysis on the valuation of investment in oligopolistic markets, as we are convinced that this line of research will not only improve our theoretical understanding of optimal investment timing, but may also contribute to modern capital budgeting practice and methodology.

APPENDIX

Proof of Proposition 2

Consider the payouts to strategic investment for firm 1. Suppose first that it invests at time 0 while firm 2 does not. If demand next period is so low that late entry by firm 2 is not profitable, ie $\theta < 2K$, firm 1 will be a zero marginal cost monopolist. If $\theta > 2K$, firm 2 enters, but firm 1 earns higher payouts as a low cost, higher market share Cournot competitor. Thus firm 1's expected payout from solitary strategic investment at 0 is:

$$E_0\left[\frac{\theta^2}{4}\middle|\theta < 2K\right] + E_0\left[\frac{(\theta + K)^2}{9}\middle|\theta \geq 2K\right] - I$$

If instead firm 2 enters as well, they will share the market as Cournot competitors, and their expected payout equals $E[\theta^2/9] - I$.

Consider now the payout to not investing. If firm 1 does not invest and firm 2 does, firm 1's expected payout equals the sum of zero when future entry is unprofitable ($\theta < 2K$) and its profits as a high cost entrant otherwise:

$$0 + E_0\left[\frac{(\theta - 2K)(\theta + K)}{9}\middle|\theta \geq 2K\right]$$

Finally, if neither firm enters at 0, their *ex post* payouts are zero if $\theta < K$ and equal to the high cost, symmetric Cournot equilibrium payouts otherwise:

$$0 + E_0\left[\frac{(\theta - K)^2}{9}\middle|\theta \geq K\right]$$

Note next that the relative payout to investment by either firm increases monotonically with θ_0, independently of the competitor's timing of investment. To see this, assume the other firm does not invest; the net gain from entering is

$$E_0\left[\frac{\theta^2}{4}\middle|\theta < 2K\right] + E_0\left[\frac{(\theta + K)^2}{9}\middle|\theta \geq 2K\right] + E_0\left[\frac{(\theta - K)^2}{9}\middle|\theta \geq K\right] - I$$

which can be re-written as:

$$E_0\left[\frac{\theta^2}{4}\middle|\theta < K\right] + E_0\left[\frac{5\theta^2 - 4K^2 + 8K\theta}{36}\middle|K \leq \theta \leq 2K\right] + E_0\left[\frac{4K\theta}{9}\middle|\theta \geq 2K\right] - I$$

In this expression, each term is positive and increasing in θ_0. If the other firm invests as well, the symmetric Cournot payout is clearly increasing in θ_0.

The net gain equals zero at $\theta_0 = \theta^R$, the threshold value such that neither firm will invest. To verify that neither firm wishes to invest if the other firm is not expected to invest, let $y = 0$. Firm 1 will not invest if its net gain from strategic investment is negative. This occurs when

$$\phi_2 + \phi_3 - I < \phi_5$$

which is satisfied for $\theta_0 < \theta^R$, since the net gain is monotonic in θ_0. The reasoning for firm 2 is analogous, so $y = x = 0$ is an equilibrium in this range. When θ_0 is below the threshold, neither firm invests even if it were certain to gain a market advantage. A similar comparison of payouts can be made when firm 2 does invest. The net benefit (loss) of strategic investment rises (falls) with θ_0 and equals zero exactly when θ_0 equals θ^S. When θ_0 is above this threshold, both firms invest.

To verify this, let y = 1. Firm 1 invests today with certainty if:

$$\phi_1 - I \geq \phi_4$$

which is satisfied for all $\theta_0 > \theta^S$.

When θ_0 is in the intermediate region, both firms invest with positive probability. Let $x(\theta_0)$ be the probability of entry by firm 1 and $y(\theta_0)$ the corresponding probability for firm 2. Firm 1 will randomise between investing and not investing as long as it is indifferent between the two choices.

This occurs if:

$$yE_0\left[\frac{\theta^2}{9}\right] + (1 - y)E_0\left[\frac{\theta^2}{4}\middle|\theta \leq 2K\right] + E_0\left[\frac{(\theta + K)^2}{9}\middle|\theta \geq 2K\right] - I$$

$$= yE_0\left[\frac{(\theta - 2K)^2}{9}\middle|\theta > 2K\right] + (1 - y)E_0\left[\frac{(\theta - K)^2}{9}\middle|\theta \geq K\right]$$

In the game with symmetric beliefs, the equilibrium value of $y(\theta_0)$ is equal to $x(\theta_0)$ as described in Proposition 2. It is also increasing in θ_0. To verify this, notice that the numerator of $y(\theta_0)$ as defined above is always negative in range $\theta^S > \theta_0 \geq \theta^R$, and equals the negative of the net payout to strategic investment when the competitor does not invest, which we have shown to be increasing in θ_0. The denominator is also negative and larger in absolute value, since it equals the numerator plus $\phi_1 - I - \phi_4$ (the net gain to strategic investment when the competitors invest) which in this region is negative and decreasing in θ_0.

1 Treatments of the investment decision described costs of postponement in terms of the higher adjustment cost of rapid construction of capacity. Hartman (1972) shows that under convex adjustment costs, higher uncertainty increases the optimal amount of investment even under perfect competition. See also Abel (1984) and Pindyck (1988, 1991).

2 While considerable work has been done on the role of asymmetric information (see Tirole (1988)), the only treatment of uncertain market conditions we are aware of is the partial equilibrium approach of Appelbaum and Lim (1985), who focus on production rather than investment as a form of precommitment. Their ad hoc revenue function drives entirely their results.

3 Our view of investment in growth opportunities is fairly general. Analogous results would obtain if the investment would result in greater quality or consumer appeal.

4 This is consistent with the theoretical work by Kreps and Scheinkman (1983) on capacity choice followed by price competition.

5 We find it useful to decompose the growth option gained by strategic investment in two components. First, it results in a lower "unit exercise price" for future expansion. In addition, the optimal output Q_1^I, "the number of unit production options that are optimally exercised", also increases, as other firms choose to limit their own output to make room for the stronger competitor.

6 Note that for a range of demand beyond this point the relative gain to investment is higher than for the monopoly case. This reflects the fact that not investing leads to a lower payout because of loss of market share to competition, so in this range there is a greater incentive for investment. When demand is quite high the monopolist benefits comparatively more from the cost advantage, while the relative market share gain declines for the investing firm.

7 It is not essential that there are only two firms; however, if we continue to increase the number of potential investors the expected value of the strategic investment option and not investing would be zero.

8 See Cox, Ingersoll and Ross (1985). For instance, in a single factor asset pricing model $\lambda = (\rho_{\theta,M}/\sigma_M)(r_M - r)$, where $\rho_{\theta,M}$ is the correlation coefficient between θ and the market portfolio, and r_M and σ_M are, respectively, the rate of return and its standard deviation.

BIBLIOGRAPHY

Abel, Andrew B., 1983, "Optimal Investment under Uncertainty", *American Economic Review* 73, pp. 228–33.

Appelbaum, Elie and Chin Lim, 1985, "Contestable Markets Under Uncertainty", *Rand Journal of Economics* 16(1), pp. 28–40.

Cox, John, Jonathan Ingersoll and Stephen Ross, 1985, "An Intertemporal General Equilibrium Model of Asset Prices", *Econometrica* 53, pp. 363–84.

Dixit, Avinash, 1989, "Entry and Exit Decisions under Uncertainty", *Journal of Political Economy* 97 (June), pp. 620–38.

Dixit, Avinash, 1980,"The Role of Investment in Entry Deterrence", *Economic Journal* 90, pp. 95–106.

Dixit, Avinash, 1979, "A Model of Duopoly Suggesting a Theory of Entry Barriers", *Bell Journal of Economics* 10(1), pp. 20–32.

Dixit, Avinash, K. and R.S. Pindyck, 1994, *Investment Under Uncertainty*, Princeton University Press.

Gilbert, Richard J., 1989, "Mobility Barriers", in Schmalensee, R. and R. Willig (eds), *Handbook of Industrial Organization*, North-Holland.

Hartman, Richard, 1972, "The Effects of Price and Cost Uncertainty on Investment", *Journal of Economic Theory* 5, 258–66.

Kulatilaka, Nalin and Enrico Perotti, 1991, "Strategic Investment Timing Under Uncertainty", Working paper, London School of Economics.

McDonald, Robert and Daniel Siegel, 1986, "The Value of Waiting to Invest", *Quarterly Journal of Economics* 101, pp. 707–27.

Pindyck, Robert, 1988, "Irreversible Investment, Capacity Choice, and the Value of the Firm", *American Economic Review* 78(5), pp. 969–85.

Tirole, Jean, 1988, *The Theory of Industrial Organization*, MIT Press, Cambridge, Massachusetts.

Rothschild, Michael and Joseph Stiglitz, 1970, "Increasing Risk I: A Definition", *Journal of Economic Theory* 2, September, pp. 225–43.

2

A Real Options and Game-Theoretic Approach to Corporate Investment Strategy Under Competition*

Han T.J. Smit and L.A. Ankum

Erasmus University Rotterdam, University of Amsterdam

An investment strategy encompasses a sequence of tactical investment projects, of which several may yield a low return when considered in isolation. Often the value of such an investment consists of the option to invest in the future growth of the firm. For example, the value of a pilot project or an R&D investment does not derive so much from the exact cash inflows, but rather from the option to invest in future commercial exploitation. Standard forecasting of the expected cash inflows implicitly assumes investing in the follow-up project. Therefore, the traditional discounted cashflow (DCF) method has serious shortcomings in analysing projects when information concerning future investment decisions is not yet known. The application of option theory can be used as an analytical tool to evaluate such projects and to support the overall operating and investment strategy. Brennan and Schwartz (1985) examine the operational policy of a copper mine. Myers (1987), Trigeorgis (1986), McDonald and Siegel (1986), Majd and Pindyck (1987) and Myers and Majd (1990) provide other examples of flexible investment strategies.

We consider the option to defer investment in production facilities analogous to a call option on a dividend-paying stock. In this investment strategy, decisions involving the creation of capacity may be postponed so that

*This chapter was first published in *Financial Management* Autumn, 1993, pp. 241–50 and is reprinted with the permission of Financial Management Association International. The authors would like to thank Lenos Trigeorgis, Winfried Hallerbach, Tony Bouwan, Henk Berkman, Michiel Zwijnenburg and two anonymous referees for their helpful comments. Any remaining errors are the responsibility of the authors. Financial support from the Netherlands Organisation for Scientific Research (NWO) is gratefully acknowledged.

management can decide not to invest if market demand turns out to be unfavourable. On the other hand, deferral also has disadvantages since during the postponement period the firm misses the net operating cash inflows. The analogy with a call option, of course, is not exact. One difference between financial call options and future investment possibilities is the exclusiveness of the latter. Kester (1984) and Trigeorgis (1991) have shown that competition in the market may force the company to invest early, which may erode the flexibility value of a deferred investment strategy.

This chapter casts the real options approach for project timing in a microeconomic framework. We use microeconomic tools to analyse aspects of competition in the investment strategy in forecasting the cash inflows. In this connection, we focus particularly on the concept of economic rents, ie, the excess profit above the opportunity cost of capital. Economic rents attract new entrants to the market so that if the investment opportunity is not exclusive new entrants will diminish returns until expected and required returns are equal. Under rivalry, economic rents only exist if the firm has a specific competitive advantage in the realisation of the project. The firm, therefore, needs to identify those markets in which it has a temporary or permanent competitive advantage and concentrate investment in these areas. If the competitive advantage is temporary, expected economic rents will decline over time and postponement will erode the value of the project. Based on the firm's strength in relation to its competitors and the value of the project in relation to market uncertainty, we develop various investment tactics.

The chapter is organised as follows. We first discuss the options approach to investment timing. We study the concept of economic rents for strategic planning purposes, as well as the influence of competition on the expected economic rents over time. Later, we use an options approach for timing investment strategy in different microeconomic market contexts. We finish with some concluding remarks and directions for further research.

THE OPTION TO DEFER INVESTMENT

Some investment projects can actually be seen as the first links in a chain of subsequent investment decisions. With projects of this type, the firm essentially acquires an option to invest in a potential follow-up project. For example, an R&D project, the development of a new technology, or entry into a new geographical market may create future investment opportunities. In strategy, these projects are often compared with options for future company growth (eg, see Myers, 1987). We consider the option to defer a project (see Kester (1984) for a qualitative approach, and Trigeorgis (1986) or Kemna and Vorst (1987) for a quantitative treatment). The deferment of a capacity expansion programme caused by uncertain market demand is seen to create flexibility for management. The timing of investment in a production facility is analogous to the timing of exercising of a call option on a dividend-paying stock.

Figure 1 Analogy of a deferred project with a call option

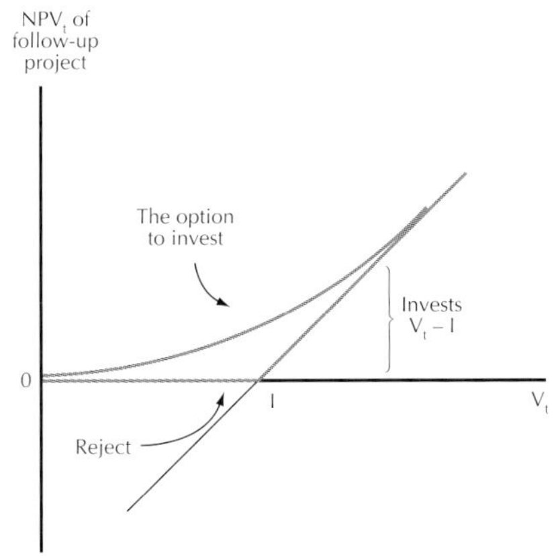

Figure 1 illustrates this analogy. The underlying value is the present value of the net operating cash inflows, V_t. The (present value of the) investment outlay in the plant, I, is equivalent to the exercise price. If, in time, market demand develops favourably, the firm would invest and the net value of the project would equal its NPV = V_t – I. If the project does not prove to be lucrative and the net present value turns out to be negative, management may decide not to invest and the value is zero. Besides this advantage to wait and see, deferment also has disadvantages. For example, if we consider a project with an infinite life, management misses the net operating cash inflows when the plant is not operative. The missed net operating cash inflow during the deferment period is equivalent to a dividend.

Based on this analogy, we can use option valuation techniques to value a deferred project and support the investment strategy. The binomial option valuation method of Cox, Ross and Rubinstein (1979) is a useful tool in this regard. We subsequently express the missed cash inflow during the postponement period as a constant proportion of the state project value. The dynamics of market demand for the product result in a series of possible project values, $V_{t,s}$, in each state (s) over time (t). The investment timing problem is solved recursively, starting with future values and working backward. The investment opportunity value is equal to the maximum of $V_{t,s}$ – I when management invests, zero when management decides not to invest, and the option value when management defers the project. The essence of the investment timing strategy is to find for every decision moment a critical project value V_t^*. At this critical value, the missed net operating cash inflow equals the flexibility value from postponing one more period. As long as the value of the project is below this critical value, the project will be deferred. Should the project value exceed the critical value, management would invest.

In asset valuation, we create shareholders' value by investing in projects for which the financial market value (of the cash inflows) exceeds the investment outlays. Therefore, we try to determine what the project would be worth if it were traded in a financial market. Option valuation models are often based on arbitrage arguments with an option equivalent. An equivalent portfolio strategy, consisting of a position in the underlying asset partly financed with a risk-free loan, is constructed so that it has in every state the same payoff as the option, and should therefore have the same value. In the real options case, the financial market valuation of the postponed project is determined by a project equivalent in the financial market. We could justify the portfolio equivalent approach if a financial instrument is traded with exactly the same risk characteristics as the project. If the underlying state variable is taken to be the price of a commodity, for example, an equivalent portfolio could be constructed with futures in the commodity (eg, see Brennan and Schwartz, 1985, for the valuation of a copper mine). In the case of postponing a plant for consumer products, however, the underlying asset is an aggregate variable. If an identical plant is not explicitly traded, application of the option valuation method seems questionable. However, if the financial markets are complete, in that the securities traded are sufficient for dynamic spanning of the underlying asset, the option valuation method can still be applied (see Mason and Merton, 1985). In complete financial markets, there would exist portfolios of securities that replicate the dynamics of the present value of the project caused by the rate of change of prices and market demand.

As noted, the analogy between a call option and a deferred project is not exact. A major difference is the degree of exclusiveness of the latter, so the call option analogy must be seen in the context of market structure (eg, see Kester, 1984 and Trigeorgis, 1986). A project in a monopoly situation is more analogous to a call option, since it involves an exclusive right to invest. A project under perfect competition, on the other hand, is like a "public good" of the whole industry. In this case, there is a loss in value from postponement caused by the entry of competitors. Emerging competition or rivalry may thus create an incentive to invest early, as postponement of the project may result in project value erosion. This is particularly so if early investment would pre-empt competitive entry (eg, Kester, 1984, Trigeorgis, 1991).

CORPORATE STRATEGY AND ECONOMIC RENTS

In an efficient financial market, the prices of all traded securities adjust rapidly to reflect all currently available information. If information about a firm's profits or new projects arrives in the financial market, investors bid up the prices until the return equals that on investments with comparable risk. Real markets are often less than perfectly competitive, so it is possible for a firm to consistently earn excess returns that exceed the opportunity cost of capital (see Shapiro, 1991). We estimate the expected net operating cash inflow as the opportunity cost of capital plus the expected economic rent:

$$\overline{CF}_t = (I)i + \overline{ER}_t, \qquad t = 1, 2, 3, \ldots, \infty \tag{1}$$

where

$\overline{CF}_t$ = the expected net operating cash inflow
I = the investment cash outflow
i = the opportunity cost of capital
$\overline{ER}_t$ = the expected economic rent

In equation (1), (I)i reflects the yearly opportunity cost of capital invested in a project with an infinite life.

Various microeconomic models can help in forecasting the excess economic rents or expected profit. In a competitive market characterised by costless entry and exit and homogeneous products, an early investment can produce only temporary economic rents. The expected economic rents can have several sources; for example, they can result from a new product introduction. Eventually, however, potential competitors will catch up and enter the industry. When the industry settles into long-run competitive equilibrium, all projects are expected to just earn their opportunity cost of capital and their expected net present value will just be zero. Economic rents are not always temporary. If, for example, the firm has a proprietary technology protected by enforceable patents, the economic rents could be persistent. As long as barriers to entry remain, the firm has an exclusive or proprietary investment opportunity. Appendix A describes the modelling of economic rents under perfect competition, monopoly and oligopoly.

Under rivalry, economic rents would exist if the firm has a competitive advantage in realising the project. Barriers to entry or a distinct competitive advantage over existing competitors (eg, economies of scale and scope, absolute cost advantages or product differentiation) are the real source of economic rents. The firm therefore needs to identify those markets in which it has a temporary or permanent competitive advantage, and concentrate investment in these areas. Understanding potential barriers to entry helps identify these markets and potential value-creating investment opportunities. One strategy, for example, will be directed towards increasing volume because of economies of scale or broadening when economies of scope are important in the market (see Hall, 1980).

The present value of a project with an infinite life and constant expected cash inflows, when the cost of capital is i, equals $\overline{V}_t = \overline{CF}_t / i$. At the end of the period, the present value equals $V_t^{inc} = (1 + i)(\overline{CF}_t / i)$. The difference between these present values reflects the net operating cash inflow during the deferment period, $CF_{t,s}$. We subsequently express this cash inflow as a proportion of the state project value in the binomial option model, $\delta_{t,s} V_{t,s}^{inc}$, so the present value, $V_{t,s}$, of the project equals $(1 - \delta_{t,s}) V_{t,s}^{inc}$. Since there is no decline in expected economic rent and expected cash inflow under monopoly, a postponed project in monopoly is analogous to a call option on a stock with a constant dividend payout ratio, $\delta_{t,s}$:

$$\delta_t = \frac{CF_t}{V_t^{inc}} = \frac{i}{(1 + i)'}, \qquad t = 1, 2, 3, \ldots \infty \tag{2}$$

Under perfect competition, the expected net operational cash inflows will change over time until the project earns just the opportunity cost of capital.[1] The deferred project under perfect competition is thus analogous to a call option on stock with a changing dividend payout ratio:

$$\delta_{t,s} = \frac{CF_{t,s}}{V_{t,s}^{inc}} = \frac{(I)i + \left\{1 - \frac{1}{(1 + i)} e^{-d}\right\}\{V_{t,s}^{inc} - (1 + i)I\}}{V_{t,s}^{inc}}$$

$$t = 1, 2, 3, \ldots \infty \tag{3}$$

In equation (3), (I)i reflects the yearly opportunity cost of capital, while $\{1 - (1/(1 + i))e^{-d}\}\{V_{t,s}^{inc} - (1 + i)I\}$ reflects the economic rent in each state of nature. Note that when the industry settles into competitive long–run equilibrium, the net present value of the project becomes zero ($V_{t,s}^{inc}/(1 + i) = I$), and $\delta_{t,s}$ equals $i/(1 + i)$ as with constant expected cash inflows. If the projects have a positive net present value, competitors will enter and $\delta_{t,s} > i/(1 + i)$. If instead projects have negative net present values, competitors will exit and $\delta_{t,s} < i/(1 + i)$.

The number of firms active in an industry characterises the intensity of competition. In monopoly, only one firm operates in the market, whereas in perfect competition, the number of firms is very large. Oligopoly is between these two extremes. In the case of duopoly, two firms are operating so that the behaviour of each individual competitor directly influences the value of the project. The investment timing strategy of an incumbent firm in duopoly is coupled to the timing tactics of the rival. In the case of an early investment, it is sometimes possible to avoid a decrease in value by pre-empting the competitor. The threat of pre-emption is primarily determined by the strength of the firm and the intensity of competition in the industry (eg, see Tirole, 1990).

Applying basic principles from game theory, we can develop different investment tactics under each market structure. In duopoly, the timing of a project will be influenced by the development of market demand and the behaviour of the competitor. Instead of the maximum of $(V_{t,s} - I, 0)$ and the deferment value in the binomial option model, the state value would now be equal to the (Nash equilibrium) outcome of a simultaneous investment subgame. Figure 2 presents an example of this subgame in the normal form (see Figure A1 for the option pricing model in extensive form). There are four possible situations: (i) both firm A and firm B decide to invest; (ii) A decides to invest but B waits; (iii) A waits but B invests; and (iv) both A and B wait. The first number in each cell indicates the investment opportunity value for firm A, while the second number is the value of a competing project of firm B. The investment opportunity net value is estimated as the net present value, $V_{t,s} - I$, for a direct investment, or the option value, $C_{t,s}$, for a postponed investment. Note that each value depends on the tactic of the other firm (wait or invest), particularly if an early investment preempts a competitor or leads to a cost advantage:

Figure 2 Simultaneous investment subgame

		Firm B: Invest	Firm B: Defer
Firm A		I	II
	Invest	$V_{t,s} - I \quad V_{t,s} - I$	$V_{t,s}^{leader} - I \quad C_{t,s}^{follower}$
		III	IV
	Defer	$C_{t,s}^{follower} \quad V_{t,s}^{leader} - I$	$C_{t,s} \quad C_{t,s}$

$$(V_{t,s}^{leader} > V_{t,s}, \text{ or } C_{t,s}^{follower} < C_{t,s})$$

INVESTMENT TIMING STRATEGY AND ECONOMIC RENTS

A crucial strategic choice for competing in emerging industries is the timing of investment in capacity. We next utilise a simple numerical example to illustrate the timing strategy, first under perfect competition and then under monopoly. The firm in this example has invested in a pilot project to get an early foothold in the market. If the market opens up and creates expansion opportunities, the firm will increase capacity significantly by investing in a follow-up project. This pioneering strategy involves high risk, but also involves low competition in the early stage of the market leading to substantial expected economic rents. Here we assume:

- An industry portfolio is traded in the financial market with exactly the same risk characteristics as the project, yielding an expected return of 16%. The dynamics in the present value of the follow-up project imply a 25% increase ($u = 1.25$) or a 20% decline ($d = 0.8$) per year. This represents a standard deviation in project present value of 22.315% per year. The real rate of interest is 5% per year ($r = 1.05$).
- The expected economic rent of the follow-up project in the first year is 10. Due to anticipated entry, the cash inflows will decline exponentially at a 30% annual rate until the project earns the opportunity cost of capital of 16%.
- The production facility has an infinite physical life. The required investment outlay is 500.

Given the declining expected economic rents and the opportunity cost of capital of 16%, the present value of the expected operating cash inflows from the production facilities is estimated to be 524.[2] The net present value

(NPV) therefore is 524 – 500 = 24. Based on the net present value rule, management should invest immediately.

However, when we consider the option to defer capacity expansion and wait to see how the market develops further, management should postpone. The dynamics of market demand for the product result in a series of possible project values over time. Based on the binomial option pricing model, the value of the deferred project may increase by a factor of $u(1 - \delta_{t,s})$ per period if market demand turns out to be favourable, or it may decrease by a factor of $d(1 - \delta_{t,s})$ per period if market demand turns out to be unfavourable, ie,

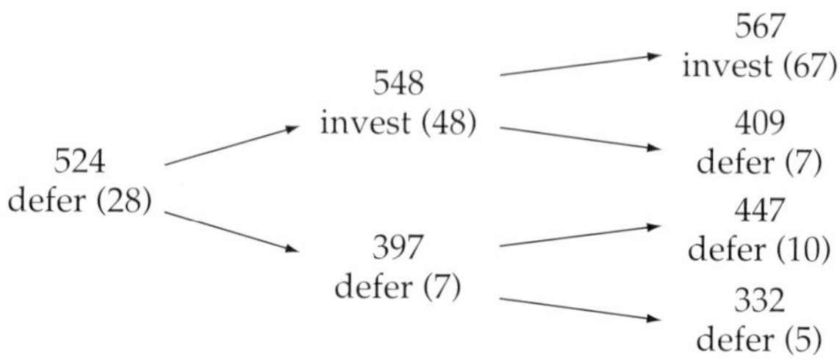

The net value (inside the parentheses) at each state is equal to the maximum of:

$$V_{t,s} - I \qquad \text{invest}$$
$$0 \qquad \text{stop}$$
$$C_{t,s} = \frac{pC_{t+1,u} + (1 - p)C_{t+1,d}}{r} \qquad \text{defer}$$

where $p = (r - d)/(u - d)$ is the risk-neutral probability.[3]

The value of the option to invest in the follow-up project is 28, which is greater than the net present value of 24. The optimal strategy, therefore, is to defer the follow-up project until the market develops favourably.

If we consider a project with the same economic rent, ER_1, in a monopoly, the expected net operating cash inflows are constant. The present value of the production facilities is now 563, and the NPV equals 563 – 500 = 63. When we consider the option to defer capacity expansion, management should invest immediately:

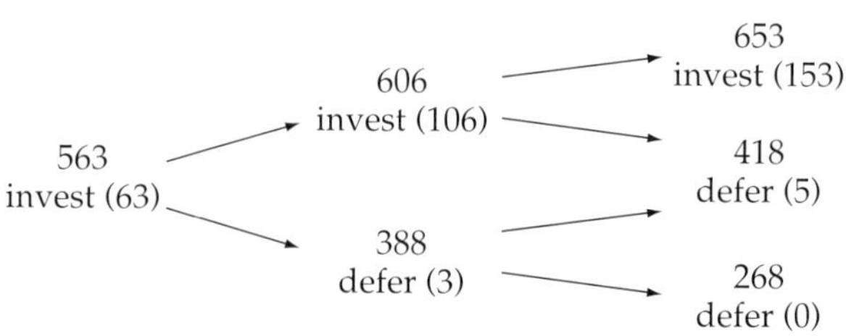

Figure 3 Investment timing strategy under perfect competition and monopoly

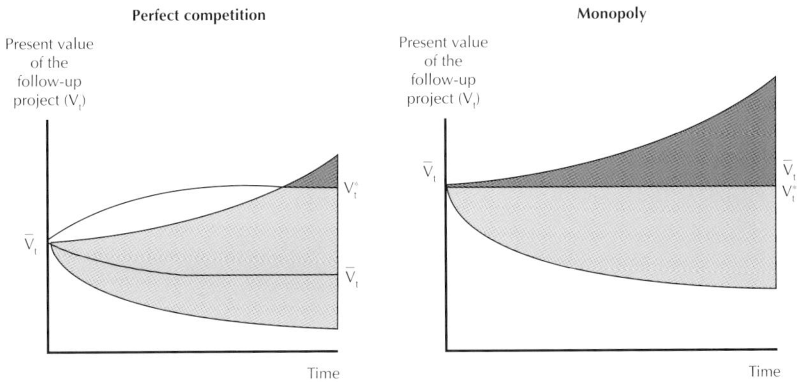

Figure 3 illustrates how the dynamics in present value, V_t, and the critical value above which it becomes optimal to invest, V_t^*, relate to each other. In the case in which the firm shares the investment opportunity with competitors, anticipated entry causes a decline in the expected present value of the project, V_t, and a low critical value, V_t^*. The absence of a structural competitive advantage may result in a tendency to invest early to preclude erosion in the value of the project. The figure on the right-hand side of Figure 3 shows how the dynamics in present value and the critical value relate to each other in the case of an exclusive investment opportunity. Under monopoly, postponement of the project does not necessarily imply a decrease in expected value. Assuming the project has an infinite life, the critical value would be constant over time. In comparison to perfect competition, there is a stronger tendency to postpone projects under monopoly with relatively low net present values. However, if the project has a relatively high NPV due to permanent economic rents, there would be an incentive to invest early.

Now consider the timing of the follow-up project in a duopoly market. The expected net operating cash inflows are constant until a competitor invests in a competing project. However, in the case of an early investment, it is sometimes possible to avoid a decrease in value by pre-empting the market. Instead of the maximum of ($V_{t,s}$ – I, 0) or the deferment value in the binomial option model, the state value would now equal the (subgame perfect equilibrium) outcome of a simultaneous investment subgame.

Consider an example with two identical firms pursuing similar tactics for investment in the follow-up project.[4] Figure 4 illustrates the net values of a two-period game in extensive form. The actions to invest (I) or defer (D) are reflected by the branches. When both firms A and B invest, the game ends; when both firms defer, nature (N) moves and the game is repeated; and when one firm invests first, nature moves and the follower

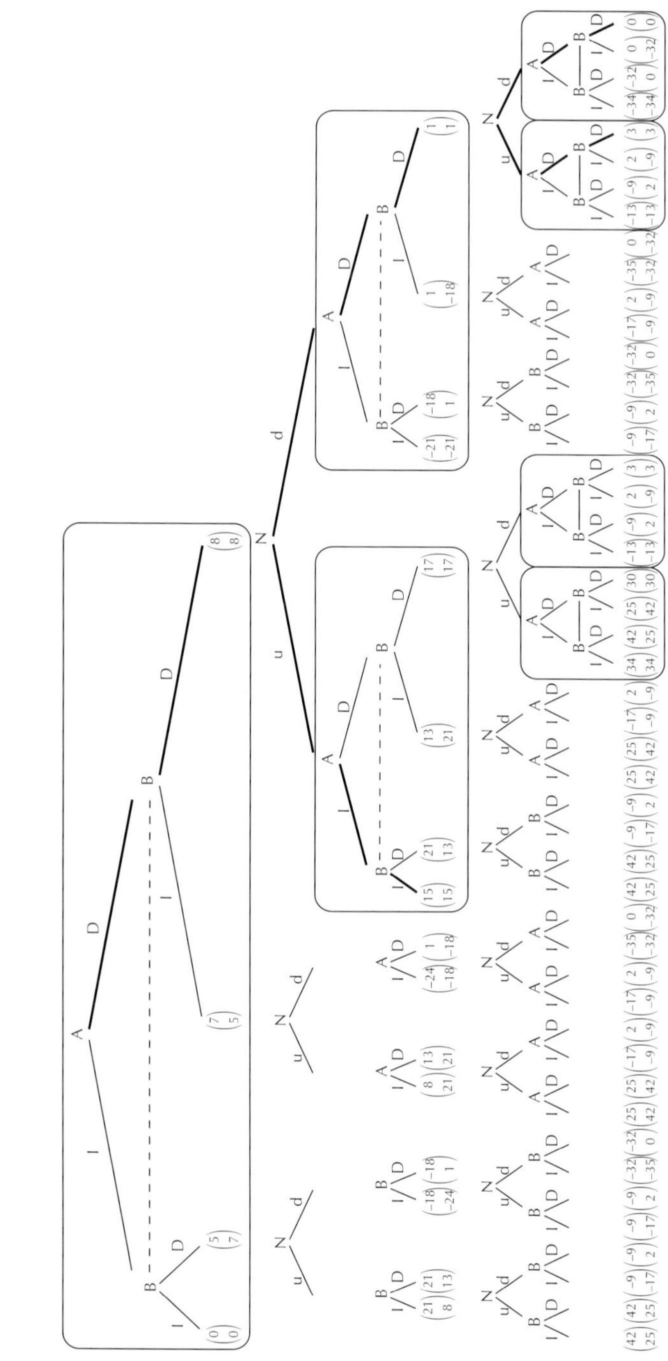

Figure 4 Project value outcomes of a game in extensive form when two competitors have symmetrical market power

may then decide to invest later. The strategy of each firm consists of mapping the information set about the competitor's actions and the development of market demand (N moves) to an action by the firm. The subgame perfect equilibrium set of strategies can be found by backward induction, starting from the future project values for both firms and working back along the tree. The bold branches along the tree reflect the optimal actions along the equilibrium path. In this example, both firms have an identical strategy, ie, deferring (D) the follow-up project and investing (I) if, in the early stage, market demand turns out to be favourable (ie, N moves up); or deferring the project further if market demand turns out to be unfavourable (ie, N moves down) and continuing deferment in the later stage of the market (N moves down, up or down, down).

Consider next the subgame following initial deferral (D) by both firms and favourable demand (N moves up) in which both firms follow a strategy to defer (D) and market demand turns out to be favourable (N moves up). In this case, a decision to invest has a higher net value for the follow-up project, compared to deferral, regardless of whether the competitor decides to invest or not (34 > 25 and 42 > 30). Both firms have a strictly dominant strategy to invest, resulting in net present values of (34, 34) for both follow-up projects. If the market develops unfavourably in the last stage (N moves down), the dominant strategy for both firms is to defer, leading to (3, 3) values. Calculating the state values recursively along the node results in (17, 17) option values of the investment opportunities when both defer (D) in the earlier stage of the market.[5] The dilemma of this subgame (A, B defer, N moves up) is known as the prisoner's dilemma. Firm A's dominant strategy is to invest whether the competitor decides to invest or not. The dominant strategy for firm B is also to invest immediately (15 > 13 and 21 > 17). Thus, both firms will choose to invest. The paradox is that the outcome (15, 15) is worse for both, compared with the situation when both defer (17, 17). If both firms coordinated their investment strategies, they would be better off in the latter situation. If market demand develops favourably (N moves up), both firms would invest and the NPV of the projects are (15, 15), whereas, if the market develops unfavourably (N moves down), both firms would defer and the option values of the investment opportunities are (1, 1). Using these outcomes in the backward valuation process results in (8, 8) option values for the investment opportunities when both firms defer in the first stage.

In the previous example, we assumed identical tactics and equal market power for the two competitors. Of course, the strengths of each firm in an industry are not always identical. Subsequently, we focus on the influence of asymmetrical market power on a firm's investment tactics. A dominant position in the industry may give a competitive advantage to the firm in the realisation of project value. Consider, for example, the case where backward induction leads to the asymmetrical game between a dominant and a weak firm shown in Figure 5.[6]

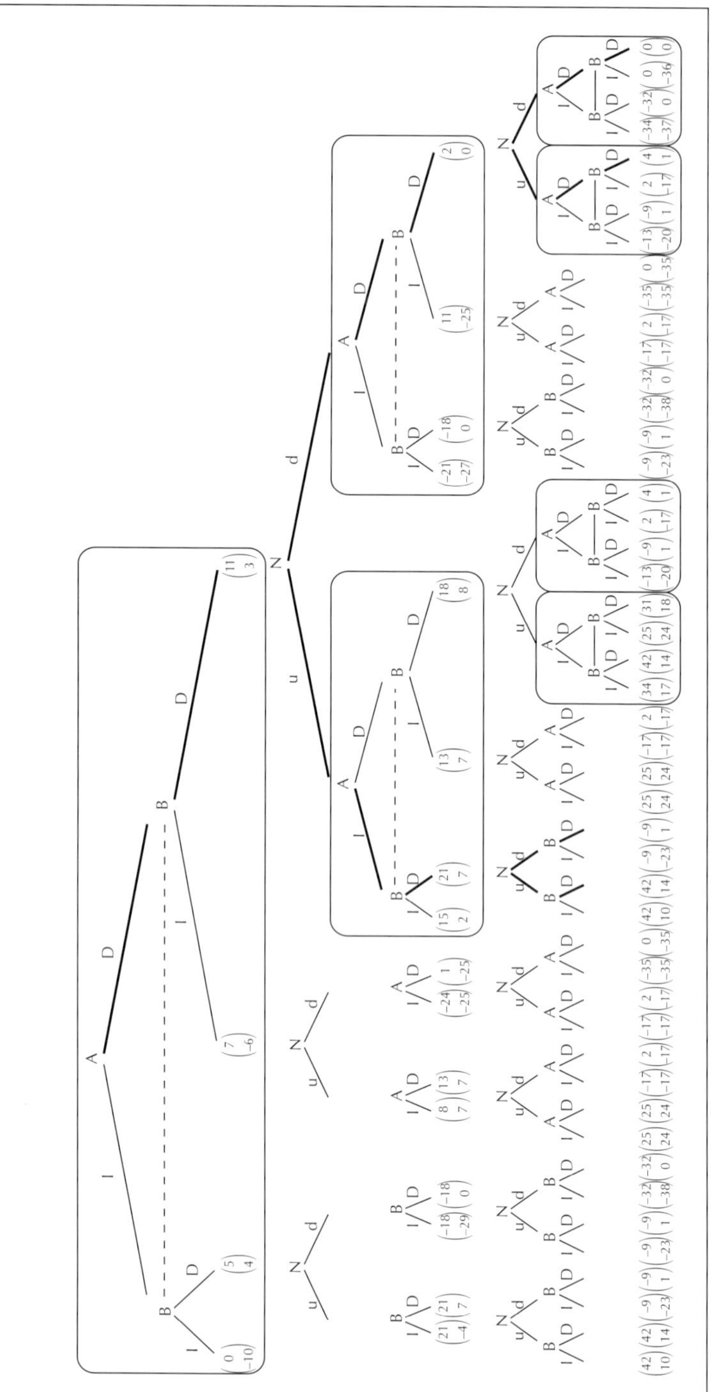

Figure 5 Project value outcomes of a game in extensive form when two competitors have asymmetrical market power

When both firms defer in the early stage (D) and market demand develops favourably (N moves up), the competitive advantage of firm A results in a higher net present value from immediately investing in the project and a lower value of waiting. Thus, for firm A, the dominant strategy is to invest early, irrespective of the strategy of competitor B ($15 > 13$ and $21 > 18$). Because a comparable project will have a lower value for the weaker competitor, the competitor will not invest first. Knowing that firm A will invest, the dominant strategy for firm B is to defer ($7 > 2$). In this example, the two firms pursue different strategies. Dominant firm A invests and pre-empts the market (receiving 21) if, in the early stage, market demand turns out to be favourable, or further defers if market demand turns out to be unfavourable. The preferred strategy for the weak firm is in each case to defer the project, waiting until the market develops sufficiently before entering with its project.

In general, market power can influence the threat of pre-emption. Typically, projects wit low present value from immediate investment would be postponed. The low project value of weaker competitors gives them little ability to pre-empt a stronger competitor with high project value. The market power of a firm may also influence the value erosion associated with pre-emption of a competitor. By investing early, a weak competitor may only gain a small share of the total market, but the potential loss in the value of the project would be minimal. The loss in value, if the firm becomes a follower, will be larger as the market power of the leader grows, because then an early investment can take significant market share away.

Based on the market strength of a firm in relation to its competitors and the value of the project in relation to market uncertainty, we can distinguish the following investment tactics:

- Projects with relatively small net present values from immediate investment in uncertain markets have relatively larger decision flexibility value favouring postponement. If the firm has a dominant position in the industry, there is little threat of complete pre-emption by competitors. The firm can safely postpone the project, and invest only if the market develops favourably or if the weaker competitor invests first. It is even possible that a dominant firm can delay investing even when a weaker competitor invests. This could be the case if the competitor can test the market without taking away a significant market share.
- Projects with relatively large net present values in stable market environments have relatively small flexibility values. The opportunity cost of deferment in the form of missed cash inflows during the deferment period is high, and so the dominant firm (firm A in the previous example) will invest early.

- ❑ If the firm has a weak position in the market and the project does not have a large net present value, it may be better to wait until the market develops sufficiently in the future to start the project (firm B in the previous example).
- ❑ If the firm has a weak market position but the project appears to have a positive net present value, it would invest immediately if early investment can pre-empt a competitor or create a cost advantage. However, because of the weak position of the company, a stronger competitor may come in and erode its net value (even turn it negative.)

CONCLUSIONS

The net present value (NPV) method has serious shortcomings in analysing projects when future decisions are contingent on intermediate developments in an uncertain environment. Option theory provides a better analytical tool to evaluate such projects. One important difference between a real investment opportunity and a call option is the exclusiveness of the latter. If we use the analogy with a call option to value an investment opportunity, we have to take this difference into account.

Using simple numerical examples, we have illustrated the influence of competition on project value and investment timing. Postponement under perfect competition implies a loss in the expected value of the project due to anticipated competitive entry. Absence of a structural competitive advantage may thus result in a tendency to invest early if the firm can preclude this erosion of value. An investment opportunity in a monopoly is exclusive, so during the postponement period there is no expected loss in value due to competition. In comparison with perfect competition, there is a stronger tendency under monopoly to postpone projects with relatively low net present values. On the other hand, an exclusive project with large present value creates a tendency to invest early.

Oligopoly lies between these two extremes. Here, a few firms are operating in the industry with individual market power. In duopoly, it may be attractive for both firms to defer investment when there is low project value and uncertain market demand. If the firms can coordinate their investment, both will postpone. However, as soon as one of the firms invests, the other firm will follow suit with a similar project. If competitive rivalry is intense and there is no coordination, both firms will invest immediately, which may be suboptimal. In the case where firms in the industry have asymmetric market power, we have distinguished various investment tactics based on the firm's strength in relation to its competitors, and the net present value from immediate investment in relation to market uncertainty.

In general, a firm will postpone projects when net present value is low, market demand is uncertain, and interest rates are high. In contrast to a financial option, the degree of non-exclusiveness of the investment opportunity influences the investment strategy. When there are many competitors, each with a negligible market power, anticipated entry can erode the value of the project. With a few competitors having individual market power, there may be a threat of complete pre-emption. Future research can be directed toward applying the option valuation approach in real cases. Extension of the theory can be directed toward modelling option pricing in combination with dynamic game theory and incomplete information. However, with higher mathematical complexity, much of the intuition of viewing projects as real options may be sacrificed.

APPENDIX A

FORECASTING CASH INFLOWS IN DIFFERENT MARKET STRUCTURES

This appendix relates cash inflows to economic rents. We estimate the expected net operating cash inflow as the opportunity cost of capital plus the expected economic rent, according to equation (1). We assume that (i) the financial market is efficient; and (ii) the project has an infinite physical life.

Perfect competition

In perfect competition, economic rents are expected to decline exponentially, as follows:

$$\overline{ER}_t = (\overline{ER}_0)e^{-dt} \qquad t = 1, 2, 3, \ldots\infty \qquad \textbf{(A1)}$$

Monopoly

In monopoly, the expected economic rents are constant.

$$\overline{ER}_t = \overline{ER}_1 \qquad t = 1, 2, 3, \ldots\infty \qquad \textbf{(A2)}$$

Duopoly

In oligopoly, the timing of a project will be influenced by the behaviour of a competitor. Instead of the maximum of ($V_{t,s}$ – I, 0) and the deferment value in the original version of the binomial model, a state value equals the outcome of a simultaneous subgame. The investment timing problem is then solved by backward induction leading to a subgame perfect equilibrium. The details are shown in Figure A1.

$V_{t,s}$ = present value of operating cash inflows as if there is no competition
θ = proportion of value left when the firm pre-empts the market
ν = proportion of value left when both firms invest simultaneously
δ_t = proportion of value expressing net operating cash inflow
p = risk-neutral probability

Figure A1 The game in extensive form

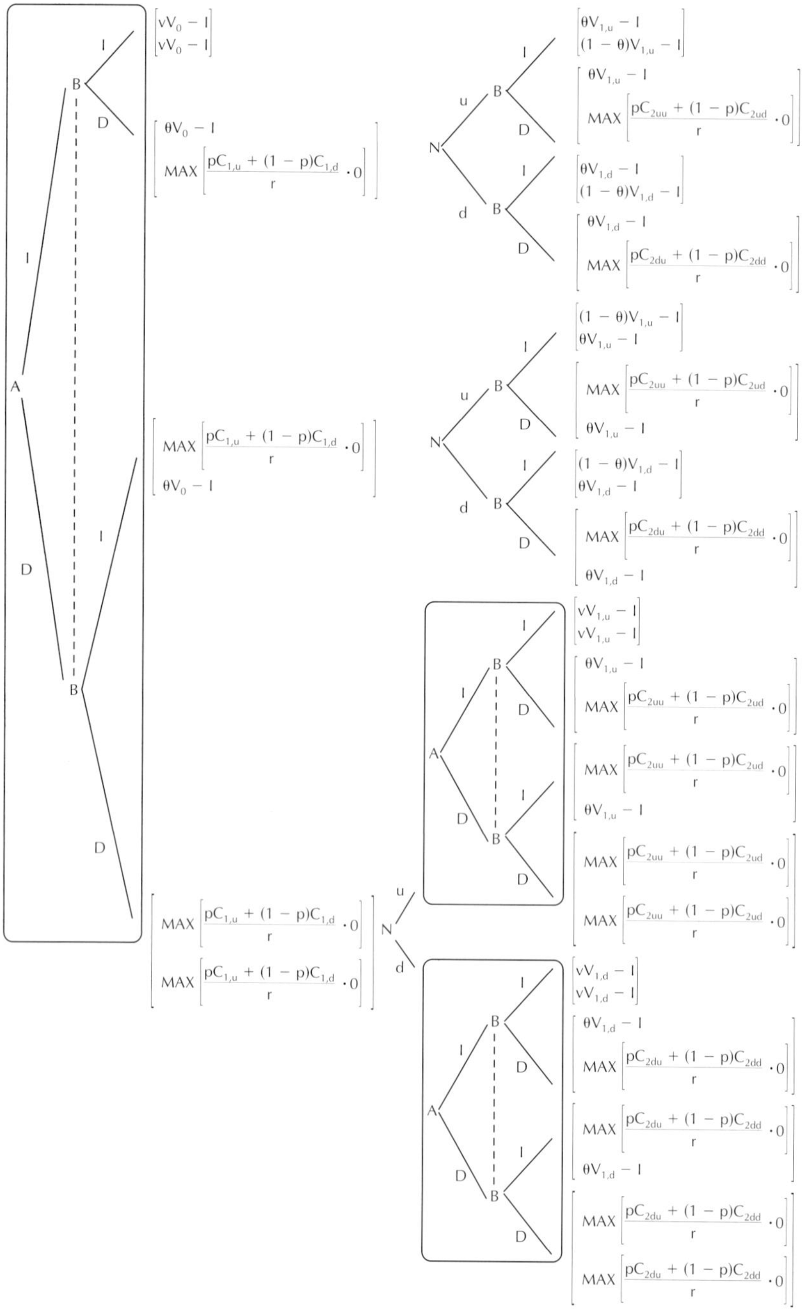

1 Assuming that an individual competitor does not take the value of waiting into account.

2

$$524 = \frac{10 + 0.16(500)}{1.16} + \frac{7.4 + 0.16(500)}{(1.16)^2} + \frac{5.5 + 0.16(500)}{(1.16)^3} + \ldots \frac{0 + 0.16(500)}{(1.16)^\infty}$$

3 p is the probability that would prevail if investors were risk-neutral. This valuation principle is based on arbitrage arguments with a replicating portfolio.

4 In the symmetrical duopoly example, we use $u = 1.5$, $d = 0.66$, $r = 0.1$, $p = 0.52$, $\theta = 0.55$, $v = 0.5$, $V_0 = 100$ and $I = 50$. The values in Figure 4 are rounded and based on more periods.

5

$$\frac{p(34) + (1 - p)(3)}{r} = 17$$

6 In the asymmetrical duopoly example, we use the same input variables as in the symmetrical game, except for the present value of the investment opportunity of the weak firm B. For firm B's project, we use $V_0 = 80$ and $I = 50$ instead.

BIBLIOGRAPHY

Brennan, M.J. and E.S. Schwartz, 1985, "Evaluating Natural Resource Investments", *Journal of Business* April, pp. 135–57.

Cox, J.C., S.A. Ross and M. Rubinstein, 1979, "Option Pricing: A Simplified Approach", *Journal of Financial Economics* September, pp. 229–63.

Hall, W.K., 1980, "Survival Strategies in a Hostile Environment", *Harvard Business Review* September–October, pp. 75–85.

Kemna, A.G.Z. and T.F.C. Vorst, 1987, "Options and Optimal Investment Timing", *Finance and Investment State of the Art*, pp. 107–23.

Kester, W.C., 1984, "Today's Options for Tomorrow's Growth", *Harvard Business Review* March–April, pp. 153–60.

Majd, S. and R. Pindyck, 1987, "Time to Build, Option Value, and Investment Decisions", *Journal of Financial Economics* March, pp. 7–27.

Mason, S. and R.C. Merton, 1985, "The Role of Contingent Claims Analysis in Corporate Finance", in *Recent Advances in Corporate Finance*, E.I. Altman and M. Subrahmanyan (eds.), Homewood, IL, R.D. Irwin.

McDonald, R.L. and D.R. Siegel, 1986, "The Value of Waiting to Invest", *Quarterly Journal of Economics* November, pp. 707–27.

Myers, S.C., 1987, "Finance Theory and Financial Strategy", *Midland Corporate Finance Journal* Spring, pp. 6–13.

Myers, S.C. and S. Majd, 1990, "Abandonment Value and Project Life", *Advances in Futures and Options Research*, pp. 1–21.

Shapiro, A.C., 1991, *Modern Corporate Finance*, Macmillan Publishing Company.

Trigeorgis, L., 1986, "Valuing Real Investment Opportunities: An Options Approach to Strategic Capital Budgeting", unpublished doctoral dissertation, Harvard University.

Trigeorgis, L., 1991, "Anticipated Competitive Entry and Early Preemptive Investment in Deferrable Projects", *Journal of Economics and Business* May, pp. 143–56.

Tirole, J., 1990, *The Theory of Industrial Organization*, MIT Press.

Part II

Real Options and Game-Theoretic Industry Equilibria

3

*The Strategic Exercise of Options: Development Cascades and Overbuilding in Real Estate Markets**

Steven R. Grenadier

Graduate School of Business, Stanford University

This chapter develops an equilibrium framework for strategic option exercise games. I focus on a particular example: the timing of real estate development. An analysis of the equilibrium exercise policies of developers provides insights into the forces that shape market behaviour. The model isolates the factors that make some markets prone to bursts of concentrated development. The model also provides an explanation for why some markets may experience building booms in the face of declining demand and property values. While such behaviour is often regarded as irrational overbuilding, the model provides a rational foundation for such exercise patterns.

In the case of traditional financial options, optimal exercise strategies can be derived without consideration of the strategic interactions across option holders. Most financial options represent widely held side-bets between agents external to the firm, and therefore their exercise does not influence the characteristics of the underlying security or the options themselves. A notable exception is the case of warrants or convertible securities. In the case of warrants, exercise results in the firm issuing new shares of common stock, thereby influencing the underlying stock value as well as the value of the remaining warrants.

*This chapter was first published in the *Journal of Finance* 51(5), 1996 and is reprinted with the permission of Blackwell Publishers. The author would like to thank Geert Bekaert, Kent Daniel, Jim Follain, Michael Harrison, Tony Sanders, Rene Stulz, (the editor), two anonymous referees and seminar participants at Harvard University, Northwestern University, the Board of Governors of the Federal Reserve, and the Ohio State/Price Waterhouse Housing and Real Estate Economics summer workshop.

In this chapter, I demonstrate that a game-theoretic approach to option exercise can be very useful in explaining real-world investment decisions. I develop an equilibrium framework for solving option exercise strategies. In order to emphasise the applicability of such an approach, I focus on a particular real-world example: the behaviour of real estate markets. This analysis of the strategic equilibrium exercise policies of real estate developers (ie, the exercise of the "option to build") provides a potential explanation for several puzzling real estate market phenomena. For example, some real estate markets have been prone to pronounced bursts of development activity, while others have been characterised by smooth patterns of development over time. Thus, one can use the model to examine the conditions that influence the time between construction starts (ie, exercise). Similarly, some real estate markets have been prone to lengthy periods of overbuilding, where new development increases in the face of significant declines in demand and building values. The model provides a potential rational explanation for this phenomenon. Developers, fearing pre-emption by a competitor, proceed into a "panic" equilibrium in which all development occurs during a market downturn. The model is also extended to the case of a perfectly competitive equilibrium, and the impact of market structure on equilibrium exercise strategies is examined.

Research on the strategic exercise of warrants and convertible securities provides an important background for the present analysis. Emanuel (1983) demonstrates that if a warrant holder has a sufficient degree of monopoly power, it is optimal to exercise warrants sequentially, rather than simultaneously. Extending the analysis to a competitive market structure, Constantinides (1984) analyses the case in which a continuum of competitive warrant holders choose their equilibrium exercise strategies. Spatt and Sterbenz (1988) expand on these results in several directions. First, they demonstrate that for oligopolistic markets, the equilibrium set of exercise strategies may lead to warrant valuations that may be even lower than those of the competitive case. Essentially, fear of pre-emptive exercise by competitors can lead to a simultaneous rush to early exercise. Second, they demonstrate that optimal exercise strategies depend on the issuing firm's policy regarding the use of the warrant-exercise proceeds. For example, when firms reinvest the proceeds in their underlying assets, the underlying risk of the equity shares increases, leading to an increase in the value of the unexercised warrants. This can then lead to sequential exercise. Alternatively, if the proceeds are invested in riskless bonds, the underlying equity risk is not increased, thus leading to a simultaneous exercise equilibrium.

This chapter's strategic option exercise framework has numerous applications in real asset markets. This area of research has been relatively untapped. Williams (1993) is a significant exception. Williams derives an equilibrium set of exercise strategies for real estate developers, where equilibrium development is symmetric and simultaneous. In Williams' equilibrium, all developers build at the maximum feasible rate whenever income rises above a critical value. In contrast, equilibrium development in this model may arise endogenously as either simultaneous or sequential. In addition, the present model allows for the possibility of pre-emptive equilibria in which development bursts occur following a fall in demand.[1] The framework provided in this chapter also has implications for corporate investment policy. For example, expenditures on research and development are analogous to exercising the option to invest in future growth opportunities. With a limited number of competitors, each firm's timing and quantity of research effort must take into consideration the equilibrium exercise strategies of its rivals. Related applications to real options have included the study of capital budgeting in settings with pre-emption and learning (Spatt and Sterbenz (1985)), investment with strategic competition (Kulatilaka and Perotti (1995)), and the adoption of new technology (Dybvig and Spatt (1983)). Porter and Spence (1978) provide an interesting actual example of a real option in which exercise was simultaneous due to pre-emption in a study of capacity additions in the corn wet milling industry. An early application of option pricing analysis to real estate development is Titman (1985). Grenadier (1995a) uses an option pricing approach to provide a rational explanation for boom-and-bust cycles in real estate markets.

In this chapter I develop an equilibrium model for the case of real estate development. Two building owners lease their existing properties in a local real estate market. In addition, each holds the option to develop a new, superior building. The option to develop is a call option with an exercise price equal to the cost of construction, and the underlying security is a new building. However, the exercise of the development option by one building owner has repercussions on the value of both options. The first to build (the Leader) will pay the construction cost earlier, but benefit by being able to lease the superior space without competition. The other developer (the Follower) will see the value of its existing building affected: it is rendered (relatively) obsolete by the presence of a new building. If and when the Follower exercises his development option, he will gain the value from renting a new building. The Leader will see his monopoly rentals discontinued, and must now compete in a duopoly. The exercise strategy is complicated, yet made more realistic by the inclusion of a time-to-build feature.

I identify and analyse subgame perfect equilibria for this strategic development option game. Depending on the initial conditions and parameter values, either simultaneous or sequential exercise may take place. The methodology is that of stochastic stopping-time games; in particular, the results of Dutta and Rustichini (1991) underlie the approach. The approach in this chapter, however, is more on a heuristic level. For a more rigorous development of the theory of equilibrium in stochastic stopping-time games, see Dutta and Rustichini (1991) or Huang and Li (1991). Smets (1993) provides an application in the area of international finance.

This first section provides an overview of the model. The second derives the value of pursuing either the Leader or Follower strategy and the third section determines the equilibrium set of exercise strategies for both developers. It considers the model's implications on the propensity for development cascades. The chapter goes on to apply the model to analysing building booms in the face of declining demand. Finally it generalises the model to the case of a perfectly competitive market setting.

A SUMMARY OF THE MODEL

The strategic exercise of options is illustrated through a specific case: the option to develop real estate. At the outset of the model, a local real estate market consists of two buildings, owned by distinct individuals. Each building is identical and rents at a rate of R per unit time. Both owners have an opportunity to redevelop their properties into new, superior buildings. These buildings can earn potentially greater rentals. Thus, each owner holds an option to develop. The option to develop has an exercise price equal to I, the cost of construction.

A new building will yield a rental rate according to a downward-sloping inverse demand function. The demand function will be subject to continuous shocks. The inverse demand function is of the following form[2]:

$$P(t) = X(t) \cdot D[Q(t)] \tag{1}$$

where P(t) is rent flow at time t, Q(t) is the supply of improved buildings at time t, and $D(\cdot)$ is a differentiable function with $D' < 0$. This market is characterised by evolving uncertainty in the state of demand. Even demand at the next instant is uncertain today. X(t) represents a multiplicative demand shock, and evolves as a geometric Brownian motion:

$$dX = \mu X\, dt + \sigma X\, dz \tag{2}$$

where dz is an increment of a standard Wiener process. The constant μ is the instantaneous conditional expected percentage change in X per unit time and the constant σ is the instantaneous conditional standard deviation per unit time. One can easily envision potential real-world candidates for such market shocks. For office development, such shocks may take the form of business profitability. For apartment development, the shocks may be due to job growth. For industrial markets, the shocks may be due to industrial production.

The exercise of the development option does not yield immediate payoffs. An important aspect of real estate development is the time-to-build. The construction period for real estate varies widely across property types. For office buildings, the average length of time between the initiation and completion of construction is two and a half years (Wheaton

(1987)). For some very large office projects in highly regulated cities, the construction lags could extend to five years or more. In the case of industrial buildings, the lag is much shorter, perhaps on the order of six months (Wheaton and Torto (1990)). Developers must use current information on demand and supply in order to determine the optimal time to begin construction. I model this aspect of real estate development by assuming a time-to-build of δ years. Thus, if an owner begins construction (exercises the option) at time τ, the buildling will not begin receiving rentals until time $\tau + \delta$.

In equilibrium, one owner will begin construction first and become the Leader. The other will build later and become the Follower. Of course, it may be the case that each wishes to enter simultaneously and become the Leader. A slight technical assumption is imposed. It is assumed that if each tries to build first, one will randomly (ie, through a toss of a coin) win the race.[3] The loser of the race may then, if he wishes, enter a split second later. This will be indistinguishable from simultaneous development.

The strategic nature of the dual option exercise problem will now be made clear. The exercise of the development option will result in repercussions on both the option exerciser as well as the other building owner. First, consider the exercise of the development option by the Leader. The Leader pays an initial construction cost today, and loses current rentals on the existing building. In return, at the end of construction in δ years, the Leader begins receiving monopoly rents of $X(t) \cdot D(1)$ on the newly developed building. This option exercise also affects the fortunes of the Follower. The competitor's construction of an improved building lessens the demand for the existing building. For example, the existing tenants in the older building may demand rent concessions or improvements to their space, in return for not moving to the newer building at the end of the lease. I account for the degree of "harm" to the old building's operations caused by the Leader's exercise by assuming that the old building's rentals fall to $(1 - \gamma) \cdot R$, with $\gamma \in (0, 1)$.[4]

Now, consider the impact of the Follower's exercise of the development option on both owners. The Follower will pay the cost of construction, lose current rent, and at the end of construction begin receiving rent on the new, improved building. The Leader is also affected because he now faces competition in the leasing of improved space. Each building will lease at a duopoly rental of $X(t) \cdot D(2)$.

In the following sections I derive a set of equilibrium exercise strategies. Specifically, I derive a pair of symmetric, Markovian exercise strategies that form a subgame perfect equilibrium. Simply put, at each point of the game, each owner's exercise strategy is optimal conditional on the other's exercise strategy. Depending on the initial conditions, some equilibria will be characterised by sequential development, while others will be characterised by simultaneous development. The starting point of the game can arise in various ways. One possibility is that an exogenous shock, such as a zoning change allowing commercial development in an area, instigates developer interest. Another possibility is that the game has an infinite history and developers merely wait for the local market to grow to a sufficient size before beginning development.

A brief comment about the option valuation framework is in order. In traditional financial option pricing models, the approach to valuation is based on arbitrage arguments, where one uses the fact that one can instantaneously trade the underlying asset and a riskless asset so that the option is precisely replicated. In the case of real estate, where one cannot short sell the building, transactions costs are high, and assets cannot be traded in infinitesimal increments, such an arbitrage technique is highly unrealistic. The alternative valuation technique used here is an equilibrium approach. This approach is now standard in much of the literature on real options. For simplicity, I assume risk neutrality.[5] This seemingly restrictive assumption can easily be relaxed by adjusting the drift rate μ to account for a risk premium in the manner of Cox and Ross (1976).

THE VALUE OF THE LEADER AND FOLLOWER

In the model, one individual will develop first and become the Leader, and the other will become the Follower. In this section, I derive the value of pursuing each strategy. For the moment, I simply take the roles as given. In the next section, I derive the equilibrium set of exercise strategies that determine the identity of the Leader and Follower.

First, I value the payoff of choosing to be the Follower. As is standard in solving dynamic games, I work backward in a dynamic-programming fashion. Thus, in this section it is assumed that one owner has already begun (or completed) construction. Since the Leader has already exercised his option, the Follower can construct his exercise strategy without fear of pre-emption. Therefore, this is a standard option pricing problem in which the equilibrium Follower strategy is to maximise his option's value.

The value of the Follower can be characterised as a portfolio containing the existing building, plus an option to exchange the existing building for a new building.[6] The existing building, conditional on the Leader having begun development, yields a rent flow of $(1-\gamma) \cdot R$. Since this payment is certain, its value is $(1-\gamma) \cdot R/r$. The value of the Follower also emanates from the value of the option to exchange the old building for a new building. Upon exercise, the Follower pays a construction cost of I, and exchanges the old building for a new building δ years from completion. Denote the value of this option to exchange as W(X). The total value of the Follower, F(X), will be the portfolio value: $F(X) = (1-\gamma) \cdot R/r + W(X)$.

Consider the dynamics of W(X) in the region where the Follower does not exercise. By Itô's Lemma, the instantaneous change in W(X) can be written as:

$$dW(X) = [\tfrac{1}{2}\sigma^2X^2W''(X) + \mu XW'(X)]dt + \sigma XW'(X)dz \quad \textbf{(3)}$$

Since this option yields no current cashflow, the total instantaneous return comes in the form of capital gains, $dW(X)$. Let α_F denote the expected return per unit time on the Follower's option value. Applying the expectations operator to equation (3) and dividing by W, the expected return can be written as:

$$\alpha_F = \frac{1}{W(X)} \cdot \left[\frac{1}{2}\sigma^2X^2W''(X) + \mu XW'(X)\right] \quad \textbf{(4)}$$

In equilibrium, the expected return on the Follower's option must equal the risk-free rate r. Setting α_F equal to r yields the following differential equation, which must be satisfied in equilibrium:

$$0 = \tfrac{1}{2}\sigma^2X^2W''(X) + \mu XW'(X) - rW(X) \quad \textbf{(5)}$$

Equation (5) must be solved subject to specific boundary conditions. Upon exercise of the development option, the Follower exchanges the cashflows on the old building, $(1 - \gamma)R$, for the rentals on a new building, $XD(2)$, which begin in δ years. As in American option pricing problems of this sort, there will be a trigger value at which it will be optimal to exercise. Specifically, there will exist a trigger value, X_F, such that the Follower will exercise the option to build the first time that $X(t)$ equals or exceeds X_F.[7] Mathematically, one can write the optimal entry time of the Follower as $T_F \equiv \inf[t \geq 0\colon\ X(t) \geq X_F]$. Therefore, I impose two boundary conditions which serve to define the optimal exercise policy:

$$W(X_F) = \left[\frac{D(2)}{r - \mu}e^{-(r-\mu)\delta}\right] \cdot X_F - \frac{(1 - \gamma)R}{r} - I$$

$$W'(X_F) = \left[\frac{D(2)}{r - \mu}e^{-(r-\mu)\delta}\right] \quad \textbf{(6)}$$

The first boundary condition is commonly termed the "value-matching" condition. It simply reflects the fact that, upon exercise, the payoff of the option is the exchange of a new building for the older building (δ years into the future), minus the cost of construction.[8] The second boundary condition is known as the "high-contact" or "smooth-pasting" condition.[9] Essentially, this condition ensures that X_F is the trigger that maximises the value of the Follower's option.

A final boundary condition is imposed: $W(0) = 0$. This is the result of the fact that zero is an absorbing barrier of the rent process; if $X(t)$ ever reaches zero, it will remain there forever. Thus, if X hits zero, the option will never be exercised and will become worthless.

Solving the equilibrium differential equation in (5) subject to the above boundary conditions results in the following value of the Follower's option (conditional upon the Leader having entered):

$$W(X) = \begin{cases} \left(\dfrac{I + (1-\gamma)R/r}{\beta - 1}\right) \cdot \left(\dfrac{X}{X_F}\right)^{\beta} & \text{if } X < X_F \\[2ex] \left[\dfrac{D(2)}{r-\mu} e^{-(r-\mu)\delta}\right] \cdot X - \dfrac{(1-\gamma)R}{r} - I & \text{if } X \geq X_F \end{cases} \tag{7}$$

where:

$$\beta = \frac{-(\mu - \frac{1}{2}\sigma^2) + \sqrt{(\mu - \frac{1}{2}\sigma^2)^2 + 2r\sigma^2}}{\sigma^2} > 1$$

$$X_F = \left(\frac{\beta}{\beta - 1}\right)\left(\frac{r-\mu}{D(2)}\right) e^{(r-\mu)\delta}\left[I + \frac{(1-\gamma) \cdot R}{r}\right]$$

Finally, the value of being the Follower equals the sum of the value of the existing building plus the value of the option to exchange. Thus,

$$F(X) = \frac{(1-\gamma) \cdot R}{r} + W(X) \tag{8}$$

Proposition 1 summarises the optimal Follower strategy.

Proposition 1. *Conditional on the Leader having begun construction, the optimal Follower strategy is to begin construction the first moment that* $X(t)$ *equals or exceeds the trigger value* X_F*, as defined in equation (7). That is, the optimal entry time of the Follower,* T_F*, can be written as:*

$$T_F = \inf\left\{t \geq 0 : X(t) \geq \left(\frac{\beta}{\beta - 1}\right)\left(\frac{r-\mu}{D(2)}\right) e^{(r-\mu)\delta}\left[I + \frac{(1-\gamma) \cdot R}{r}\right]\right\}$$

Now, consider the value of the Leader, conditional on the Follower pursuing the optimal Follower strategy. Suppose the Leader has already exercised his development option and now has τ years until completion, where $\tau \in [0, \delta]$. Over the next τ years, the Leader receives no rentals. Once completed, the Leader earns a monopoly rental rate of $X(t) \cdot D(1)$. However, since the Follower enters at the stopping time T_F, the Leader will begin receiving the duopoly cashflow, $X(t) \cdot D(2)$, at time $T_F + \delta$. It is important to note that it is the Follower's exercise strategy, T_F, which impacts the Leader's valuation.

The value of the Leader can be replicated by the following portfolio of options:

1 Purchase a call option on an asset which pays a perpetual dividend rate of $X(t) \cdot D(1)$. This call option has a zero exercise price, and a fixed expiration date of τ.
2 Purchase a call option on an asset which pays a perpetual dividend rate of $X(t) \cdot [D(2) - D(1)]$. This call option has a zero exercise price, and a stochastic expiration date of $T_F + \delta$.

The value of this portfolio can be derived using the same approach as in the case of the Follower (except the smooth-pasting condition is no longer needed). Denote the value of the Leader, prior to the exercise of the Follower, $L(X, \tau)$. X is the current level of demand, and τ is the remaining time before completion, $\tau \in [0, \delta]$. The solution for $L(X, \tau)$ can be expressed as:

$$L(X, \tau) = \begin{cases} \dfrac{e^{-(r-\mu)\tau}}{r-\mu} XD(1) + \dfrac{\beta}{\beta - 1} \dfrac{D(2) - D(1)}{D(2)} \left(I + \dfrac{(1-\gamma)R}{r} \right) \left(\dfrac{X}{X_F} \right)^{\beta} & \text{if } X < X_F \\ \dfrac{e^{-(r-\mu)\tau}}{r-\mu} XD(1) + \dfrac{Xe^{-(r-\mu)\delta}}{r-\mu} \left[D(2) - D(1) \right] & \text{if } X \geq X_F \end{cases} \tag{9}$$

Finally, consider the relative values of the Leader and Follower at the moment the Leader begins construction. The Leader receives a payoff of $L(X, \delta) - I$ and the Follower receives $F(X)$. Depending on the initial entry time (which is determined by equilibrium considerations and derived in the next section), the Leader's payoff may be greater or less than that of the Follower. The following proposition describes the relative valuations.

Proposition 2. *There exists a unique point,* $X_L \in (0, X_F)$, *with the following properties:*

$$\begin{aligned} L(X, \delta) - I &< F(X) \quad \text{for } X < X_L \\ L(X, \delta) - I &= F(X) \quad \text{for } X = X_L \\ L(X, \delta) - I &> F(X) \quad \text{for } X_L < X < X_F \\ L(X, \delta) - I &= F(X) \quad \text{for } X \geq X_F \end{aligned}$$

Proof. See Appendix.

Proposition 2 demonstrates that there is a unique value of $X \in (0,X_F)$, called X_L, at which the payoffs to both Leader and Follower are equal. At any point below X_L, each developer would prefer being the Follower. At any point above X_L, each developer would prefer (or be indifferent to) being the Leader. At X_L, the benefits of a potentially temporary monopoly just equal the costs of paying the construction costs earlier.

THE EQUILIBRIUM EXERCISE STRATEGIES

Given the value of being the Leader or Follower, I now work back to the beginning of the game, where the two developers choose their equilibrium entry strategies. Each is currently leasing the previous generation of buildings at the rent flow of R. I construct a pair of symmetric, subgame perfect equilibrium entry strategies in which each developer's exercise strategy, conditional upon the other's exercise strategy, is value-maximising.

I focus on two classes of equilibria that occur depending on the initial state of demand, X(0). If $X(0) < X_F$, equilibrium exercise will be sequential; once one agent begins development, a period of time passes before the other begins development. If $X(0) \geq X_F$, equilibrium exercise will be simultaneous; one exercises instantly after the other.

Equilibrium with sequential exercise: $X(0) < X_F$

If the development game begins with an initial level of X that is lower than X_F, then we will observe a sequential equilibrium: the Leader will begin development the first moment that X(t) equals or exceeds X_L, and the Follower will wait and begin development the first moment that X(t) rises to the level X_F. I state this more formally in the following definition.

Sequential exercise equilibrium

Suppose the initial level of X is below X_F. A pair of symmetric, subgame perfect equilibrium exercise strategies is for each agent to act as follows:

> If your competitor has not yet begun construction, then initiate development the first moment that X(t) equals or exceeds X_L. If your competitor has already begun construction, then wait until X(t) rises to X_F before beginning construction.

Given the above strategies, the sequential equilibrium will appear as follows. If $X(0) < X_L$, one developer will wait until the trigger X_L is reached, and the other will wait until the trigger X_F is reached. The developers will be indifferent between leading or following. If $X(0) \in [X_L, X_F)$, each will race to build immediately. The (random) winner of the race will then build, and the loser will wait until the trigger X_F is reached.

It is simple to demonstrate that this set of strategies forms an equilibrium. From Proposition 2, entry before X hits X_L leads to a value strictly less than that of following. Thus, if your competitor follows the equilibrium strategy, the best that you can do is also to pursue the equilibrium strategy by attempting to enter the first moment that X equals or exceeds X_L. In addition, Proposition 1 demonstrates that the optimal strategy for whomever builds second is to enter when the trigger X_F is reached.

Equilibrium with simultaneous exercise: $X(0) \geq X_F$

Now, consider the range where $X(0) \geq X_F$. In this range, any equilibrium will be characterised by simultaneous exercise. If either developer begins construction at a level of X greater than X_F, the other will enter immediately thereafter. This is due to Proposition 1, which depicts the optimal Follower strategy of entering the first moment that X_F is equalled or exceeded. Thus for any exercise strategy over this region, the equilibrium will be characterised by simultaneous entry: one enters an instant after the other.

There will be an infinite number of equilibrium strategies over this region. I focus on a class of equilibria that can be characterised by a band of values, $(X_F, \hat{X})$, where both agents begin development the first time X(t) exits this range. It is simple to demonstrate that such a set of strategies forms an equilibrium. First, we know that if X ever falls to the level X_F, each will be forced to enter immediately. This is due to the threat of pre-emption. Should the value of X fall below X_F, then the firm that enters first will have a value strictly greater than the laggard's value, as in Proposition 2. Thus, it is in each firm's interest to enter immediately. Second, for any $\hat{X} > X_F$, it is an equilibrium for both to enter when X(t) first equals or exceeds $\hat{X}$. That is, given that one firm will wait until $\hat{X}$ is reached, it is optimal for the other firm to do likewise. This follows from Propositions 1 and 2: the value of leading is strictly greater than following, and the Follower's optimal response to losing the race is to enter immediately.

Therefore, there is an infinite set of Markovian, subgame perfect equilibrium exercise strategies characterised by immediate construction upon the first moment that X(t) falls outside the range $(X_F, \hat{X})$, $\forall \hat{X} \geq X_F$. From this set of simultaneous equilibria, I focus on one in particular: the Pareto optimal one.[10] That is, given that the two developers will simultaneously enter the first moment that X(t) reaches $\hat{X}$ (provided X(t) does not first fall to X_F), then the Pareto optimal equilibrium will set $\hat{X}$ such that both values are maximised.

I now solve for the Pareto optimal set of simultaneous exercise strategies. Denote the value of each developer as G(X). Using an analogous line of reasoning to that of the second section, over the range at which each developer is idle, their values will satisfy the following equilibrium differential equation:

$$0 = \tfrac{1}{2}\sigma^2 X^2 G''(X) + \mu X G'(X) - rG(X) + R \qquad \textbf{(10)}$$

The boundary conditions will determine the optimal simultaneous exercise region. Denote the optimal joint exercise trigger as X_J. Thus, if X rises to the trigger X_J, each will exercise their development option, and their compound option values will be maximised. However, if X falls to the level X_F, each will exercise in order to avoid pre-emption. These boundary conditions are:

$$G(X_F) = \frac{1}{r - \mu} e^{-(r-\mu)\delta} X_F D(2) - I$$

$$G(X_J) = \frac{1}{r - \mu} e^{-(r-\mu)\delta} X_J D(2) - I \qquad (11)$$

$$G'(X_J) = \frac{1}{r - \mu} e^{-(r-\mu)\delta} D(2)$$

The first boundary condition reflects the fact that if X falls to the level of X_F, each will enter immediately. Even though one firm will win the race, the other will find it optimal to enter in the next instant. The second boundary condition is a "value-matching" condition: at the optimal joint-entry trigger, each firm enters immediately. The final boundary condition is a "smooth-pasting" condition which ensures that the joint-entry trigger X_J is indeed the one that maximises value.

The solution to differential equation (10) can be written as:

$$G(X) = \begin{cases} A_1 X^{\alpha} + A_2 X^{\beta} + \dfrac{R}{r} & \text{for } X \in (X_F, X_J) \\[2ex] \dfrac{e^{-(r-\mu)\delta} X D(2)}{r - \mu} - I & \text{for } X \notin (X_F, X_J) \end{cases} \qquad (12)$$

where:

$$\alpha = \frac{-(\mu - \frac{1}{2}\sigma^2) - \sqrt{(\mu - \frac{1}{2}\sigma^2)^2 + 2r\sigma^2}}{\sigma^2} < 0$$

$$\beta = \frac{-(\mu - \frac{1}{2}\sigma^2) + \sqrt{(\mu - \frac{1}{2}\sigma^2)^2 + 2r\sigma^2}}{\sigma^2} > 1$$

and A_1, A_2 and X_J are the solutions to the following system of three non-linear equations:

$$A_1 X_F^{\alpha} + A_2 X_F^{\beta} + \frac{R}{r} = \frac{e^{-(r-\mu)\delta} X_F D(2)}{r - \mu} - I$$

$$A_1X_J^{\alpha} + A_2X_J^{\beta} + \frac{R}{r} = \frac{e^{-(r-\mu)\delta}X_J D(2)}{r-\mu} - I \tag{13}$$

$$\alpha A_1X_J^{\alpha-1} + \beta A_2X_J^{\beta-1} = \frac{e^{-(r-\mu)\delta}D(2)}{r-\mu}$$

While this system does not simplify to a closed-form solution, it is easily solved numerically.

I now summarise the simultaneous equilibrium in the following definition.

Simultaneous exercise equilibrium

Suppose the initial level of X is at or above X_F. A symmetric, subgame perfect equilibrium set of exercise strategies is for each owner to act as follows:

> Begin development the first moment that X(t) falls to X_F, or equals or exceeds X_J. If your competitor wins the race to build first, then enter instantaneously thereafter.

Given the above strategies, the sequential equilibrium will appear as follows. If $X(0) \in (X_F, X_J)$, then there will be an initial period of inactivity. Both agents will begin construction the first moment that either X_F or X_J is reached. If $X(0) = X_F$, or $X(0) \geq X_J$, then both begin construction immediately.

DEVELOPMENT CASCADES: THE RAPIDITY OF CONSTRUCTION STARTS

In this section, I analyse the amount of time between the exercise of options. The goal of this analysis is to determine the average time between construction starts. That is, once development occurs in a market, when is it likely to be followed by additional development? Here I focus on the sequential exercise equilibrium, where development options are exercised at distinct stopping times. Depending on the underlying conditions of the market, equilibrium construction may occur in short bursts or more evenly over time. I refer to a rapid succession of exercise strategies as a "development cascade".[11]

A measure of the rapidity of construction starts

I focus on the case in which the market begins with a period of inactivity followed by a sequence of construction starts. Thus, in this section I analyse the case in which the market begins with $X(0) < X_L$.[12]

Briefly, recall the nature of the sequential equilibrium setting. The real estate market is dormant until demand rises to the trigger level X_L. At that time, one developer begins construction. The other developer will hold off on construction until demand rises to the second trigger level, X_F. Therefore, the time between construction starts is the first passage time of the stochastic process X(t) reaching the level X_F, when starting at the level X_L.

Define the time between construction starts as T_s. Thus, T_s can be expressed as:

$$T_s \equiv \inf\{t \geq 0: X(t) = X_F, X(0) = X_L\} \quad \textbf{(14)}$$

Using equation (1.11) in Harrison (1985), and the application of a simple change in variables, the cumulative distribution function of the first passage time T_s can be written as:

$$P[T_s \leq t] = \Phi\left[\frac{-\ln(X_F/X_L) + (\mu - \frac{1}{2}\sigma^2)t}{\sigma\sqrt{t}}\right]$$

$$+ \left(\frac{X_F}{X_L}\right)^{2(\mu-(1/2)\sigma^2)/\sigma^2} \Phi\left[\frac{-\ln(X_F/X_L) - (\mu - \frac{1}{2}\sigma^2)t}{\sigma\sqrt{t}}\right] \quad \textbf{(15)}$$

To provide a measure of the average time between construction starts, one might first consider the expected value of T_s. Unfortunately, for the case in which $\mu < \frac{1}{2}\sigma^2$, the expectation is undefined. Essentially, if the variance in demand is sufficiently high, the probability that the upper trigger is never reached is large enough to make the expectation infinite. To overcome this problem, I choose to measure the average time between starts by the median of T_s. The median time between construction starts, M_s, is the solution to the following nonlinear equation:

$$\Phi\left[\frac{-\ln(X_F/X_L) + (\mu - \frac{1}{2}\sigma^2)M_s}{\sigma\sqrt{M_s}}\right]$$

$$+ \left(\frac{X_F}{X_L}\right)^{2(\mu-(1/2)\sigma^2)/\sigma^2} \Phi\left[\frac{-\ln(X_F/X_L) - (\mu - \frac{1}{2}\sigma^2)M_s}{\sigma\sqrt{M_s}}\right] = \frac{1}{2} \quad \textbf{(16)}$$

Comparative statics

I now examine the effect of changes in underlying parameters on the median time between construction starts. In particular, I consider three potential influences on M_s: the volatility of demand (σ), the time-to-build (δ), and the degree of current obsolescence (γ).

Consider the effect of market volatility on the time between construction starts. There are two competing forces at work here. First, a well-established result from the theory of real options is that increasing uncertainty leads to an increased willingness to postpone investment.[13] Thus, beginning from any initial level of demand, this result would tend to delay the exercise of any future development option. However, a second force at work here is that increasing volatility may also make it more likely that a

Figure 1 The effect of demand volatility on the median time between construction starts

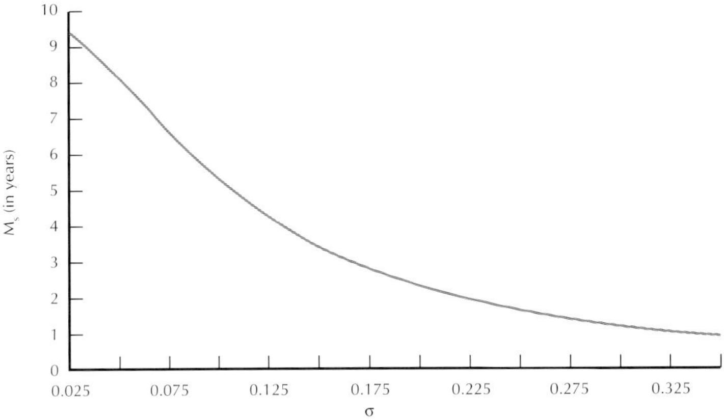

This figure plots the effect of demand volatility, σ, on the median time between construction starts, M_s. From the instant that the first development option is exercised, M_s is the median number of years before the second development option is exercised. The greater the market's volatility, the greater the propensity for development to occur in bursts. An existing building, earning a rent of R, can be redeveloped into a new building earning a stochastic rent flow of X(t)D[Q(t)], where the demand state variable X is a geometric Brownian motion with drift μ and volatility σ, and D(·) is a downward-sloping inverse demand function. γ is the fractional loss in rent to an older building when a new building is constructed. The cost of construction is I, and development requires a construction period of δ years. The risk-free rate is r. For this figure, the following parameters are fixed: R = 0.1, μ = 0.02, D(1) = 1, D(2) = 0.9, γ = 0.8, I = 1, δ = 1 and r = 0.04.

substantial increase in demand is reached in a shorter period of time. Therefore, changing σ alters the optimal exercise strategies as well as the stochastic properties of the underlying demand.

Figure 1 demonstrates that the net effect of increasing demand volatility is a decrease in the median time between starts.[14] For example, given two markets differing only in the degree of demand volatility, the one with the greater volatility will be the one more prone to development cascades. Notice that for σ = 0.025, the median time between starts is almost 9.5 years. For σ = 0.15, the median time between starts falls to 3.5 years. For σ = 0.35, the median time between starts falls to less than one year. That is, there is a 50% likelihood that the time between starts will be less than one year.

The intuition for this result is as follows. As volatility increases, the option to wait increases. Of course, this option to wait affects both the Leader and Follower. Thus, even though the Leader will delay entering, so too will the Follower. Therefore, the "option to wait" factor will have little effect on the time between starts. On the other hand, the increased volatility makes it more likely that the time it takes for demand to rise from X_L to X_F will fall. Therefore, it is this effect which dominates.

I now consider the effects of changes in the time-to-build and degree of current obsolescence on the propensity for development cascades. Proposition 3 provides a mathematical result that leads to a striking comparative static.

Proposition 3. *The ratio of the two trigger values, X_F/X_L, is independent of the time-to-build parameter δ and the degree of current obsolescence γ:*

$$\frac{\partial(X_F/X_L)}{\partial\delta} = \frac{\partial(X_F/X_L)}{\partial\gamma} = 0$$

Proof. Proposition 2 defines X_L implicitly as the solution to $L(X_L, \delta) - I - F(X_L) = 0$. Using this identity, and the implicit function theorem, one may obtain an expression for $\partial X_L/\partial\delta$ and $\partial X_L/\partial\gamma$. In addition, one can directly differentiate the expression for X_F given in equation (7). Finally, simple algebraic manipulation yields the stated result. QED.

Proposition 3 leads directly to the following result: changes in the time-to-build or degree of current obsolescence have no effect on the time between construction starts. In fact, not only will such changes have no effect on the median time between starts, but they will also have no effect on the entire probability distribution of the time between starts. That is, not only is the median time between starts unaffected, but so to is the mean, variance, skewness and other moments of the distribution. This is apparent by considering the cumulative distribution function of the time between starts given in equation (15). The distribution function for the time between starts, $P[T_s \leq t]$, depends on the parameters δ and γ only through the ratio (X_F/X_L). Proposition 3 demonstrates that this ratio is independent of these two parameters.

The reason for this result is that both the Leader and Follower adjust their equilibrium exercise triggers proportionately to the changes in the underlying parameters. Unlike the case of changes in the underlying volatility, these parameters have no effect on the underlying state variable process. Therefore, changes in δ and γ do not affect the underlying stochastic environment, but simply the consequences of competitive exercise. Since the exercise strategy response is symmetric and proportional, there will be no change in the distribution of the time between exercise.

Empirical implications

The model predicts that demand volatility increases the likelihood of construction cascades. This implies that both regions and property types with more volatile demand would be the most prone to concentrated bursts of development.

One implication of this result is that cities with diversified economies would be less prone to development cascades than cities with significant dependence on a single economic factor.[15] Consider the cases of the Denver and Houston office markets. These cities have historically been heavily dependent on the oil industry. Over the thirty-year period from 1960 through 1990, over half of all office construction was completed in a four-year interval: 1982–85.[16] Contrast these two markets' experiences with those of Chicago and Minneapolis, both historically more diversified than the oil-price dependent cities. Over no four-year period did Chicago have more than one-quarter of its completions, and over no four-year period did Minneapolis have more than one-third of its completions. Relatively speaking, Chicago and Minneapolis saw development occur more gradually, rather than through short bursts. Thus, we do find that less diversified economies may be more prone to development concentrated over a short span of time.[17]

In addition to providing predictions about the cross-sectional differences across cities, the model also points to timing distinctions across property types. For example, office markets appear more volatile than industrial space markets. The Russell-NCREIF database provides a time-series of net operating income by property type. Over the period from 1978 to the first-half of 1994, the annualised standard deviation of the growth rate in office earnings was 0.1587. For warehouse space the number was 0.0892. Thus, office earnings were twice as volatile. Figure 2a displays the patterns of development for the office and industrial property markets. Note that industrial development occurred fairly evenly throughout the 1978–93 period, while office development was more erratic. In terms of square footage, Birch (1990) estimates that 43% of all office space ever built in the US was built during the 1980s, and 60% was built in the 1970s and 1980s.

The model also predicts that the tendency toward development booms should be unrelated to time-to-build or obsolescence. For example, high-rise office towers will take considerably longer to build than small, one-storey office buildings. If one assumes that the demand volatility is relatively equal between these two building types, the model predicts no difference in their tendencies toward development cascades. As another example, the model would predict that, all things being equal, markets with dilapidated buildings would be just as prone to development cascades as those with relatively modern buildings. Unfortunately, a test of such propositions cannot be accomplished until appropriate data become available.

A RATIONAL UNDERPINNING FOR "IRRATIONAL" CONSTRUCTION: RECESSION-INDUCED CONSTRUCTION BOOMS

In this section, I examine the properties of the simultaneous exercise equilibrium. In this equilibrium, both options are immediately exercised the first moment that X(t) either *rises* to an upper trigger, or *falls* to a lower

Figure 2 Office and industrial property market behaviour (1978–93)

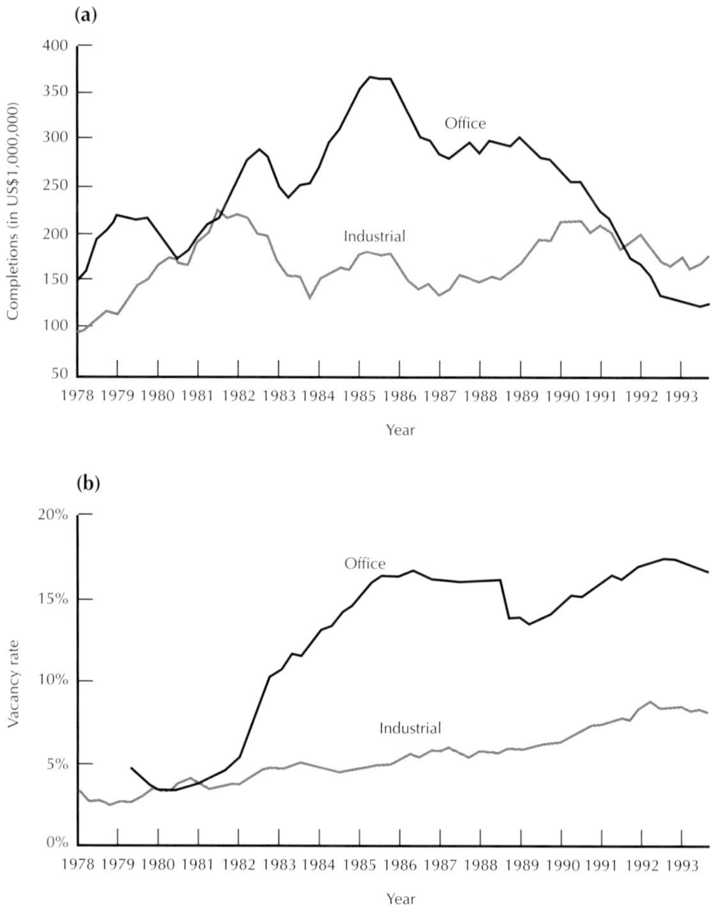

(a) Plots the real value of office and industrial property completions over time. The completion series are obtained by deflating the Commerce Dept Construction put in place series by the implicit construction cost. Thus, all completions are in constant (1993) dollars. **(b)** Plots the office and industrial property vacancy rates. The source of the vacancy rate series is CB Commercial.

trigger. Obviously, both developers would prefer exercising after a run-up in demand, but it is still rational for each of them to rush to exercise should demand fall. I refer to this phenomenon, simultaneous development following a decline in demand, as a "recession-induced construction boom" (RCB).

Periods of construction booms in the face of rising vacancies and plummeting demand are a recurring phenomenon in real estate markets. This is often termed "overbuilding" and almost universally attributed to some form of irrationality on the part of developers, lenders, or both.[18] The model in this chapter presents an alternative, rational explanation for RCBs. Developers see the market begin to erode. They realise that if it erodes any further, and if any of their competitors begins to build, they will be shut out of the market. Before this comes to fruition, each builds simultaneously in an attempt to avoid pre-emption. While building in a downturn is harmful to developers, it is less harmful than the alternative of becoming a Follower in a down market.

Another potential rational explanation for building in the face of high and rising vacancies is a Tobin's q type of argument: even though the values of new buildings were falling, the building costs were falling even faster. On a national level, this does not appear to be the case. During the past decades, national indexes of construction costs (eg, the Department of Commerce Implicit Price Deflator for Construction) were consistently rising, even during periods when commercial property values were falling dramatically. However, such figures ignore the circumstances of local markets. Labour and material market conditions in different cities may have led to rational building during periods of declining rents and prices.

A. Definition of a recession-induced construction boom

Recall from the construction of an equilibrium in the third section that there are conditions under which rational, value-maximising developers will simultaneously build in the presence of declining demand. This is true if the initial level of demand satisfies $X_F \leq X(0) \leq X_J$. The equilibrium set of strategies is for both owners to commence construction simultaneously the first moment that X(t) reaches either X_F or X_J. I now analyse the likelihood that the lower trigger is reached first.

Let t denote the current time and suppose that neither owner has started construction. Thus, $X(t) \in [X_F, X_J]$. The following definitions will be useful:

$$\begin{aligned} T_1(x) &= \inf\{\omega \geq 0: X(t+\omega) = X_F, X(t) = x, x \in [X_F, X_J]\} \\ T_2(x) &= \inf\{\omega \geq 0: X(t+\omega) = X_J, X(t) = x, x \in [X_F, X_J]\} \\ T(x) &= \min[T_1(x), T_2(x)] \end{aligned} \tag{17}$$

Thus, $T_1(x)$ is the first passage time for X hitting X_F starting from the level x, and $T_2(x)$ is the first passage time for X hitting X_J starting from the level x. In equilibrium, both will begin construction at the minimum of these two stopping times, T(x).

Mathematically, a RCB occurs if $T_1(x) < T_2(x)$. Denote the probability of a RCB, conditional on starting at demand level x, as $\Psi(x)$, where $\Psi(x) = \Pr[T_1(x) < T_2(X)]$.

Using the techniques in Harrison (1985), Chapter 3, this probability can be expressed as follows:

$$\Psi(x) = \frac{x^{-2(\mu-(1/2)\sigma^2)/\sigma^2} - X_J^{-2(\mu-(1/2)\sigma^2)/\sigma^2}}{X_F^{-2(\mu-(1/2)\sigma^2)/\sigma^2} - X_J^{-2(\mu-(1/2)\sigma^2)/\sigma^2}} \quad \text{when} \quad \mu \neq \frac{1}{2}\sigma^2 \qquad \textbf{(18)}$$

and

$$\Psi(x) = \frac{\ln(x) - \ln(X_J)}{\ln(X_F) - \ln(X_J)} \quad \text{when} \quad \mu = \frac{1}{2}\sigma^2$$

Comparative statics

I now investigate two factors that influence a market's propensity for RCBs. First, consider the influence of time-to-build on the propensity for RCBs. For a given level of demand, x, I vary δ and investigate the effect on $\Psi(x)$. Clearly, $\Psi(x)$ will always be decreasing in x; the higher the demand, the less likely it will be for demand to hit the lower trigger prior to the upper trigger. Simulation results are presented in Figure 3a, for three different levels of demand. Notice that for each level of demand, the likelihood of a RCB is monotonically increasing in the time-to-build parameter. For example, consider the curve with $x = 0.10$. Suppose two real estate markets were exactly equivalent, except one took six months to build and the other took 6.5 years to build. They will display substantially different propensities for RCBs. The market with the shorter lag will have a 50% probability of experiencing a RCB; the market with the longer lag will have a 75% probability of experiencing a RCB.

In Figure 3b, the impact of demand volatility, σ, is displayed. For the most part, the probability of a RCB is increasing in the volatility of demand. For example, consider the curve with $x = 0.14$. Suppose two real estate markets were exactly equivalent, except one had little volatility ($\sigma = 0.05$) and the other had substantial volatility ($\sigma = 0.16$). The market with low volatility will have virtually no chance of a RCB occurring. The more volatile market will have a 50% chance of a RCB occurring. However, for higher levels of initial demand, the probability begins to decline as σ rises to higher levels. This is seen in the curve for which $x = 0.16$. This effect appears quite modest.

Empirical implications

The model predicts that markets with greater time-to-build will display a greater propensity toward RCB. This has implications on the difference between the office and industrial property markets. For office space, an estimate of the average length of time between the initiation and completion of construction is 2.5 years. The construction time for industrial buildings (which are predominately one storey, and basically shells of buildings) is

Figure 3 The probability of a recession-induced construction boom (RCB)

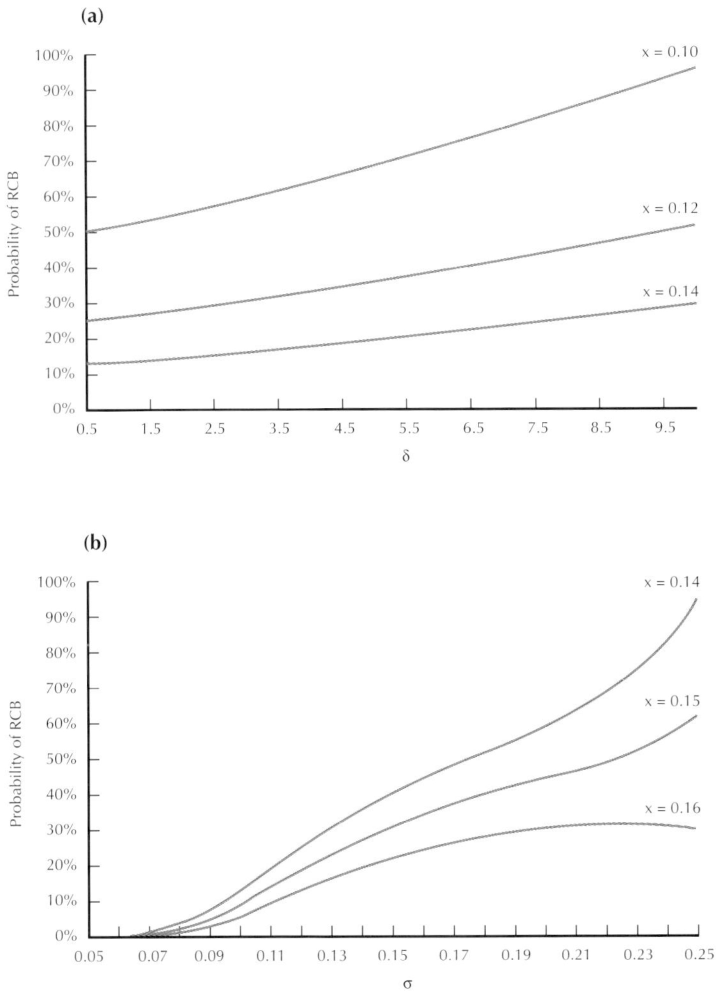

A RCB is the event in which both owners exercise their options to build simultaneously, following a downturn in demand. Such a simultaneous exercise equilibrium results from a rational fear of preemption. Each of the two graphs displays the sensitivity of the probability of a RCB to changing parameter values, for three levels of initial demand, x. **(a)** Displays the effect of the time-to-build, δ, on the probability of a RCB. The effect is monotonically increasing; markets in which there are large construction lags will be more prone to overbuilding. **(b)** Displays the effect of demand volatility, σ, on the probability of a RCB. For most ranges of demand, the effect is monotonically increasing; volatile real estate markets will be more prone to building booms in the face of declining demand. An existing building, earning a rent of R, can be redeveloped into a new building earning a stochastic rent flow of $X(t)D[Q(t)]$, where the demand state variable X is a geometric Brownian motion with drift μ and volatility σ, and $D(\cdot)$ is a downward-sloping inverse demand function. γ is the fractional loss in rent to an older building when a new building is constructed. The cost of construction is I, and r is the risk-free rate. For this figure, the following parameters are fixed (unless varied along the horizontal axis): $R = 0.1$, $\mu = 0.02$, $\sigma = 0.1$, $D(1) = 1$, $D(2) = 0.9$, $\gamma = 0.8$, $I = 1$, $\delta = 1$ and $r = 0.04$.

significantly less, perhaps on the order of six months. The model would predict a greater tendency for RCBs in the office market than in the industrial market. Figure 2b displays the vacancy rate patterns for the office and industrial property markets. When combined with the construction patterns displayed in Figure 2a, the tendency for office markets to experience RCBs is highlighted. Even as office markets experienced dramatically high and rising vacancy rates, new development continued to arrive and be initiated.

The model also predicts that greater volatility will contribute to the propensity toward RCBs. On a locational basis, this would imply that cities which are less economically diversified should be more prone to overbuilding. This would help explain the cases of the oil-patch cities, such as Denver and Houston, described earlier. During the time of their concentrated burst of construction in 1982–85, each had vacancy rates near 30%. In fact, by 1983, vacancy rates were already above 27% in each of these cities. In terms of property types, the volatility effect would also make office markets more prone to overbuilding than industrial markets. Recall from the earlier section on empirical implications that the standard deviation of the growth rate in office earnings was twice that of industrial earnings. This would tend to reinforce the time-to-build effect which implies that office markets should be especially prone to overbuilding.

ANALYSIS OF COMPETITIVE EQUILIBRIUM EXERCISE STRATEGIES

The model characterises the equilibrium set of exercise strategies in the case of a symmetric duopoly. In this section I derive the equilibrium exercise strategies of developers in a competitive equilibrium formulation. This is analogous to Constantinides' (1984) construction of competitive equilibrium warrant exercise strategies. In this case, developers ignore the impact of their entry on the underlying rent process. In addition, competitive pressures ensure that the value of development, at any point of entry, equals the cost of construction. I show that the exercise strategies are very similar to those derived in the duopoly case. The one significant difference is in the case for RCBs. Under perfect competition, there can be no such phenomenon.

The presentation of the competitive equilibrium approach will be mostly heuristic. A more detailed description of the technique can be found in Grenadier (1995b) and Leahy (1993). For simplicity of presentation, I derive the equilibrium development strategies by assuming zero construction lags.[19]

Derivation of competitive equilibrium exercise triggers

Consider a competitive real estate market where a large number of owners lease their buildings. Some owners have exercised their development option and earn a rent flow of P(t), while others have not exercised their development option and earn a rent flow of $R \cdot (1 - \gamma)$. At any time, an owner can exercise his development option by paying a construction cost of I. Assume that the number of redeveloped buildings can be represented as a continuum whose mass at time t is Q(t). Therefore, the setup is the same as in the duopoly model, with the demand function, $P(t) = X(t) \cdot D[Q(t)]$. However, Q should now be considered a continuous variable rather than only taking the values 0, 1, or 2.

Using Itô's lemma, the evolution of the equilibrium rent process can be written as[20]:

$$dP = \mu P\,dt + \sigma P\,dz + \frac{D'[Q(t)]}{D[Q(t)]} P dQ \tag{19}$$

Over the range in which no new development occurs ($dQ = 0$), equilibrium rents follow a geometric Brownian motion process. Note that while developers are idle, the instantaneous mean and variance of the rent process are independent of the level of supply. At moments at which new development occurs, the level of the rent processes will shift downward.

Let G(P) denote the value of a building prior to exercise of the development option, and H(P) the value of a building after development. These valuations will not depend explicitly on Q. This can be seen by the following two points. First, when supply is unchanging, the rent process is independent of Q, as is clear from equation (19), with $dQ = 0$. Second, even when supply is increasing due to development, the free entry condition will ensure that the payoff from exercising the development option is simply its cost, I, independent of Q.

Using the same techniques as in the second section, the values of buildings before and after exercise must satisfy the following differential equations in equilibrium:

$$\begin{aligned} 0 &= \tfrac{1}{2}\sigma^2 P^2 G'' + \mu P G' - rG + R \cdot (1 - \gamma) \\ 0 &= \tfrac{1}{2}\sigma^2 P^2 H'' + \mu P H' - rH + P \end{aligned} \tag{20}$$

These differential equations must be solved simultaneously. They are subject to two forms of boundary conditions. The first type are the standard individual optimisation conditions. These are merely the value-matching and smooth-pasting conditions that ensure rational exercise strategies. The second type is the competitive equilibrium condition of free entry. Even though individual owners select the optimal moments at which to develop, competitive forces ensure that new development will eat away all excess profits. These boundary conditions can be written as:

$$\begin{aligned} G(P^*) &= H(P^*) - I \\ G'(P^*) &= H'(P^*) \\ H(P^*) &= I \end{aligned} \tag{21}$$

where P^* is the level of rent that triggers new development.

The first boundary condition of equation (21) is the value-matching condition: the value of a building at the moment of development equals the value of a developed building minus the cost of construction. The second boundary condition is a smooth-pasting condition that ensures that agents exercise their development options optimally. The third condition is the free-entry condition ensuring zero net present value of development. Finally, I impose the regularity conditions, $G(0) = H(0) = 0$ to reflect the absorbing barrier at zero.

By solving differential equations (20) subject to boundary conditions (21), the solution to the optimal exercise trigger P^* can be expressed as:

$$P^* = \left(\frac{\beta}{\beta - 1}\right)(r - \mu)\left[I + \frac{(1-\gamma)R}{r}\right] \quad \textbf{(22)}$$

Using the inverse demand function (1), competitive equilibrium construction is triggered whenever X(t) rises to a function of existing supply Q(t). Define the level of X that triggers competitive equilibrium entry as $X^C(Q)$. From equations (22) and (1), we have:

$$X^C(Q) = \left(\frac{\beta}{\beta - 1}\right)\frac{(r - \mu)}{D(Q)}\left[I + \frac{(1-\gamma)R}{r}\right] \quad \textbf{(23)}$$

Note that this is precisely the same exercise trigger value for the Follower in the case of the duopoly (with no lags), except that it is now a function of Q.

The following proposition summarises the competitive equilibrium development exercise strategies.

Proposition 4. *In a competitive equilibrium setting, the supply of developed properties increases to the level* Q *the first moment that* X(t) *equals or exceeds the trigger function* $X^C(Q)$*, as defined in equation (23). That is, the entry time of the Qth increment of new development,* T_Q *can be written as:*

$$T_Q = \inf\left\{t \geq 0: X(t) \geq \left(\frac{\beta}{\beta - 1}\right)\frac{(r - \mu)}{D(Q)}\left[I + \frac{(1-\gamma)R}{r}\right]\right\} \quad \textbf{(24)}$$

Analysis

The nature of the competitive equilibrium exercise strategies is as follows. At any time other than the initial moment that the market opens, supply increases will be sequential. That is, given a supply of developed buildings Q(t) at t, the next increment of supply will be developed the next instant that X rises to the level $X^C(Q + dQ)$. Similarly, supply will rise to the level $\bar{Q}$, for any $\bar{Q} > Q(t)$, the first moment that X reaches $X^C(\bar{Q})$.

This is very similar to the sequential equilibrium of the duopoly model. Suppose the supply of developed buildings has just increased to Q. Then, define $T_C(Q, \bar{Q})$ as the time it takes for the supply of developed buildings to increase from Q to $\bar{Q}$, for any $\bar{Q} > Q$. Thus, $T_C(Q, \bar{Q})$ can be expressed as:

$$T(Q, \bar{Q}) = \inf\{t \geq 0: X(t) = X^C(\bar{Q}), X(0) = X^C(Q)\} \qquad \textbf{(25)}$$

However, unlike the duopoly case, there will not be any chance of a recession-induced construction boom. The only time that there can be simultaneous exercise is at the moment the market opens (which could have been in the distant past), and this is not triggered by a decline in demand. When the market opens, enough new development occurs such that Q(0) satisfies $X(0) = X^C[Q(0)]$. Following that, only incremental exercise ensues.

Therefore, the basic model's specification for conditions which lead to RCBs depends on the structure of the market. As the market structure approaches that of perfect competition, the potential for RCBs goes away.[21] The structure of the market, however, cannot be defined solely by the number of competitors, since the real estate "product" is highly differentiated by location. For example, while a major city may have a large number of potential developers and sites, there are likely to be many submarkets differentiated by such attributes as distance from the central business district, zoning restrictions and the quality of amenities.

CONCLUSION

Real-world situations do not always allow one to formulate an optimal exercise strategy in isolation. While one person may be solving a partial-differential equation with its associated smooth-pasting condition, a competitor may be doing likewise. In the example of real estate development, these calculations cannot be conducted separately, but must be done as part of a strategic equilibrium. The model demonstrates that development options might be exercised sequentially or simultaneously, depending on the underlying conditions of the market.

The solution to this option exercise game was not conducted for its own sake, but rather to provide an underlying theory from which one may begin to understand actual market behaviour. Real estate development options are not always exercised in a smooth pattern; sometimes markets sit idle for years and then blast off in a surge of construction. The model is able to isolate the conditions that make such a phenomenon more or less likely. In addition, sometimes these bursts of construction occur when the underlying demand for space is falling. While this is often regarded as irrational overbuilding, the model provides a rational equilibrium foundation for such exercise patterns.

The model of strategic exercise can be applied to a variety of real option settings. Consider a pharmaceutical industry competing over the search for a new drug. Beginning a research programme is costly and yields uncertain results. However, each firm must formulate its research strategy fully cognizant of the development strategies of its competitors. Consider also the decision of a firm to introduce a new product to consumers. The initial advertising and distribution expenses must be paid now in order to reap the uncertain benefits of sales in the future. If the firm faces competitors able to produce a similar product, then the timing of the new product option exercise is a strategic exercise game.

APPENDIX

Proof of Proposition 2

Proposition 2. *There exists a unique point, $X_L \in (0, X_F)$, with the following properties:*

$$\begin{aligned} L(X,\delta) - I &< F(X) \quad \text{for} \quad X < X_L \\ L(X,\delta) - I &= F(X) \quad \text{for} \quad X = X_L \\ L(X,\delta) - I &> F(X) \quad \text{for} \quad X_L > X < X_F \\ L(X,\delta) - I &= F(X) \quad \text{for} \quad X \geq X_F \end{aligned}$$

Proof. Define the function $D(X) = L(X, \delta) - F(X) - I$, where $L(X, \delta)$ was derived in equation (9), by setting $\tau = \delta$, and $F(X)$ in equations (7) and (8).

First, I establish the existence of a root for $D(X)$ in the interval $(0, X_F)$. Evaluating at $X = 0$ yields $D(0) = -[I + (1 - \gamma)R/r] < 0$. Similarly, evaluating at $X = X_F$ yields $D(X_F) = 0$. Finally, some algebraic manipulation yields $\lim_{X \to X_F^-} D'(X) < 0$. Therefore, $D(X)$ must have at least one root in the interval $(0, X_F)$.

To prove uniqueness, one merely needs to demonstrate strict concavity over the interval. Differentiating $D(X)$ twice yields:

$$D''(X) = -\frac{\beta}{X^2}\left[1 + \beta\left(\frac{D(1) - D(2)}{D(2)}\right)\right]\left[1 + \frac{(1-\gamma)R}{r}\right]\left(\frac{X}{X_F}\right)^{\beta} < 0$$

Thus, the root is unique.

Finally, from Proposition 1, if $X \geq X_F$, the optimal response of the Follower is to begin construction immediately. Thus, both Leader and Follower will have the same value over this range. QED.

1 Several key assumptions account for the differing nature of the Williams equilibrium from the present model's. The Williams model may be seen as an approximation of a steady-state equilibrium, in that the supply of undeveloped assets is permitted to increase at the same rate as the supply of developed assets. In addition, in the Williams model development can

be done in infinitesimal increments and with no construction lag. In the present model, development is done in finite increments, there is time-to-build, and there is no automatic increase in land holdings following development.

2 This demand structure is similar to that in Williams (1993) and Grenadier (1995b).

3 A potential rationale for this assumption is that development requires approval from the local government. Approval may depend on who is first in line or arbitrary considerations.

4 It would be more realistic to assume that the deterioration in the rent to the old building depends on the stage of the Leader's completion. The assumption that the harm is immediate is done for tractability. However, none of the basic results of the model is altered under a time-varying harm function.

5 Or, equivalently, that the process dz is uncorrelated with aggregate wealth.

6 See Margrabe (1978) for an analysis of the option to exchange one risky asset for another.

7 See the appendix to Chapter 4 in Dixit and Pindyck (1994) for a discussion of optimal stopping regions.

8 The term $\{[D(2)/(r-\mu)]e^{-(r-\mu)\delta}\}X_F$ equals the present value of a perpetual rent flow of $X(t) \cdot D(2)$, beginning in δ years, and conditional on the current level of X being X_F.

9 See Merton (1973) for a derivation of the high-contact condition.

10 Fudenberg and Tirole (1985) argue that if one equilibrium Pareto dominates all others, then it is the most reasonable to expect.

11 Caplin and Leahy (1994) provide an example of a development cascade in a specific location: Lower Sixth Avenue in New York City. Caplin and Leahy posit an information-spillover explanation for the rapid growth of a previously deserted retail area.

12 Of course, a similar analysis could be performed for other ranges of initial starting points. However, this is the only case in which the time between starts is independent of the (arbitrary) initial starting level of demand.

13 In particular, see McDonald and Siegel (1986).

14 In this case, an analytical solution to the comparative static result is not obtainable. However, equation (16) is easily solved using numerical procedures. While the results in Figure 1 were derived under a set of initial parameter values, the same qualitative results hold under a wide range of assumed parameter values.

15 An additional factor that may influence the timing of development in a city (or submarket) is zoning restrictions. Such restrictions can slow down or speed up development and are subject to changes over time (eg, changes in the zoning commission).

16 This data was provided by William Wheaton. This proprietary database contains square-footage office completions, by city, semi-annually over the period from 1960–91.

17 One might possibly explain such behaviour by looking for bursts in office demand. However, this does not appear to be the case. Office employment comes predominantly from two sectors of the economy: the Finance-Insurance-Real Estate and the Services sectors (DiPasquale and Wheaton (1996)). The growth rate in these two sectors of employment over the period from 1982 to 1985 was the opposite of what one would expect, given the rates of construction growth. Denver grew by 12%, Houston grew by 9.7%, Chicago grew by 14.3%, and Minneapolis grew by 16.1%. Of course, since real estate is a very long-lived asset, development reflects expectations of demand growth over a longer period. However, even demand growth over the period from 1985 to 1990 fails to explain the differing development rates.

18 See Wheaton and Torto (1986) for a discussion of the popular explanations for overbuilding.

19 In order to accommodate a time-to-build feature, the supply of completed units Q must be modified to include construction in progress, as in Bar-Ilan, Sulem and Zanello (1992).

20 There are no second order terms in Q because the supply process is variation finite. If this were not the case, the economy would be incurring infinite costs of adjustment.

21 Of course, in the case of a monopolist, there is also zero probability of pre-emption. Thus, a monopolist will only build when demand rises to an upper trigger, and there will be no analogue to a RCB.

BIBLIOGRAPHY

Bar-Ilan, A., A. Sulem and A. Zanello, 1992, "Time to Build and Capacity Choice", Working paper, Dartmouth College and Tel-Aviv University.

Birch, David, 1990, *America's Future Office Space Needs,* National Association of Industrial and Office Parks, Washington, DC.

Caplin, Andrew and John Leahy, 1993, "Miracle on Sixth Avenue: Information Externalities and Search", Discussion paper no. 1665, Harvard University.

Constantinides, George M., 1984, "Warrant Exercise and Bond Conversion in Competitive Markets", *Journal of Financial Economics* 13, pp. 371–97.

Cox, J.C. and S.A. Ross, 1976, "The Valuation of Options for Alternative Stochastic Processes", *Journal of Financial Economics* 3, pp. 145–66.

DiPasquale, Denise and William C. Wheaton, 1996, *The Economics of Real Estate Markets,* Prentice-Hall, Englewood Cliffs, NJ.

Dixit, Avinash K and Robert S. Pindyck, 1994, *Investment Under Uncertainty,* Princeton University Press, Princeton, NJ.

Dutta, Prajit K and Aldo Rustichini, 1991, "A Theory of Stopping Time Games with Applications to Product Innovation and Asset Sales", Discussion paper no. 523, Columbia University.

Dybvig, Philip H. and Chester S. Spatt, 1983, "Adoption Externalities as Public Goods", *Journal of Public Economics* 20, pp. 231–47.

Emanuel, David C., 1983, "Warrant Valuation and Exercise Strategy", *Journal of Financial Economics* 12, pp. 211–35.

Fudenberg, Drew and Jean Tirole, 1985, "Preemption and Rent Equalization in the Adoption of New Technology", *Review of Economic Studies* 52, pp. 383–401.

Grenadier, Steven R., 1995a, "The Persistence of Real Estate Cycles", *Journal of Real Estate Finance and Economics* 10, pp. 95–119.

Grenadier, Steven R., 1995b, "Valuing Lease Contracts: A Real-Options Approach", *Journal of Financial Economics* 38, pp. 297–331.

Harrison, J. Michael, 1985, *Brownian Motion and Stochastic Flow Systems,* Wiley, New York, NY.

Huang, Chi-fu and Lode Li, 1991, "Entry and Exit: Subgame Perfect Equilibria in Continuous-time stopping games", Working paper no. 54, Yale University.

Kulatilaka, Nalin and Enrico C. Perotti, 1995, *What is Lost by Waiting to Invest,* mimeo, Boston University.

Leahy, John V., 1993, "Investment in Competitive Equilibrium: The Optimality of Myopic Behavior", *Quarterly Journal of Economics* 108, pp. 1105–33.

Margrabe, William, 1978, "The Value of an Option to Exchange one Asset for Another", *Journal of Finance* 33, pp. 177–98.

McDonald, Robert and Daniel Siegel, 1986, "The Value of Waiting to Invest", *Quarterly Journal of Economics* 101, pp. 707–27.

Merton, Robert C., 1973, "Theory of Rational Option Pricing", *Bell Journal of Economics and Management Science* 4, pp. 141–83.

Porter, Michael E. and Michael Spence, 1978, "The Capacity Expansion Process in a Growing Oligopoly: The Case of Corn Wet Milling", Working paper no. HBS78-61, Harvard University.

Smets, Frank R., 1993, "Essays on Foreign Direct Investment", unpublished dissertation, Yale University.

Spatt, Chester S. and Frederic P. Sterbenz, 1985, "Learning, Preemption, and the Degree of Rivalry", *Rand Journal of Economics* 16, pp. 84–92.

Spatt, Chester S. and Frederic P. Sterbenz, 1988, "Warrant Exercise, Dividends, and Reinvestment policy", *Journal of Finance* 43, pp. 493–506.

Titman, Sheridan, 1985, "Urban Land Prices Under Uncertainty", *American Economic Review* 75, pp. 506–14.

Wheaton, William C., 1987, "The Cyclic Behavior of the National Office Market", *AREUEA Journal* 15, pp. 281–99.

Wheaton, William C. and Raymond Torto, 1986, "The National Office Market, History and Future Prospects, III: Federal Taxes and the Supply of Office Space", unpublished manuscript, Massachusetts Institute of Technology.

Wheaton, William C. and Raymond Torto, 1990, "An Investment Model of the Demand and Supply for Industrial Real Estate", *AREUEA Journal* 18, pp. 530–47.

Williams, Joseph T., 1993, "Equilibrium and Options on Real Assets", *Review of Financial Studies* 6, pp. 826–50.

4

*Equilibrium and Options on Real Assets**

Joseph T. Williams

University of British Columbia

In aggregate, options on real and financial assets can have very different properties. Typically, the good or service produced by a real asset has a finite elasticity of demand, and developers have finite capacities. Also, the supply of options can be limited, and developers can be less than perfectly competitive. In a subgame-perfect Nash equilibrium with these properties, the optimal exercise policy and resulting values of developed and undeveloped assets are calculated explicitly. The novel comparative statics are discussed in detail.

To date, options on real assets, like real estate, natural resources, and capital assets, have been analysed only in partial equilibrium.[1] When viewed from the perspective of a single developer with a single undeveloped asset, real options have many characteristics in common with financial options. With real estate and other real assets, the option is an undeveloped property, the underlying asset is a developed property, the exercise price is the cost of development, and the maturity is generally infinite. In previous articles special characteristics of real options have been emphasised. These include the time to build, the often stochastic cost of construction, the controllable quantity or density of development, and the option to abandon either developed or undeveloped property.[2]

When viewed in aggregate, options on real and financial assets can have very different properties. At least four differences are potentially significant. Exercising options to develop real assets affects the aggregate supply

*This chapter was first published in *Review of Financial Studies* 6(4), 1993, pp. 825–50, and is reprinted with the permission of Oxford University Press. The author wishes to thank Chester Spatt for suggesting this problem, and Chester Spatt (the editor), Suresh Sundaresan (the referee), and the participants in seminars at UBC, UCI, and WFA for their helpful comments on the original paper.

of developed assets and thereby the price in equilibrium of the output from each developed asset. In turn, this affects both a developer's optimal exercise policy and the value of his option to develop. Also, if contractors have increasing costs or limited capacities, then the cost of development depends on the aggregate demand for development. This not only alters each developer's optimal exercise policy but also may preclude simultaneous exercise of all outstanding options.[3] In addition, the supply of undeveloped assets can be constrained, possibly by the supply of developed assets. Again, this affects exercise policies and option values. Finally, with monopolistic or oligopolistic ownership of undeveloped assets, the action of any one owner can affect the values of other assets in his portfolio, as well as the actions of other owners. How each owner responds in equilibrium depends partly on the concentration of ownership. These properties of real options do not apply to most financial options, including calls and puts on stocks and bonds. With the exception of warrants and convertible bonds, financial options are exercised by trading paper that is held in zero net supply by a large number of autonomous individuals.[4] For the latter securities there is no downward-sloping demand by consumers, no aggregate constraint on the rate of exercise, no limited supply of options and no monopolistic or oligopolistic exercise by holders.

How significant are these distinguishing characteristics of real options? The significance, as measured by the magnitude of the effect on optimal exercise policies and the resulting market values, depends partly on the elasticity of demand for products from the real asset, the capacities of developers, the supply of undeveloped assets and the degree of concentration among developers. To illustrate possible magnitudes, consider one category of real assets: single-family houses within the metropolitan area of Phoenix, Arizona, between 1970 and 1990. Despite the area's rapid growth during this period, the annual construction of new houses never exceeded 3% of the housing stock. Because the average time for housing permits remained roughly constant at less than two years, the supply of building lots never exceeded 6% of the stock.[5] In any year, half of the houses were constructed by no more than 10 developers and the remainder by no more than 200 developers.[6] Although no reliable estimates of the price elasticity of demand for housing services are available for Phoenix, most estimates for the United States have been less than 1.0.[7]

In this chapter options to develop real assets are analysed in an equilibrium with the aforementioned properties. By assumption, the demand for the product from the real asset depends on its current price, as well as an exogenous factor that evolves as a geometric Wiener process. The aggregate demand function has a constant elasticity that includes as a special, limiting case the infinite elasticity from most previous models of real options.[8] Also, all developers have the same limited capacity to construct developed assets using perfectly competitive contractors. This limited capacity precludes simultaneous exercise of all real options and thereby constrains the aggregate rate of development. In addition, the aggregate supply of undeveloped assets is proportional to the supply of developed assets, with smaller constants of proportionality representing tighter restrictions on growth. Although development increases the supplies of developed and undeveloped assets at the same rate over time, and thereby adds to developers' supplies of undeveloped assets, the limited supply at any one time of options to develop affects their optimal rate of exercise. Finally, ownership of undeveloped assets is divided equally among a fixed number of identical investors, ranging from a monopolist, to oligopolists, to perfect competitors. Owners choose their rates of development in a Markov perfect equilibrium.

Under these assumptions the optimal exercise strategy and resulting values of developed and undeveloped assets are calculated explicitly. In equilibrium, development optimally occurs at all levels of current income from a developed asset that exceed or equal a lower bound. This bound is a switching point between no development and development at the maximum feasible rate. It replaces the single point of optimal exercise from previous articles on real options. With this rate of development, the ratio of income to value for the developed asset is everywhere increasing in income. The ratio converges to a constant discount rate as the elasticity of demand increases to infinity. In turn, the value of the undeveloped asset depends on both the optimal rate of development and the resulting value of the developed asset. In this sense, the market values of the developed and undeveloped assets are interdependent, much like stocks and warrants but unlike stocks and their traded options. Other properties of the solution are similar to previous results on options.

The novel economic insights come from the new comparative statics. For example, the optimal switching point for development, as described earlier, is increasing in three parameters of the developed asset: the elasticity of demand for its good or service, its maximum rate of development and its rate of depreciation. It is also decreasing in both the supply of undeveloped relative to developed assets and the number of identical owners of undeveloped assets. The maximum rate of development, relative supply of undeveloped assets, and ownership of undeveloped assets are entirely new parameters. Also, all the comparative statics have new interpretations. In standard models of options with an exogenous pricing process for the associated security, optimal exercise is deferred by changes in the values of parameters that increase the value of the option. In this model, optimal exercise is also deferred by changes in parameters that either decrease the value of the developed (optioned) asset or increase the effect of exercise on the value of the remaining undeveloped assets (options) owned by a monopolist or an oligopolist. Undeveloped assets are always held in equilibrium because the optimal rate of development is finite.

The chapter is organised as follows. The model is presented in the first section, and the equilibrium is displayed in the following one. The numerical computations are then discussed, and the results are summarised. All technical details appear in the Appendix.

MODEL

Consider a category of real assets, such as single-family houses within a metropolitan area. For notational simplicity, all properties within the category are assumed to be identical, after adjusting for measurable differences, such as the number of square feet in a house or its distance from the central business district. Also, the category is sufficiently broad so that the supply and demand for other categories of real assets can be ignored, except as specified later. For this category, q is the current aggregate supply of developed assets; x is the current value of an exogenous variable affecting the demand for the good or service produced by each developed asset; and y is the current income from that asset. For example, with single-family houses the exogenous variable x summarises such factors as the current ratio of buyers to sellers, as well as the current income and net worth of buyers, and the income y equals the current rent or implicit rent minus operating costs.

The good or service from the developed asset can be sold in a perfectly competitive, continuous spot market. At each instant of time, atomistic consumers maximise their utility and thereby generate an aggregate demand for the good or service that has a constant price elasticity. This constant elasticity holds with an instantaneously perishable commodity, such as rental services, if, for example, all consumers have identical, time-additive, isoelastic utilities. In addition, the output from each developed asset is produced with an identical, convex, Cobb–Douglas cost function. Finally, control of the developed assets is sufficiently dispersed that producers of the good or service are perfect competitors in the spot market. As in all option pricing models, these assumptions are introduced mainly for analytical convenience. When combined with subsequent assumptions, they permit an explicit solution in the next section.

Under the foregoing assumptions the clearing condition in the spot market has an analytically convenient form. Specifically, the aggregate demand for the good or service, the optimal output from each developed asset and the resulting income per developed asset y are proportional to power functions of the current price of the good or service. Thus, the aggregate demand for the good or service and its supply by each developed asset are proportional to power functions of the income y. As a result, the exogenous factor x can be rescaled to capture the constants of proportionality, and the requirement that aggregate demand always equals aggregate supply in the spot market can be simplified to the clearing condition:[9]

$$q = (x/y)^{\alpha} \tag{1}$$

with constant α. The elasticity α depends on two constant elasticities for the good or service: one in the aggregate demand by consumers and another in the cost of production. Details, including the derivation of (1), appear in the Appendix. Given (1), the income per unit of developed assets is $y = xq^{-1/\alpha}$. In most previous models of real options, income has been assumed to be exogenous: $y = x$. Exogenous income is a special case of (1) in which demand is perfectly elastic: $\alpha = \infty$.

The exogenous factor affecting demand changes stochastically through time. By assumption, it evolves continuously in a proportional random walk. Also, the mean and variance of its growth rate are constant per unit of time. In this case, the exogenous variable x follows a geometric Wiener process:

$$dx/x = \mu\, dt + \sigma\, dz \qquad \textbf{(2)}$$

In (2) the variables t and z are time and a standard Wiener process, and the constants μ and σ^2 are the mean and variance of the growth rate in the exogenous demand x. The stochastic increments in (2) can be correlated with changes in other state variables affecting the demands for other categories of real assets. Thereby, the rates of development to be derived here can be correlated in equilibrium for different categories of real assets.

All owners of undeveloped assets are identical. At all times the number of such owners is constant at ν, with $\nu \in \{1, \ldots, \infty\}$. These identical owners have equal numbers of undeveloped assets.[10] When developing an asset, an owner acts as his own developer and hires a contractor in the perfectly competitive market for contractors. Hence, owners of undeveloped assets are also called developers. To produce a developed asset, each owner or developer must use one undeveloped asset and pay his contractor the perfectly competitive, constant cost, $\gamma > 0$. This cost is constant up to each developer's capacity of $\beta q/\nu$ assets per unit of time. For example, a developer of single-family houses in a metropolitan area with q houses can build over a very short period of length Δt at most $(\beta q/\nu)\Delta t$ houses. With this analytically convenient assumption, each developer's capacity is identical and proportional to the current aggregate supply of developed assets q.[11] Also, as in most models of real options, development is instantaneous once it is started.[12]

The aggregate supply of the real asset also changes through time, increasing with development and decreasing with depreciation. If developer n builds $b_n q$ units per unit of time, then he adds to the aggregate supply q at the rate b_n, for $n = 1, \ldots, \nu$. Development then occurs at the rate $b \equiv \Sigma_n b_n$. The aggregate building rate b cannot exceed the maximum building rate β, $0 \le b \le \beta$. Simultaneously, each developed property depreciates at the constant rate δ per unit of time. The depreciation rate δ is also less than the maximum building rate β, $0 \le \delta < \beta$. Net of depreciation the aggregate supply q then grows at the rate:

$$\dot{q}/q = b - \delta \tag{3}$$

with $0 \leq b \leq \beta$.

The values of developed and undeveloped assets depend on the aggregate supplies of developed and undeveloped assets, as well as the demand for developed assets, as reflected in the clearing condition (1). The supply and demand for developed assets are determined by the state variables, q and x. To avoid adding another state variable, we must restrict the aggregate supply of undeveloped assets to depend at most on these two variables. Also, for an explicit solution under these assumptions, the aggregate supplies of developed and undeveloped assets must be proportional. Proportional supplies are motivated later. Given proportionality, the aggregate supply of the undeveloped asset is θq, with the constant, $0 < \theta < \beta/\delta$. Smaller values of the constant θ represent tighter restrictions on growth. With proportionality the supplies of developed and undeveloped assets grow at the same rate over time. At any time the supply of undeveloped assets is finite. As shown subsequently, this finite supply affects developers' optimal exercise policies. Realistically, the restrictive assumption of proportional supplies should be viewed in the same context as the previous assumptions of proportional building capacity and proportionality in (1), (2) and (3).[13] Under the foregoing assumptions, the current state of the market for both developed and undeveloped assets is summarised by the state variables, q and x. In this case, the current values of each developed and undeveloped asset can be specified as $V^d(q, x)$ and $V^u(q, x)$, respectively.

Proportional supplies of developed and undeveloped assets may be a reasonable representation of a market in steady state. An example of developed and undeveloped assets is single-family houses and prepared building lots. Over time, as houses are constructed, building lots are withdrawn from the stock of undeveloped assets at the rate (3). This rate, $b - \delta$, is the measured net of the depreciation rate δ at which houses are returned to the undeveloped state of building lots. If the city is to stay in steady state, then building lots must be created at the urban boundary at a rate that is equal to the current net use of building lots, $b - \delta$. In practice, this could reflect a regulatory rule of planning officials or the capture of regulators by developers. Given this growth rate of building lots, the aggregate supply of undeveloped assets at time t is θq_t. These new building lots could be created at a virtually constant cost from non-urban land at the ever-expanding boundary of the city. However, the supply of raw land cannot be perfectly elastic because land at the boundary is closer to urban centres of employment and therefore more valuable than land away from the boundary.[14]

With some additional assumptions, developed and undeveloped assets can be valued by using standard techniques from the literature on option pricing. The first assumption is common to all models of real options.

All investors are either risk neutral, or the stochastic evolution in (2) of the exogenous variable x is spanned by the concurrent, instantaneous returns from a portfolio of securities that can be traded continuously without transaction costs, in a perfectly competitive capital market. With spanning, trading in liquid securities substitutes for trading in the illiquid real asset.[15] With single-family houses the substitute portfolio could include listed securities such as real estate investment trusts, master limited partnerships, and common stocks of home builders and their suppliers. For the substitute portfolio the excess mean rate of return per unit of standard deviation is assumed to be a constant λ. This constant is zero if all investors are risk neutral and positive if investors are risk averse. Also, the riskless rate of interest is constant at ι per unit of time. In this familiar case, the risk-adjusted expected rate of growth per unit of time for income y is another constant: $\rho \equiv \mu - \lambda\sigma$. Alternatively, with risk neutrality and no spanning, the risk-adjusted expected rate of growth ρ equals μ. To ensure that the values of developed and undeveloped assets are positive and finite, the risk-adjusted mean ρ must be bounded above: $\rho < \iota - \delta/\alpha$.

The description of the valuation problem begins with a developed asset. By a now-familiar argument from the literature on option pricing, the valuation function for a developed asset V^d must satisfy the partial differential equation:

$$0 = \tfrac{1}{2}\sigma^2 x^2 V^d_{xx} + (b - \delta)qV^d_q + \rho x V^d_x - (\delta + \iota)V^d + xq^{-1/\alpha} \tag{4}$$

In (4) the subscripts denote partial derivatives. Although this valuation equation is well specified for all feasible building rates b, it is solved subsequently only for the optimal building rates in equilibrium. Also, since development is irreversible, (4) must be satisfied for all feasible values of the variables q and x: $q, x \geq 0$.[16]

A valuation equation for the developed asset, such as (4), appears in other models of real options but not financial options. With financial options the analogue of the developed asset is the underlying security – for example, the stock on which a call option is exercised – and the value of this security is specified exogenously as one of the state variables. Here, as in other models of real options, the state variables q and x are not traded assets. This causes no complication with respect to the aggregate supply q because it evolves continuously and deterministically in (3). However, the exogenous securities demand x is a geometric Wiener variate in (2); hence, substitute securities or special preferences must be invoked to identify the coefficient ρ in (4). As a result, the risk-adjusted rate of growth ρ that appears in (4) as the coefficient of xV^d_x replaces the riskless rate of interest ι that appears at this position in valuation equations for financial options. In addition, developed assets depreciate at the constant rate δ, so that the sum $\delta + \iota$ replaces the riskless rate of interest ι in valuation equations for financial options. Finally, the last term in (4) is the asset's current income: $y = xq^{-1/\alpha}$.

The value of a developed asset must also satisfy two boundary conditions. If the exogenous variable affecting demand is zero ($x = 0$), then the asset's income y must be zero from that time forward, under (1) and (2). In this case, a developed asset has no value:

$$V^d(q, 0) = 0 \tag{5}$$

Also, if the developed asset is to have a finite price-earnings ratio, then its value per unit of income must be bounded above by some positive constant υ:

$$V^d(q, x) \leq \upsilon x q^{-1/\alpha} \tag{6}$$

as $x \to \infty$. The limiting condition (6) precludes bubbles in prices. Together, the valuation equation (4) and the boundary conditions (5) and (6) determine the valuation function for the developed property, V^d.[17]

Undeveloped assets are valued similarly. In a Markov perfect equilibrium, each player has a strategy that depends on the history of the game only through state variables that follow a Markov process. Moreover, these Markov strategies must yield a Nash equilibrium in every proper subgame. In this equilibrium each owner or developer n conjectures correctly that each other owner-developer currently builds at his optimal rate b^*_m, for $m = 1, \ldots, n-1, n+1, \ldots, \nu$. The optimal building rate b^*_m can depend only on the current state variables, q and x. Thereby, each developer n correctly anticipates in equilibrium that all other developers currently build at the aggregate rate: $b^*_{-n} \equiv \Sigma_{m \neq n} b^*_m$. Given this conjecture, each developer then maximises his wealth by maximising the current market value of his inventory of undeveloped assets. In this case, developer n solves the problem:

$$0 = \max_{b_n} \left\{ \frac{1}{2}\sigma^2 x^2 V^u_{xx} + \left(\frac{\nu - 1}{\nu} b^* + b_n - \delta \right) q V^u_q + \rho x V^u_x - \iota V^u + \frac{\nu}{\theta} b_n (V^d - V^u - \gamma) \right\} \tag{7}$$

subject to $0 \leq b_n \leq \beta/\nu$, for all $n = 1, \ldots, \nu$. In equilibrium each developer always builds at his optimal rate, equal to the conjectured rate b^*_n, for $n = 1, \ldots, \nu$. Development then occurs at the aggregate rate: $b^* \equiv \Sigma_n b^*_n$. Without loss of generality, each developer is assumed to build in (7) whenever he is indifferent between building and not. Because the rates of development b_n are bounded above by the maximum building rate β/ν, the valuation equation (7) must be satisfied for all feasible states: $q, x \geq 0$.[18]

This valuation equation for an undeveloped asset differs somewhat from the previous equation for a developed asset. At the solution to (7), all developers optimally build at the rate $b_n = b^*$, in which case the terms qV^d_q in (4) and qV^u_q in (7) have the same coefficient, $b^* - \delta$. Also, only developed assets depreciate, so that the constant δ appears as a coefficient of V^d in (4) but not of V^u in (7). To understand the final terms in (7), note that developer n builds $b_n q$ units per unit of time, using his inventory of undeveloped assets. Because all undeveloped assets θq are owned by developers and all ν developers are identical, all building by each developer can be restricted to his portfolio of $\theta q/\nu$ undeveloped assets. Measured per undeveloped asset, developer n then builds per unit of time the number of developed assets: $b_n q/(\theta q/\nu) = (\nu/\theta)b_n$. Finally, each developed asset yields the capital gain, $V^d - V^u - \gamma$. This gain reflects the dependence of the value of undeveloped assets V^u on the concurrent value of developed assets V^d. With monopolistic or oligopolistic developers, $\nu < \infty$, the gain may be positive, whereas with perfect competition, $\nu = \infty$, it should be zero.[19] In fact, as shown in the third section, the gain is positive with a monopoly or oligopoly and zero otherwise.

The equation for an undeveloped asset also differs from familiar valuation equations for financial options. In (7) the option is exercised on a developed asset, the value of which depends on the elasticity α. Most financial options have zero net supply, so their exercise has no effect on cashflows from real assets.[20] Second, options to develop a real asset can be exercised in (7) no more rapidly than the maximum building rate of each developer: $b_n \leq \beta/\nu$. In contrast, options on securities are exercised by exchanging paper and hence have no economically significant restrictions on rates of exercise. In addition, the supply of undeveloped assets is restricted in (7) relative to the supply of developed assets by the ratio θ. In familiar models of financial options, the supply of options is unlimited because writers of options are not required to provide collateral. Finally, the number of developers ν measures the concentration of ownership of undeveloped assets. Owners can be monopolistic, $\nu = 1$, oligopolistic, $1 < \nu < \infty$, or competitive, $\nu = \infty$. In all previous models of real options, the representative developer is endowed with one undeveloped asset. He then chooses his exercise policy to maximise his gain on that asset: $V^d - V^u - \gamma$. This matches the maximand in (7) with perfectly competitive developers, $\nu = \infty$.[21]

Finally, the boundary conditions for developed and undeveloped assets are identical. For an undeveloped asset the initial condition corresponds to (5):

$$V^u(q, 0) = 0 \tag{8}$$

Also, the limiting condition matches (6):

$$V^u(q, x) \leq \upsilon x q^{-1/\alpha} \tag{9}$$

as $x \to \infty$. This limiting condition replaces the optimal stopping conditions in previous articles on options with unbounded rates of exercise. Together, (7) to (9) determine the value of an undeveloped property.[22]

EQUILIBRIUM

The previous problem is solved in three steps. First, the region in which development is optimal is assumed to satisfy a simple switching condition. Conditional on this assumption, the value of a developed asset V^d is then calculated from (4) to (6). Next, this solution is used with (7) to (9) to derive both the interval of optimal development and the resulting value of an undeveloped asset V^u. Finally, the solution is shown to satisfy the initial switching condition under two constraints on the parameters. All results are displayed in this section and derived in the Appendix.

The analysis begins with two preliminaries. First, the variables in (4) to (9) are transformed as follows: $F(y) \equiv V^d(q, x)$ and $G(y) = V^u(q, x)$. With this transformation $F(y)$ and $G(y)$ are the current market values of a developed and an undeveloped asset, conditional only on the income per unit of the developed asset y. Next, the optimal building rate b_n^* from problem (7) is hypothesised to satisfy the switching condition:

$$b_n^* = \begin{cases} 0 & 0 \le y < y^* \\ \beta/\nu & y^* \le y < \infty \end{cases} \tag{10}$$

for some non-negative switching point y^* and all developers $n = 1, \ldots, v$. If (10) is correct, then the Markov perfect equilibrium is symmetric with the aggregate building rate: $b^* = vb_n^*$. In other words, all developers optimally build at the same rate: $b_n^* = b^*/v$. This equilibrium is characterised by no development, $b^* = 0$, at all incomes y below the switching point y^*, and building at the maximum rate, $b^* = \beta$, otherwise. As shown in the Appendix, problem (4) to (9) has a unique solution under the previous transformation that satisfies hypothesis (10) if the parameters satisfy two inequalities. The inequalities are derived in the Appendix, displayed in this section, and evaluated numerically in the next section.

The value of a developed asset follows from (4) to (6). Conditional on the previous transformation, problem (4) to (6) has the unique solution:

$$F(y) = \begin{cases} \phi_1 y - \chi_1 y^*(y/y^*)^{\zeta_1} & 0 \le y < y^* \\ \phi_2 y + \chi_2 y^*(y/y^*)^{\zeta_2} & y^* \le y < \infty \end{cases} \tag{11}$$

In (11) the new parameters are

$$\zeta_1 \equiv 1 - \left(\frac{1}{2} + \frac{\rho + \delta/\alpha}{\sigma^2}\right) + \sqrt{\left(\frac{1}{2} + \frac{\rho + \delta/\alpha}{\sigma^2}\right)^2 + \frac{2}{\phi_1 \sigma^2}} > 1$$

$$\zeta_2 \equiv \frac{1}{2} - \frac{\rho + (\delta - \beta)/\alpha}{\sigma^2} - \sqrt{\left[\frac{1}{2} - \frac{\rho + (\delta - \beta)/\alpha}{\sigma^2}\right]^2 + 2\frac{\delta + \iota}{\sigma^2}} < 0$$

$$\phi_1 \equiv \frac{1}{\delta + \iota - \rho - \delta/\alpha} > 0, \qquad \phi_2 \equiv \frac{1}{\delta + \iota - \rho + (\beta - \delta)/\alpha} > 0$$

and

$$\chi_1 \equiv \frac{\beta}{\alpha}\frac{1 - \zeta_2}{\zeta_1 - \zeta_2}\phi_1\phi_2 > 0, \qquad \chi_2 \equiv \frac{\beta}{\alpha}\frac{\zeta_1 - 1}{\zeta_1 - \zeta_2}\phi_1\phi_2 > 0$$

The value of a developed asset is displayed in Figure 1. The valuation function F is non-negative and increasing, strictly concave below the switching point y^*, and strictly convex above y^*. It is bounded above by $\phi_1 y$ and below by its asymptote $\phi_2 y$. As the income y increases, the ratio of

Figure 1 Value of a developed asset

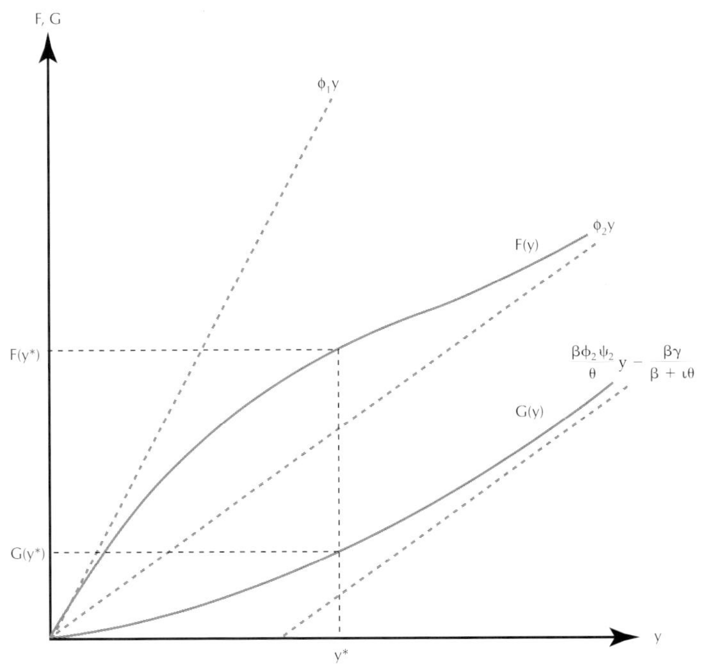

The values of a developed and undeveloped asset, F(y) and G(y), depend on the current income y from a developed asset. In equilibrium, developers do not build, $b^* = 0$, when $0 \leq y \leq y^*$, and build at the minimum rate $b^* = \beta$ when $y^* \leq y < \infty$.

income to value, $y/F(y)$, increases. At $y = 0$, this ratio, commonly called the capitalisation rate, attains its lower bound: $\delta + \iota - \rho - \delta/\alpha$. The lower bound is the risk-adjusted, discount rate for future income when no building is currently optimal: $b^* = 0$. With $\dot{q}/q = -\delta$, this bound equals the riskless rate of interest ι minus the expected growth rate of income y from (1), adjusted for both risk and depreciation, $\rho + \delta/\alpha - \delta$. Also, as $y \to \infty$, the ratio $y/F(y)$ approaches its upper bound: $\delta + \iota - \rho + (\beta - \delta)/\alpha$. This asymptote is the same risk-adjusted, discount rate but with building at the maximum rate: $b^* = \beta$. In addition, as the elasticity of demand α increases the growth rate of income y becomes less sensitive to the growth rate of aggregate supply q, and the difference between the above two discount rates decreases. In the limit the capitalisation rate converges to a constant: $y/F(y) \to \delta + \iota - \rho$ as $\alpha \to \infty$. This constant discount rate appears in previous solutions for real options with exogenous income.[23]

The value of an undeveloped asset is determined by the solution to problem (7) to (9). With (10) and the transformation above (10), the unique solution is

$$G(y) = \begin{cases} \beta\left[\dfrac{\gamma}{\beta + \iota\theta}\dfrac{\eta_2}{\eta_1 - \eta_2} + \left(\dfrac{1 - \eta_2}{\eta_1 - \eta_2}\dfrac{\phi_2\psi_2}{\theta} + \dfrac{\zeta_2 - \eta_2}{\eta_1 - \eta_2}\dfrac{\chi_2}{\beta - \delta\theta}\right)y^*\right]\left(\dfrac{y}{y^*}\right)^{\eta_1} & 0 \le y < y^* \\ \beta\left[\dfrac{\gamma}{\beta + \iota\theta}\dfrac{\eta_1}{\eta_1 - \eta_2} - \left(\dfrac{\eta_1 - 1}{\eta_1 - \eta_2}\dfrac{\phi_2\psi_2}{\theta} + \dfrac{\eta_1 - \zeta_2}{\eta_1 - \eta_2}\dfrac{\chi_2}{\beta - \delta\theta}\right)y^*\right]\left(\dfrac{y}{y^*}\right)^{\eta_2} \\ \quad + \beta\left[\dfrac{\phi_2\psi_2}{\theta}y + \dfrac{\chi_2}{\beta - \delta\theta}y^*\left(\dfrac{y}{y^*}\right)^{\zeta_2} - \dfrac{\gamma}{\beta + \iota\theta}\right] & y^* \le y < \infty \end{cases} \tag{12}$$

In (12) the new parameters are

$$\eta_1 \equiv 1 - \left(\frac{1}{2} + \frac{\rho + \delta/\alpha}{\sigma^2}\right) + \sqrt{\left(\frac{1}{2} + \frac{\rho + \delta/\alpha}{\sigma^2}\right)^2 + \frac{2}{\psi_1\sigma^2}} > 1$$

$$\eta_2 \equiv \frac{1}{2} - \frac{\rho + (\delta - \beta)/\alpha}{\sigma^2} - \sqrt{\left[\frac{1}{2} - \frac{\rho + (\delta - \beta)/\alpha}{\sigma^2}\right]^2 + 2\frac{\iota + \beta/\theta}{\sigma^2}} < 0$$

and

$$\psi_1 \equiv \frac{1}{\iota - \rho - \delta/\alpha} > 0, \qquad \psi_2 \equiv \frac{1}{\iota + \beta/\theta - \rho + (\beta - \delta)/\alpha} > 0$$

The value of an undeveloped asset is also displayed in Figure 1. Under the subsequently specified constraints on the parameters, the valuation function G is non-negative, increasing, and bounded below by its asymptote. Also, it is strictly convex on the initial interval, $0 < y < y^*$, strictly concave on some intermediate interval, $y^* < y < y^{**}$, and strictly convex on the subsequent interval, $y^{**} < y < \infty$. Depending on the values of the parameters, the intermediate interval may not exist: $y^* \leq y^{**} < \infty$. The interpretation of the asymptote is familiar, even though its parameters are new. In the asymptote the constant $\beta\gamma/(\beta + \iota\theta)$ is an exercise price. As $y \to \infty$, development occurs asymptotically at the constant rate β per unit of developed assets or, equivalently, at the rate β/θ per unit of undeveloped assets, with θ again representing the ratio of undeveloped to developed assets. Also, the development cost is γ per unit of time. To compute the present value, the riskless cost per unit of time $\beta\gamma/\theta$ must be discounted at the riskless rate $\iota + \beta/\theta$, reflecting the rate β/θ at which undeveloped assets are eliminated by development. In the asymptote the remaining term, $(\beta\phi_1\psi_2/\theta)y$, is the asymptotic value of development. As previously discussed following (11), developed properties have the asymptotic present value $\psi_2 y$ per unit. With development at the asymptotic rate β/θ per unit of time, development then yields the risky income $(\beta\phi_2/\theta)y$ per unit of time. This risky income has the present value $(\beta\phi_2\psi_2/\theta)y$, calculated with the present-value factor ψ_2 or, equivalently, the discount rate $\iota + \beta/\theta - \rho + (\beta - \delta)/\alpha$. This risky discount rate equals the riskless discount rate from above, $\iota + \beta/\theta$, minus the risk-adjusted, expected growth rate of income, $\rho - (\beta - \delta)/\alpha$. With undeveloped assets there is no adjustment for depreciation.

The solution to problem (4) to (9) also indicates the optimal switching point in (10). This switching point is calculated from (7) under the transformation above (10) as follows:

$$y^* = \gamma \frac{1 + \left(1 + \dfrac{\eta_1\theta}{\alpha\nu}\right)\dfrac{\beta}{\beta + \iota\theta}\dfrac{\eta_2}{\eta_1 - \eta_2}}{\phi_1 - \chi_1 - \beta\left(1 + \dfrac{\eta_1\theta}{\alpha\nu}\right)\left(\dfrac{1 - \eta_2}{\eta_1 - \eta_2}\dfrac{\phi_2\psi_2}{\theta} + \dfrac{\zeta_2 - \eta_2}{\eta_1 - \eta_2}\dfrac{\chi_2}{\beta - \delta\theta}\right)} \tag{13}$$

Under the constraints on the parameters to be identified, the switching point (13) is positive and finite. Below the switching point y^*, no developer builds in the symmetric Nash equilibrium: $b^* = 0$ for all $0 \leq y < y^*$. Above the switching point all developers build at the maximum rate: $b^* = \beta$ for all

$y^* \leq y < \infty$. The switching point (13) replaces the single point of optimal exercise from previous models of real options in partial equilibrium because the capacity constraints of developers preclude simultaneous exercise of all options at one level of income y. This switching point has economically plausible comparative statics, as illustrated numerically in the next section.

The symmetric Nash equilibrium is completely characterised by (11) through (13) if assumption (10) is correct. This assumption is correct if (10) satisfies the first-order condition from (7) under the transformation above (10). As shown in the Appendix, (10) and (13) uniquely satisfy the first-order condition if the switching point (13) satisfies the upper bound in the following inequalities:

$$\frac{\eta_2\gamma/(\beta + \iota\theta)}{(\eta_2 - 1)\phi_2\psi_2/\theta + (\eta_2 - \zeta_2)\chi_2/(\beta - \delta\theta)} \leq y^* < \frac{\gamma\eta_1}{(\eta_1 - 1)\phi_1 + (\zeta_1 - \eta_1)\chi_1} \tag{14}$$

With the lower bound in (14), the valuation function (12) is non-negative everywhere. Because the switching point (13) depends only on the parameters, the combination of (13) and (14) constrains the values of the parameters. These constraints are too complicated to evaluate analytically. Fortunately, (14) is not binding for plausible values of the parameters, as shown numerically in the next section.

DISCUSSION

Comparative statics are discussed in this section. Because the solution, (11) to (13), is a complicated function of the parameters, most results must be calculated numerically. However, numerical computations have the advantage of revealing magnitudes that can help to answer interesting questions. For example, do realistic elasticities of demand generate significantly different exercise policies and market values than the infinite elasticity from most previous models of real assets in partial equilibrium? Also, do constraints on the rate of development and the supply of undeveloped assets significantly affect rates of growth in equilibrium? Finally, how many developers are necessary for the optimal rate of development to approximate the competitive solution?

Numerical values of the optimal switching points and resulting values of the developed and undeveloped assets are displayed in Table 1 for various values of all parameters. In each vertical triplet the top number is the optimal switching point y^*; the middle number is the resulting value of the developed asset $F(y^*)$; and the bottom number is the value of the undeveloped asset $G(y^*)$. The column heading is the value of the parameter in question, and all other parameters are the values from the base case. The constraints (14) are binding nowhere in the table.

Table 1 Optimal switching point y^* and the resulting values of developed and undeveloped assets $F(y^*)$ and $G(y^*)$

Parameter values for the base case

$\alpha = 1.0$	$\beta = 0.03$	$\gamma = 5.0$
$\delta = 0.02$	$\theta = 1.0$	$\iota = 0.10$
$\nu = 10$	$\rho = 0$	$\sigma^2 = 0.05$

α elasticity of demand

1.0	2.0	∞
0.5448	0.5921	0.6464
5.116	5.216	5.386
0.0939	0.1908	0.3013

β maximum building rate

0.02	0.03	0.04
0.5415	0.5448	0.5454
5.186	5.116	5.028
0.1510	0.0939	0.0227

γ construction cost per unit

4.0	5.0	6.0
0.4358	0.5448	0.6538
4.093	5.116	6.139
0.0751	0.0939	0.1126

δ depreciation rate

0.01	0.02	0.03
0.5437	0.5448	0.5507
5.077	5.116	5.198
0.0609	0.0939	0.1629

θ relative supply of undeveloped asset

0.5	1.0	1.5
0.5500	0.5448	0.5438
5.165	5.116	5.107
0.1306	0.0939	0.0873

ι riskless rate of interest

0.08	0.10	0.12
0.4448	0.5448	0.6443
5.117	5.116	5.109
0.0957	0.0939	0.0872

ν number of developers

1	10	∞
0.5792	0.5448	0.5421
5.440	5.116	5.091
0.1301	0.0939	0.0911

ρ risk-adjusted, expected rate of growth in exogenous demand

–0.02	0.00	0.02
0.6449	0.5448	0.4446
5.071	5.116	5.186
0.0551	0.0939	0.1563

σ^2 variance of growth rate in exogenous demand

0.04	0.05	0.06
0.5373	0.5448	0.5517
5.082	5.116	5.147
0.0651	0.0939	0.1197

In each vertical triplet, the top number is the optimal switching point y^* above which development is optimal, the middle number is the value of a developed asset $F(y^*)$, and the bottom number is the value of an undeveloped asset $G(y^*)$. The column heading is the value of the indicated parameter, and values of all other parameters are indicated in the base case.

Of all the results reported in the table, the most interesting comparative statics relate to the optimal switching point and the parameters that do not appear in models of financial options. These parameters are the elasticity of demand α, the maximum building rate β, the depreciation rate δ, the relative supply of undeveloped assets θ and the number of developers ν. As indicated in the table, the optimal switching point y^* is increasing in α, β and δ, but decreasing in θ and ν. Moreover, these comparative statics hold for all possible combinations of the three different values of each parameter in the table.

The comparative statics are economically plausible. With larger elasticities α, income y is less sensitive in (1) to the aggregate supply of developed assets q, so the option to develop is more valuable in (12). Thus, the exercise of that option is optimally deferred in (13). Similarly, with larger maximum building rates β developers are less constrained from building only when building is most profitable. As a result, development is again deferred in (13). Also, with larger depreciation rates δ the developed asset is less valuable in (11) and development is retarded in (13). In contrast, with larger relative supplies θ the undeveloped asset is relatively less abundant and thereby less valuable in (12), so development is accelerated in (13). Finally, with more developers ν each has a smaller inventory of undeveloped assets. Hence, each developer is less concerned in (7) about the effect of his development on the values of the remaining undeveloped assets, and development is accelerated in (13).

For a deeper understanding of these results, some preliminaries are necessary. For each developer n the maximand in (7) is linear in the control b_n. Under the transformation above (10), the coefficient in (7) of the control b_n is proportional to the value H(y) from the function:

$$H \equiv F - G - (\theta/\alpha\nu)yG' - \gamma \tag{15}$$

for all $y \geq 0$. Let π represent a generic parameter, and consider parameter values that satisfy the inequalities in (14). In this case, the switching point, y^* in (13), is the unique root of the function (15): $H(y^*|\pi) = 0$. Moreover, this root satisfies the inequality, $H'(y^*|\pi) > 0$. Under these conditions the switching point y^* increases with small increments in the parameters π if and only if $H(y^*|\pi)$ decreases similarly in π.

Conditions under which the function H is decreasing in a parameter can be seen by examining (15). In (15) the first term F is the value of the developed asset, and the second term G is the value of the undeveloped asset or, equivalently, the option to develop. The third term, $(\theta/\alpha\nu)yG'$, is the effect of current construction by each developer on the income y and thereby the values of his remaining undeveloped assets. In models with an infinite elasticity of demand, $\alpha = \infty$, or perfectly competitive developers, $\nu = \infty$, the third term is zero. However, with finite elasticity and imperfect competition, this term is positive. In standard models of financial options other than warrants and sinking fund bonds, the first term is independent of all parameters and the third term is zero. Hence, for financial options a larger value of the parameter π increases the switching point y^* and thereby defers optimal exercise of the option if and only if it also increases the value of the option $G(y^*|\pi)$. This is a familiar result. With real options and either infinite elasticity of demand or perfect competition among developers, a larger value of the parameter π increases the switching point y^* if and only if it also increases the difference between the values of the undeveloped and developed assets, $G(y^*|\pi) - F(y^*|\pi)$. Finally, with real options, finite elasticity, and imperfect competition, a larger value π defers optimal exercise if and only if it increases the weighted difference, $(\eta_1\theta/\alpha\nu)G(y^*|\pi) - F(y^*|\pi)$, from (13) and (14).

Although these results can be used to interpret all the comparative statics in the table, for brevity the discussion is limited to the new parameters, α, β, δ, θ, and ν. From a numerical evaluation of the valuation functions for developed and undeveloped assets, F and G, the following conclusions can be drawn. Increments in the elasticity α do not affect the developed value $F(y^*)$ but do increase the undeveloped value $G(y^*)$ sufficiently to offset the decreased weight $\eta_1\theta/\alpha\nu$ in (15) and, thereby, increase the switching point y^*. Increments in either the maximum building rate β or the depreciation rate δ also increase the switching point y^*, but do so by decreasing the developed value $F(y^*)$ without affecting the undeveloped value $G(y^*)$. In contrast, increases in the relative supply of undeveloped assets θ do not change the developed value $F(y^*)$ but greatly decrease the undeveloped value $G(y^*)$ and, thereby, decrease the switching point y^*. Finally, an increased number of developers ν affects neither the developed nor the undeveloped values, $F(y^*)$ or $G(y^*)$, but decreases the weight $\eta_1\theta/\alpha\nu$ and thereby decreases the switching point y^*.

Conclusions concerning numerical magnitudes can also be drawn from the table. If the true elasticity of demand is unity, $\alpha = 1$, but income is falsely hypothesised to be exogenous, $\alpha = \infty$, then the value of the option to develop at the point of first exercise $G(y^*)$ is falsely overstated by the multiple: $0.3013/0.0939 = 3.2$. With smaller elasticities α, the error is even greater. What are realistic values for the elasticity α? Consider two constants: the elasticity of aggregate demand with respect to the price of the good or service, $\xi > 0$, and the exponent from the convex, Cobb–Douglas cost function for output from each developed asset, $\kappa > 1$. As shown in the Appendix, the elasticity in the clearing condition (1) is $\alpha = \xi + (1 - \xi)/\kappa$. Since $\kappa > 1$, the elasticities α and ξ have the relationship $\alpha \leq 1$ and if $\xi \leq 1$, with $\alpha = 1$ whenever $\xi = 1$. For housing services most estimates of the elasticity ξ are concentrated in a small interval bounded above by 1.[24] For houses a realistic elasticity of demand α is then slightly less than or equal to 1. Also, as illustrated in Table 1 increments in the maximum building rate β have a substantial effect on the value of the option $G(y^*)$ when $\alpha = 1$. With exogenous income, $\alpha = \infty$, this effect disappears completely. Finally, imperfect competition among developers can greatly decrease the value of the option to develop at the point of first exercise $G(y^*)$. However, effects of major magnitude require a very small number of developers ν. For example, with duopolists, $\nu = 2$, the switching point, $y^* = 0.5574$, and resulting option value, $G(y^*) = 0.1071$, are much closer to the corresponding values in Table 1 for perfect competition, $\nu = \infty$, than the values for monopoly, $\nu = 1$.

CONCLUSION

In this chapter real options are valued under conditions that are characteristic of real assets but not most financial assets. Here, real assets have four distinguishing characteristics. Each real asset produces goods or services that consumers demand with a finite elasticity. The rate at which assets can be developed is limited by developers' capacity, so all options to develop cannot be exercised simultaneously. Also, the supply of undeveloped assets is limited relative to the current supply of developed assets. Finally, the ownership of undeveloped assets is monopolistic, oligopolistic, or competitive. For analytical convenience, the elasticity of demand is constant, the cost of development is constant up to a capacity constraint that is proportional to the supply of developed assets; the supplies of developed and undeveloped assets are proportional; and the number of owners or developers is fixed. Under these assumptions owners develop their assets in a Markov perfect equilibrium.

For this equilibrium the optimal exercise policy of developers and the resulting values of developed and undeveloped assets are calculated explicitly. In equilibrium, development is optimal at all values of income per developed asset above a critical value or switching point. Below this optimal value no developer builds; above the switching point all developers build at the maximum feasible rate. This region of optimal exercise replaces the single point of optimal exercise in all previous models of real options. Also, unlike previous models the ratio of earnings to value for developed assets depends on the income per asset. Here, the ratio decreases as income increases. Finally, the solution has both new competitive statics and new interpretations of familiar results. The comparative statics are computed numerically and discussed in detail.

Some straightforward extensions with explicit solutions are possible. Assumptions can be introduced from models of single developers in partial equilibrium. For example, construction can take time; construction costs can evolve in a geometric Wiener process; and development can occur at an optimal density or scale with Cobb–Douglas cost functions. Also, rents on real estate can depend on such factors as commuting time to the central business district. In these cases explicit solutions are possible with either Wiener or geometric Wiener processes if the option to abandon property is ignored. Other explicit solutions are possible if the aggregate demand in (1) is replaced by a linear demand, the geometric Wiener process (2) is replaced by a Wiener process, and the assumptions of proportional capacity and proportional supplies are replaced with additive capacity and additive supplies. With numerical solutions still other extensions are possible.

Explicit solutions are also possible with different ownership structures of either developed or undeveloped assets. In this chapter, developers own equal numbers of all undeveloped assets but no developed assets.

Instead, assume that developers retain some of their developed assets. This has two effects on the previous analysis if the number of developers ν is finite. First, the spot market for the good or service from a developed asset is no longer perfectly competitive. With imperfect competition in the product market, the clearing condition (1) must be changed. Also, the representative developer no longer chose his building rate to maximise the market value of his portfolio of undeveloped assets. Instead, the residual term in his maximand (7) must be altered to include the impact of his building rate b_n on his portfolio of developed assets. These changes affect the optimal switching point between no development and development y^* and thereby the market values of developed and undeveloped assets, but do not alter the structure of the previous solutions.

Endogenising the ownership of either developed or undeveloped assets by allowing the number of developers to be determined in equilibrium is much more difficult. For example, assume again that developers own equal numbers of all undeveloped assets but no developed assets. In the numerical solutions from the previous section, the profit from development increases as the ownership of undeveloped assets becomes more concentrated. As shown in Table 1, the direct gain from development when development begins, $F(y^*) - G(y^*) - \gamma$, decreases in the number of developers or owners ν. It is positive with imperfect competition, $\nu < \infty$, and zero with perfect competition, $\nu = \infty$. Also, numerical evaluation of the valuation function F and G verifies that the gain, $F - G - \gamma$, decreases in the number of developers ν for all fixed incomes $y \geq y^*$ and increases in the income y for all incomes $y \geq y^*$. As a result, sellers of undeveloped assets receive the highest price from other owners of undeveloped assets, and ownership becomes more concentrated over time. In this model, as in other applications of option pricing, portfolio diversification is not an issue because, by assumption, all investors are either risk neutral or the capital market is complete. For example, with risk neutrality the concentration of ownership is determined in equilibrium by the distribution of wealth. For this reason realistic predictions about the ownership of real options in equilibrium require a very different model with risk aversion and an incomplete capital market.

Another interesting issue is the relationship in equilibrium between earnings and the ratio of price to earnings for developed assets. In previous articles on real assets, this ratio is a constant, independent of earnings per developed asset; here, it is increasing in income or earnings. In particular, higher earnings per developed asset increase the probability of future earnings above the critical value at which construction occurs. This decreases the value or price of developed property and thereby the ratio of price to earnings. In fact, casual observation suggests the reverse relationship: with higher average operating income for developed assets, which is characteristic of cyclical peaks versus troughs, the ratio of price to earnings is typically higher, not lower. Clearly, this procyclical relationship between P/E ratios and earnings cannot be explained by the behaviour of rational, value-maximising investors in an equilibrium with cyclical returns on real assets. If returns are recognised by all investors to be cyclical, then P/E ratios should be countercyclical, not procyclical. Instead, explaining procyclical P/E ratios seems to require a model with agency and/or asymmetric information. Indeed, this perplexing procyclicality appears from casual observation to be associated with the persistent entry into markets late in business cycles of relatively uninformed, institutional investors and lenders.

APPENDIX

Derivation of (1)

In the perfectly competitive spot market, the consumption good c has current price p. With respect to this price, the aggregate demand by consumers is $(x/p)^{\xi}$, with constant elasticity $\xi > 0$. Also, the cost of producing c is proportional to c^{κ} with constant exponent $\kappa > 1$. For each unit of the developed asset, the profit-maximising production for c is proportional to $p^{1/(\kappa-1)}$, and the resulting income y is proportional to $p^{\kappa/(\kappa-1)}$. Thus, the aggregate supply of the product from the developed asset is proportional to $qp^{1/(\kappa-1)}$ or, equivalently, $qy^{1/\kappa}$. Similarly, the aggregate demand is proportional to $(x/y^{1-1/\kappa})^{\xi}$. Equating aggregate supply with aggregate demand and rescaling the geometric Wiener variable x yields (1) with the constant, $\alpha \equiv \xi + (1-\xi)/\kappa$.

Hypothesis

With the transformation above (10) and the function H from (15), the unique optimal control in (7) is

$$b^* = \begin{cases} 0 & H(y) < 0 \\ \beta/\nu & H(y) \geq 0 \end{cases} \tag{A1}$$

Thus, the unique subgame perfect Nash equilibrium is symmetric with the aggregate building rate: $b^* = \nu b_n^*$. Now, suppose that the function H satisfies the inequalities: $H(y) < 0$ for all $0 \leq y < y^*$ and $H(y) > 0$ for all $y^* < y < \infty$. In this case, (10) and (A1) are equivalent. That is, there is a unique switching point y^* in (10) satisfying $H(y^*) = 0$. For this case, the indicator function I is defined as follows: $I(y) = 0$ on $0 \leq y < y^*$ and $I(y) = 1$ on $y^* \leq y < \infty$.

Derivation of (11)

With (10) and the transformation above (10), the valuation function (4) simplifies to

$$0 = \frac{1}{2}\sigma^2 y^2 F'' + \left(\rho + \frac{\delta - \beta I}{\alpha}\right) yF' - (\delta + \iota)F + y \tag{A2}$$

Also, the initial condition (5) becomes

$$F(0) = 0 \tag{A3}$$

Finally, the limiting condition (6) simplifies to

$$F(y) \le \upsilon y \tag{A4}$$

as $y \to \infty$.

The differential equation (A2) has a unique solution of the form

$$F(y) = \begin{cases} a_1 y^{\zeta_1} + a_1' y^{\zeta_1'} + \phi_1 y & 0 \le y < y^* \\ a_2 y^{\zeta_2} + a_2' y^{\zeta_2'} + \phi_2 y & y^* \le y < \infty \end{cases} \tag{A5}$$

The parameters ζ_1 and ζ_2 are defined below (11). The remaining constants are

$$\zeta_1' = \frac{1}{2} - \frac{\rho + \delta/\alpha}{\sigma^2} - \sqrt{\left(\frac{1}{2} - \frac{\rho + \delta/\alpha}{\sigma^2}\right)^2 + 2\frac{\delta + \iota}{\sigma^2}} < 0$$

and

$$\zeta_2' \equiv 1 - \left[\frac{1}{2} - \frac{\rho + (\delta - \beta)/\alpha}{\sigma^2}\right] + \sqrt{\left[\frac{1}{2} - \frac{\rho + (\delta - \beta)/\alpha}{\sigma^2}\right]^2 + \frac{2}{\phi_2 \sigma^2}} > 1$$

The initial condition (A3) requires that $a_1' = 0$, and the limiting condition (A4) requires that $a_2' = 0$. Also, F is continuously differentiable on $y > 0$, so the constants, a_1 and a_2, in (A5) are determined by the conditions:

$$F(y^* + \epsilon) - F(y^* - \epsilon) \to 0, \qquad F'(y^* + \epsilon) - F'(y^* - \epsilon) \to 0 \tag{A6}$$

$\epsilon \to 0$. Thereby, (A5) simplifies to (11).

Derivation of (12) and (13)

With assumption (10) and the transformation above (10), the valuation equation (7) simplifies to

$$0 = \frac{1}{2}\sigma^2 y^2 G'' + \left(\rho + \frac{\delta - \beta I}{\alpha}\right) y G' - \left(\iota + \frac{\beta}{\theta} I\right) G + \frac{\beta}{\theta}(F - \gamma) I \tag{A7}$$

since $b_n = b^*/v = \beta/v$. Again, the initial condition (8) becomes

$$G(0) = 0 \tag{A8}$$

Also, the limiting condition (9) simplifies to

$$G(y) \leq \upsilon y \tag{A9}$$

as $y \to \infty$.

Under assumption (10), problem (A7) to (A9) has the unique solution (12) and (13). The valuation equation (12) is derived much like (11). Again, one of the boundary conditions, (A8) and (A9), eliminates one root of the characteristic polynomial on each of the two intervals $[0, y^*)$ and $[y^*, \infty)$. This result, combined with the particular solution to the inhomogeneous equation on the upper interval, $[y^*, \infty)$ and the continuous differentiability of G at y^*, as in (A6), yields the solution (12). Finally, the switching point (13) is calculated using (11) and (12). Under assumption (10), the switching point y^* is the root of the function H from (15): $H(y^*) = 0$. With the inequalities (14) this root is positive and finite.

Verification of (10)

To complete the derivation, (10) to (13) must be shown to satisfy the switching condition (A1). The function H in (15) satisfies the corner conditions $H(0) < 0 < H(\infty)$. The latter inequality holds under the assumption above (4): $\rho < \iota - \delta/\alpha$. Also, at y^* it satisfies

$$y^* H'(y^*) = \gamma\eta_1 - [(\eta_1 - 1)\phi_1 + (\zeta_1 - \eta_1)\chi_1]y^*$$

from (11) and (12). Because the term in brackets is positive, H is increasing at y^* if the upper bound in (14) is satisfied by (13). In this case, it is sufficient to show that (13) can be the only root of the function H on the interval $[0, \infty)$. On the lower interval, $[0, y^*)$, the function H is strictly concave, given (11) and (12). Because $H'(y^*) > 0$, there is no root below y^*. On the upper interval, (y^*, ∞), the function H is bounded below:

$$H(y) > \phi_2 y - G(y) - (\theta/\alpha v) y G'(y) - \gamma \equiv B(y)$$

The function B is convex on an initial interval, (y^*, y^0), and concave on the subsequent interval, (y^0, ∞), with an inflection point y^0 satisfying $y^* \leq y^0 \leq \infty$. Also, B satisfies $B'' < H''$ for all $y^* < y < \infty$. Given this lower bound B and the initial condition $H'(y^*) > 0$, there can be no root on the upper interval (y^*, ∞) if $B(y)/y \geq 0$ as $y \to \infty$. With (12) and (15), the desired result follows if $\rho < \iota - \delta/\alpha$, as assumed above (4).

1 Options on real estate are analysed in Titman (1985), Clarke and Reed (1988), Capozza and Sick (1991), Williams (1991), and the references cited therein. Options on natural resources are the focus of Brennan and Schwartz (1985), and Paddock, Siegel and Smith (1988). Finally, options on capital assets are analysed in McDonald and Siegel (1985, 1986), Majd and Pindyck (1987), Pindyck (1988), Dixit (1989, 1991), and Sick (1989).

2 Some financial options have related characteristics, as argued in Margrabe (1978) and Stulz (1982). Also, with real estate another special characteristic is the distance or commuting time to the central business district, as developed in Capozza and Sick (1991) and references cited therein.

3 This problem has not occurred in previous articles on real options because new development has been limited to a negligible fraction of the current supply of developed assets. For example, in Brennan and Schwartz (1985) different mines have different costs, whereas in Capozza and Sick (1991) developers build only at the boundary of the urban area.

4 Warrants are studied in Constantinides (1984) and Spatt and Sterbenz (1988).

5 In fact, building lots are partially developed assets that are the product of previously incurred discretionary costs. The supply of completely raw land may be very elastic, but cannot be perfectly elastic, as argued in the first section.

6 Sources: *Phoenix Metropolitan Housing Study* and Arizona Republic, *Inside Metro Phoenix*, various years from 1971 to 1991.

7 See the survey by Rosen (1985) and references cited therein.

8 Demand curves are downward sloping in Pindyck (1988), Dixit (1989, 1991), and Bar-Ilan, Sulem and Zanello (1992).

9 Alternatively, with the additive changing condition $q = \alpha(x - y)$, the subsequent problem can be solved explicitly if the subsequent assumptions of proportionality are replaced everywhere by additive specifications.

10 This assumption permits a symmetric solution in the second section. Without the assumption owners would maximise the market value of their undeveloped assets by selling only to other owners and ownership would become concentrated over time. See the discussion in the final section.

11 This simple cost function – a constant cost γ per developed asset up to capacity – is consistent with the casual empirical observation that the real price of housing is roughly constant over time at the expanding boundaries of metropolitan areas without major physical or legal constraints on development. Under these conditions housing costs are determined largely by the highly elastic supply of new construction and not demand.

12 In models of partial equilibrium with isolated developers, instantaneous development involves no significant loss of generality. Exercising the option to develop produces a newly developing asset rather than a fully developed asset. This developing asset is priced at the risk-adjusted, expected present value of a developed asset. With downward-sloping demand, as in (1), the aggregate supply of developed assets q must be modified to include construction in progress, as in Bar-Ilan Sulem, and Zanello (1992).

13 An explicit solution is also possible with constant supply, constant capacity, linear demand in (1), a random walk in (2) and constant change in (3). This alternative appears to be less realistic than proportionality everywhere.

14 If the supply of raw land were perfectly elastic, then no parcel would have an option value. Premiums for the option to develop are commonly observed.

15 It may be possible to generalise the assumption of spanning as follows. The second term in (2) can be decomposed into a systematic component and a residual. The systematic component must be spanned by the stochastic returns on traded securities, whereas the residual is completely diversifiable by construction. In this case, the Brownian residual can be written as the limit of a sequence of Poisson variates, and the argument in Naik (1992) can be applied to the Poisson variates.

16 If developed assets could be abandoned, then the valuation equation (4) would be satisfied only at those values of the state variables q and x at which abandonment is not optimal. If developed assets could be redeveloped, then (4) must be modified to reflect the issues discussed below (7).

17 The remaining boundary condition, (6) as $q \to 0$, is superfluous under the subsequent transformation of variables.

18 In standard models with optimal exercise at a point or along a boundary, the partial

differential equation for the option must hold only on a connected region where the values of the state variables do not induce optimal exercise.

19 The latter condition, no immediate gain at the optimal stopping point, is imposed in models of real options in partial equilibrium. There, prices or values are functions only of aggregate states with exogenous growth rates, so that the representative developer is effectively a perfect competitor.

20 Some securities, such as warrants and convertible bonds, have positive net supply. Previous analyses of such securities have focused on the financial effects of positive net supply and ignored the effects on real assets. For example, in Constantinides (1984) a firm's proceeds from the exercise of its warrants are either distributed as extraordinary dividends or repurchases of stock or reinvested in projects with zero net present value.

21 Because only the competitive case is realistic with most financial options, it has been analysed almost exclusively in the literature on option pricing. Exceptions are the analyses of warrants and sinking fund bonds in Constantinides, Dunn and Spatt (1984), and the references cited herein.

22 Again, the remaining boundary condition, (9) as $q \to \infty$, is superfluous under the subsequent transformation of variables.

23 An example with real estate is Williams (1991).

24 Again, see the survey of Rosen (1985).

BIBLIOGRAPHY

Bar-Ilan, A., A. Sulem and A. Zanello, 1992, "Time to Build and Capacity Choice", Working Paper, Department of Economics, Dartmouth College and Tel Aviv University.

Brennan, M. and E. Schwartz, 1985, "Evaluating Natural Resource Investments under Uncertainty", *Journal of Business* 58, pp. 135–57.

Capozza, D. and G. Sick, 1991, "Pricing Risky Land: The Option to Develop", Working Paper, Graduate School of Business Administration, University of Michigan.

Clarke, H. and W. Reed, 1988, "A Stochastic Analysis of Land Development Timing and Property Valuation", *Regional Science and Urban Economics* 18, pp. 367–82.

Constantinides, G., 1984, "Warrant Exercise and Bond Conversion in Competitive Markets", *Journal of Financial Economics* 13, pp. 371–91.

Dixit, A., 1989, "Entry and Exit Decisions under Uncertainty", *Journal of Political Economy* 97, pp. 620–38.

Dixit, A., 1991, "Irreversible Investment with Price Ceilings", *Journal of Political Economy* 99, pp. 541–57.

Dunn, K. and C. Spatt, 1984, "A Strategic Analysis of Sinking Fund Bonds", *Journal of Financial Economics* 13, pp. 399–423.

Majd, S. and R. Pindyck, 1987, "Time to Build, Option Value, and Investment Decisions", *Journal of Financial Economics* 18, pp. 7–27.

Margrabe, W., 1978, "The Value of an Option to Exchange One Asset for Another", *Journal of Finance* 33, pp. 177–87.

McDonald, R. and D. Siegel, 1985, "Investment and the Valuation of Firms When There is an Option to Shut Down", *International Economic Review* 26, pp. 331–49.

McDonald, R. and D. Siegel, 1986, "The Value of Waiting to Invest", *Quarterly Journal Economics* 101, pp. 707–27.

Naik, V., 1992, "Optimal Trading Strategies for Hedging Jumps in Asset values and Return Volatility", Working Paper, Faculty of Commerce, University of British Columbia.

Paddock, J., D. Siegel and J. Smith, 1988, "Option Valuation of Claims on Real Assets: The Case of Offshore Petroleum Leases", *Quarterly Journal of Economics* 103, pp. 479–508.

Pindyck, R., 1988, "Irreversible Investment, Capacity Choice, and the Value of the Firm", *American Economic Review* 78, pp. 969–85.

Rosen, H., 1985, "Housing Subsidies: Effects on Housing Decisions, Efficiency, and Equity", in A. Auerbach and M. Feldstein (eds.), *Handbook of Public Economics* (vol. 1), Elsevier, New York.

Sick, G., 1989, *Capital Budgeting with Real Options*, Monograph 1989–93, Salomon Brothers Center, Stern School of Business, New York University.

Spatt, C. and F. Sterbenz, 1988, "Warrant Exercise, Dividends, and Reinvestment Policy", *Journal of Finance* 43, pp. 493–506.

Stulz, R., 1982, "Options on the Minimum or the Maximum of Two Risky Assets: Analysis and Application", *Journal of Financial Economics* 10, pp. 161–85.

Titman, S., 1985, "Urban Land Prices under Uncertainty", *American Economic Review* 75, pp. 505–14.

Williams, J., 1991, "Real Estate Development as an Option", *Journal of Real Estate Finance and Economics* 4, pp. 191–208.

5

*Investment in Competitive Equilibrium: The Optimality of Myopic Behavior**

John V. Leahy

Boston University

There is an extensive literature on the optimal irreversible investment strategy of a solitary firm facing an exogenous price process. This chapter shows that the investment strategies that this literature derives may be optimal in competitive equilibrium even though the price process is now endogenous. This provides a simple means for computing equilibrium investment strategies.

Over the past decade an extensive literature has developed analysing the effect of irreversibilities on a firm's investment decision.[1] The principal goal of this literature has been to modify the net present value rule of investment. Conventional wisdom states that a firm should invest in a project whenever the expected profits from the project exceed the cost of investment. These costs normally include costs of planning, construction and materials – all modified to take into consideration the effects of taxation and depreciation allowances. According to the recent literature, however, this list is incomplete. When investment is irreversible, the decision to invest involves a commitment. The firm gives up the option of postponing investment until market conditions change or of refraining from investment entirely. This option has value, and its loss must be included among the cost of investment. A firm should therefore invest only when expected profits exceed costs by the value of remaining uncommitted.

* This chapter was first published in *Quarterly Journal of Economics* 108(4), 1993, pp. 1105–33 and is reprinted with the permission of MIT Press Journals. The author would like to thank Giuseppe Bertola, Olivier Blanchard, Andrew Caplin, James Dana and Avinash Dixit for helpful discussions, two anonymous referees for suggestions on an earlier draft, and the Sloan Foundation for financial support. This chapter is a much revised version of the second chapter of his dissertation, "Optimality, Competitive Equilibrium and the Entry and Exit Decisions of Firms" (1990).

Most of this literature considers the firm's problem in isolation. Either the price of the firm's output is exogenous, or if the price is endogenous, the firm is a monopolist. In neither case must the firm consider the actions of others. This would appear to be a serious omission, since the value of remaining uncommitted depends on what a firm can gain by waiting, and a firm that faces competitors risks losing the market to others if it should allow price to rise too far above cost. In fact, under perfect competition the option to remain uncommitted is worthless, since competition eliminates all profits. This raises the question of how irreversibilities affect investment in competitive equilibrium.

Somewhat surprisingly, this chapter will show that the introduction of competition does not affect the timing of irreversible investment decisions at all. I shall consider the optimal policies of two types of firms. The first correctly anticipates how the strategies of other firms in the market will interact with prices and exogenous shocks. The second, like the firms in the literature on irreversible investment, completely ignores the effect that the industrywide stock of capital will remain fixed forever and that the price process will be driven solely by exogenous shocks. I shall call the policies of the second firm myopic. It is a special kind of myopia; the firm is far-sighted in the sense that it calculates present values, but it is shortsighted in the sense that the price process which it uses to calculate profits is correct only so long as no other firms invest. Such a myopic firm has static expectations regarding industry output, but rational expectations regarding other shocks that influence price in the market.

The central result of this chapter is that the prices that trigger investment by the two types of firms are identical even though the optimisation problems that they face appear very different. I shall demonstrate this correspondence in a setting with homogeneous, constant-returns-to-scale firms, linear costs of investment, downward sloping demand and uncertainty that follows a diffusion process.

The intuition underlying this result is straightforward. The decision to invest is similar to the decision to exercise an option. The investor trades the chance to own a unit of capital for the capital itself, and the strike price of this option is the cost of investment. The introduction of competition reduces the value of this option, but does so by reducing the value of the invested capital. Since competition reduces the value of actual and potential capital at the same time, the trade-off between the two is unaffected.

The correspondence between myopic and competitive investment has two practical implications. First, many of the lessons concerning the relationship between uncertainty and investment derived under the assumption of a solitary price-taking firm apply directly to competitive investment. For example, it is often noted that an increase in the variance of the price process increases the value of the option to remain uncommitted, thereby increasing the price at which a firm is willing to undertake an irreversible investment project. Similar comparative statics will be found in competitive equilibrium even though the firm's option to invest no longer has value.

Second, the equivalence between the two problems provides a simple means of calculating competitive equilibriums. Solving for competitive equilibriums in dynamic economies is made difficult by the interaction between investment strategies and the price process. Each firm's optimal strategy depends on its expectation of future prices: yet future prices are contingent on the policies that firms pursue. The fact that the myopic firm that ignores its competitors invests at the same price as a fully rational firm provides one possible way around this interdependence.

The standard means of circumventing the interaction between investment strategies and expected prices is to transform the problem of solving for a competitive equilibrium into a problem of resource allocation by a benevolent social planner. This transformation internalises all of the interactions among firms. This approach, which is taken by Lucas and Prescott (1971) and much of the literature on real business cycles, is closely related to the approach taken in this chapter. This chapter addresses the connection between the myopic firm and competitive equilibrium. Lucas and Prescott analyse the connection between competitive equilibrium and a social planner. The third link, between a myopic firm and a social planner, reflects a general correspondence between optimal stopping problems and optimal control problems that has received much recent attention in the literature on optimal control.[2] Since this third connection is largely unfamiliar to economists, I discuss it briefly in a later section. Connections between optimal stopping and optimal control problems may prove useful in many areas of economic research.

In the next two sections, I introduce the connection between the myopic and competitive policies in a certainty setting and then in an uncertainty setting. The following sections contain the main results. I demonstrate the optimality of the myopic policies in the special case in which the price at which firms invest in independent of the size of the market, and then extend the results to more general settings. The section on the social planner's problem is devoted to the connection between the myopic firm and the social planner. The chapter finishes with discussion on possible extensions, as well as limitations, of the results.

A DETERMINISTIC SETTING

To fix ideas, let us begin with a simple example of a case in which a myopic firm that ignores the presence of competitors chooses to invest at the same price as a fully rational firm. Consider an industry in which price depends only on industry supply and a time trend:

$$p(q, t) = -\alpha q + \beta t$$

Figure 1

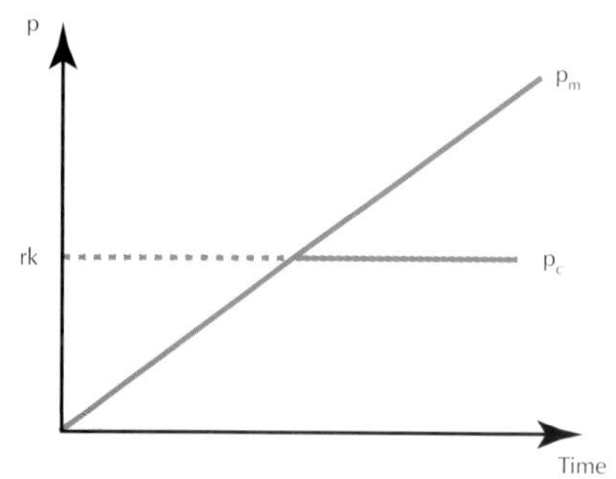

Here t represents time, q is industry output and α and β are positive constants. Suppose that firms produce output from capital with constant returns to scale; one unit of capital produces one unit of output. Suppose further that capital is infinitely divisible, and firms may add capital at a cost of k per unit. This investment is irreversible, and firms' discount future profits at the rate r.

Since firm size is indeterminant in competitive equilibrium with constant returns to scale, it is convenient to think of each infinitesimal unit of capital as a separate firm. Investment therefore consists of entry by infinitesimal firms each of which must pay an entry cost equal to k. This characterisation has no effect on competitive equilibrium and has the advantage that it keeps the expected profits of the myopic firm finite in the case of constant returns to scale.

In this example, the industry price rises at a constant rate in the absence of entry. It is clear that the competitive equilibrium is for firms to allow this price to rise until it reaches rk, the rental cost of capital, and from that point on for firms to enter in sufficient quantities to keep price constant at this level. The competitive equilibrium price path is represented by p_c in Figure 1. All firms make zero profits, and no firm has an incentive to invest earlier.

What is interesting is that the investment trigger does not depend on the endogeneity of the price process. A firm that ignored the presence of other firms and mistakenly believed that the price for its output would increase indefinitely with t, like price p_m in Figure 1, would still wait until price reached the rental cost of capital before commencing with investment. Hence the *timing* of the investment decision by a myopic and a competitive firm is identical. Of course, the *quantities* of investment may differ, as myopic firms wish to make an infinite investment.[3]

This simple example easily generalises beyond deterministic trends to arbitrary deterministic demand shocks. Consider the general inverse demand curve,

$$p_t = D(q_t, x_t), \quad D_1 < 0, \quad D_2 > 0$$

where x_t is some known function that may either rise or fall over time. The myopic firm treats q as fixed at its current level $\bar{q}$ and chooses an entry time T so as to maximise the present value of profits:

$$\max_T \int_T^{\infty} e^{-rs} D(\bar{q}, x_s) ds - e^{-rT} k$$

The first-order condition for the choice of T is simply

$$p_T = rk$$

The myopic firm will always invest at a point where price equals the annuity value of capital. The second-order condition states that the price process must cut rk from below at the time of investment. In principle, there could be many times T that satisfy the first- and second-order conditions, and the myopic firm will choose the one that maximises profits.

Now suppose that competitive firms follow the myopic policies; that is, at any point in time they behave as if industry supply will remain fixed forever at its current level and choose their entry times accordingly. Note that as investment takes place, q will in fact change, and firms' strategies will change to reflect the change in q. In spite of these changes, however, firms choose entry times as if q will remain fixed at its current level. They have static expectations regarding quantity, but rationally anticipate movements in the shock to demand.

To show that these myopic policies yield a competitive equilibrium, we need to show that the present discounted value of profits at any time of entry is equal to the cost of entry k. We know from the properties of the myopic policies that at any time of investment the price is equal to rk. It remains to determine the present value of profits between investment times. To this end, suppose that entry takes place at t_1 and t_2 but not in between. Since no investment takes place on the interval (t_1, t_2), q is constant, and the competitive and myopic price processes are the same. This implies that the present value of revenues on this interval can be no greater than $k(1 - e^{-r(t_2 - t_1)})$, which is the gain from delaying the entry cost from t_1 to t_2, or else a myopic firm would not wait until t_2 to invest. On the other hand, revenues can be no less than $k(1 - e^{-r(t_2 - t_1)})$, or some of the firms that invested at t_1 would find it optimal to postpone their entry until t_2.[4] Hence profits over the interval must be $k(1 - e^{-r(t_2 - t_1)})$. Together with the fact that firms invest at rk, this implies that the present value of profit at any investment time is equal to k. We have therefore constructed a competitive equilibrium out of the myopic strategies. Myopic behaviour is optimal in competitive equilibrium.

UNCERTAINTY

I now introduce uncertainty into the analysis by treating the demand shock as a diffusion process. Uncertainty alters the policies of both the myopic and the competitive firm. In competitive equilibrium, investment tends to limit surprise increases in the price process, while allowing price to fall. Since price generally declines following investment, firms must invest at some price trigger greater than the annuity value of capital in order to cover costs. The myopic firm, on the other hand, must consider the value of remaining uncommitted in an uncertain environment. This causes the myopic firm to wait until price has risen above the annuity value of capital before investing as well. We shall see that uncertainty causes both price triggers to increase by the same amount.

In the remainder of this section, I first extend the basic setting to include uncertainty. I then describe the evolution of prices and quantities in a competitive industry and define competitive equilibrium. In the next section I investigate the optimality of the myopic triggers.

The model

Consider a competitive industry consisting of a large number of identical firms. Each firm has a constant-returns-to-scale technology with capital as its only input; one unit of capital produces one unit of output. Unit costs are constant and equal to $c \geq 0$, and the price of a unit of output fluctuates over time so as to clear the market:

$$p_t = D(q_t, x_t) \geq 0 \tag{1}$$

where D is a time-variant inverse demand function relating price at time t to industry supply q_t and to an exogenous shock to demand x_t.[5]

At each point in time, each firm can adjust its capital stock at a cost. I generalise the previous discussion somewhat by allowing firms to disinvest, as well as invest, but I retain the assumption that the cost of adjustment is linear: increases in a firm's capital stock involve a real cost of k per unit of capital and decreases cost l. The sum of the costs $k + l$ is assumed to be positive to eliminate arbitrage. If $l = \infty$, then investment is irreversible as before.

Firms choose investment levels to maximise the present value of profits using the discount rate $r > 0$. I assume that firms employ all of their capital in production so that to reduce output they must scrap capital. They cannot let capital sit idle without disinvesting and there is no depreciation.[6]

Again, it is convenient to think of each infinitesimal unit of capital as a separate firm. In this case, investment is equivalent to entry by infinitesimal firms, and disinvestment is equivalent to exit. Each infinitesimal unit of capital faces a fixed entry cost equal to k and an exit cost equal to l. q_t represents the mass of firms in the market. Treating each unit of capital as a separate firm will allow us to compare the CRS competitive equilibrium investment strategies to the policy employed by the owner of a single investment project who ignores the effects of investment by others on the price process. I shall therefore adhere to this infinitesimal firm interpretation of the model throughout the remainder of the chapter.

The shock x is the ultimate source of uncertainty in the model. I assume that it follows a time-independent diffusion process of the form,

$$dx = \mu(x)dt + \sigma(x)dw \tag{2}$$

where w is a Wiener process.[7] This specification is slightly more general than that normally employed in models of investment under uncertainty. It covers a broad class of processes. If μ and σ are constant, then x is a Brownian motion or a continuous time random walk. If $\mu = \bar{x} - x$, then x is an Ornstein–Uhlenbeck process or a continuous time AR(1). If $\mu(x) = \mu x$ and $\sigma(x) = \sigma x$, then x is a geometric Brownian motion. Models of irreversible investment often employ this last process because the shock's average rate of growth is constant.

I assume that the inverse demand function satisfies standard regularity conditions which ensure that the price process is well behaved. In particular, $D(q_t, x_t)$ is twice continuously differentiable and strictly increasing in x, and continuously differentiable and diminishing in q. In addition,

$$E_t \int_0^{\infty} e^{-rt} |D(\hat{q}, x_t) - c| dt < \infty$$

for all fixed $\hat{q} \in \mathbf{R}_+$ so that the profits accruing to each unit of capital are finite.

I pull together the basic assumptions of the model in Assumption (*).

Assumption (*). Consider an industry with a large number of identical firms in which the following are true:

- Output is produced from capital q, with constant returns to scale.
- Investment projects are infinitely divisible. Additions to the capital shock cost k per unit and reductions cost l, where $k + l > 0$.
- Price p is represented by an inverse demand curve $D(q, x)$.
- Demand $D(q_t, x_t)$ is twice continuously differentiable and strictly increasing in x, and continuously differentiable and diminishing in q. In addition,

$$E_t \int_0^{\infty} e^{-rt} |D(\hat{q}, x_t) - c| dt < \infty$$

  for all fixed $\hat{q} \in \mathbf{R}_+$.
- The shock x follows a diffusion process,

$$dx = \mu(x)dt + \sigma(x)dw$$

  where μ and σ are continuous functions of x.

The price process and industry dynamics

One of the key determinants of a firm's investment decision is its expectation of future prices. Therefore, before introducing competitive equilibrium, it is useful to discuss the implications of Assumption (*) for the evolution of prices and quantities in the industry. An application of Itô's lemma allows us to express the evolution of the price process as a function of the shock and of the entry and exit behaviour of firms[8]:

$$dp = D_1(q_t, x_t)dq + D_2(q_t, x_t)dx + \tfrac{1}{2}\sigma(x_t)^2 D_{22}(q_t, x_t)dt \qquad \textbf{(3)}$$

The first term in equation (3) captures the direct effect of entry and exit on the price level. The remaining two terms reflect the influence of the exogenous shock for a given market size.

Since the shock is Markov, the state of the economy at any given point in time is fully characterised by the ordered pair (q_t, x_t). It is therefore natural to consider strategies that depend only on these two variables. In this case, the optimal entry and exit policies are described by regions in the (q, x) plane. Let $\overline{X}(q)$ denote the level of the shock that triggers entry when there are q firms in the market. Similarly, let $\underline{X}(q)$ denote the exit trigger. The pair of functions $\{\overline{X}(q), \underline{X}(q)\}$ describe an entry-exit policy. Note that we must have $\overline{X}(q) > \underline{X}(q)$ for $q \geq 0$, as simultaneous entry and exit is costly.

It is often more intuitive to think of entry and exit in terms of prices, rather than levels of the shock. It is possible to eliminate x_t from equation (3) by appealing to equation (1) which defines x_t as an implicit function of p_t and q_t. Setting $x_t = \phi(p_t, q_t)$ and using (2) to eliminate dx results in a general expression for the price process in terms of prices and quantities:

$$dp = \hat{\delta}(p_t, q_t)dq + \hat{\mu}(p_t, q_t)dt + \hat{\sigma}(p_t, q_t)dw \qquad \textbf{(4)}$$

where

$$\hat{\delta}(p,q) = D_1(q_t, \phi(p_t, q_t))$$

$$\hat{\mu}(p,q) = \mu(x)D_2(q,x) + \tfrac{1}{2}\sigma(x)^2 D_{22}(q,x)$$

$$\hat{\sigma}(p,q) = \sigma(x)D_2(q,x)$$

This characterisation of the price process isolates two independent effects of investment. First, investment directly influences the level of prices through its effect on the current market equilibrium: given any value for the shock, more production implies a lower equilibrium price. This effect is captured by the first term in equation (4). Second, investment may change the character of the price process by altering its infinitesimal mean and variance. A change in the size of the market alters the relationship between the shock and the market price by shifting the equilibrium to a new portion of the demand curve. This effect is captured by the latter two terms in equation (4).

An entry-exit policy $\{\overline{X}(q), \underline{X}(q)\}$ will be associated with a pair of price triggers $\{\overline{P}(q), \underline{P}(q)\}$, which are its image according to equation (1).[9] I shall refer interchangeably to strategies expressed in terms of shocks and prices.

The effect of any entry-exit strategy will be to bound the price process. The price may never rise above $\overline{P}(q)$ as then there would be firms anxious to enter the market. Their investment would either force the price down or possibly raise the entry barrier. Similarly, p_t may never fall below $\underline{P}(q)$ as long as there are still firms remaining in the market willing to exit and prevent the price from falling further. Within the bands $\overline{P}(q)$ and $\underline{P}(q)$ the price process is driven solely by the exogenous shock.[10]

Competitive equilibrium involves the simultaneous determination of the price process and the entry-exit policies of firms. I now turn to the question of how equilibrium is determined.

Competitive equilibrium

I define a competitive equilibrium as a symmetric Nash equilibrium in entry and exit strategies: given that all other firms follow a policy of entry at $\overline{P}(q)$ and exit at $\underline{P}(q)$, no individual firm can find it optimal to follow any other policy. In addition, I assume free entry so that the value of an idle firm is always equal to zero. This ensures that the value of an active firm is equal to k at the point of entry and –l at the point of exit. I present the conditions for such an equilibrium below in Definition (*).

Definition (*). An *industry equilibrium* is a diffusion process x_t and two functions $\overline{P}(q)$ and $\underline{P}(q)$ such that

- ❑ equation (1) holds;
- ❑ $p_t \in [\underline{P}(q), \overline{P}(q)]$;
- ❑ q_t increases only when $p_t = \overline{P}(q)$ and decreases only when $p_t = \underline{P}(q)$;
- ❑ entry at $\overline{P}(q)$ and exit at $\underline{P}(q)$ maximise the present discounted value of profits given that prices evolve according to (1)–(3); and
- ❑ the value of an idle firm is equal to zero.

The first three conditions ensure that p_t is the price process that would result if all firms followed policies $\overline{P}(q)$ and $\underline{P}(q)$; entry and exit at the triggers keep the price process within the triggers. The fourth condition ensures that no individual firm will rationally choose to follow any other policy. The final item is the free entry condition.

Constructing an equilibrium is equivalent to finding two functions $\overline{P}(q)$ and $\underline{P}(q)$ such that the present discounted value of profits for the price process regulated by these functions is k at the entry barrier and –l at the exit barrier. If this is the case, then at the upper barrier inactive firms will be indifferent between remaining idle or entering the market, while at the lower barrier active firms will be indifferent between remaining active or exiting the market. Furthermore, since the present discounted value of profits is strictly increasing in the price level for any given market size, no firm will find it in its best interest to follow any alternative policy.

In general, constructing such an equilibrium would appear to be an extremely difficult task due to the variety of effects that investment has on the price process. Not only does it directly affect the level of prices by altering supply, but it also affects the sensitivity of prices to demand shocks by shifting equilibrium to a new portion of the demand curve. To calculate the present discounted value of profits under a given set of policies, we have to analyse the entire evolution of prices for each possible realisation of the shock in order to determine in what ways entry and exit influenced the level, mean and variance of the price process.

In the next two sections I show that we can construct a competitive equilibrium much more simply from the policies of a myopic firm. I shall pursue this result in two steps. I first demonstrate this construction in a special case in which investment has no effect on the infinitesimal mean or variance of the price process. This restriction greatly simplifies the optimal policies and allows us to concentrate on the effect that investment has on the level of prices. I then show that more general cases can be reduced to the case in which the mean and variance of the price process are independent of the size of the market.

A SIMPLE INTERMEDIATE CASE

Let us consider cases in which the demand curve and the shock process are chosen so that entry has no effect on the mean and variance of the price process. This would be the case, for example, if the demand shock entered the inverse demand function multiplicatively and followed a geometric Brownian motion (see Dixit (1991b)). If the mean and variance of the price process depend only on the level of the price process, then prices evolve according to

$$dp = \hat{\delta}(p_t, q_t)dq + \hat{\mu}(p_t)dt + \hat{\sigma}(p_t)dw \tag{5}$$

Here q affects the level of prices through the first term in (5), but $\hat{\mu}$ and $\hat{\sigma}$ depend only on p. If the price process is of this form, then the future evolution of the price process depends only on the level of p. The optimal Markov policy is therefore independent of q, and entry and exit occur at fixed levels $\overline{P}$ and $\underline{P}$.[11] This is the principal simplifying feature of this special case.

The purpose of this section is to compare these competitive equilibrium entry and exit triggers with the optimal entry and exit policies of a myopic firm. The latter assumes that q never changes so that prices evolve according to

$$d\tilde{p} = \hat{\mu}(\tilde{p})dt + \hat{\sigma}(\tilde{p})dw \tag{6}$$

Note that the $\hat{\delta}(\tilde{p}, q)dq$ term which captures the effect of investment on the level of prices does not appear in equation (6).

Figure 2

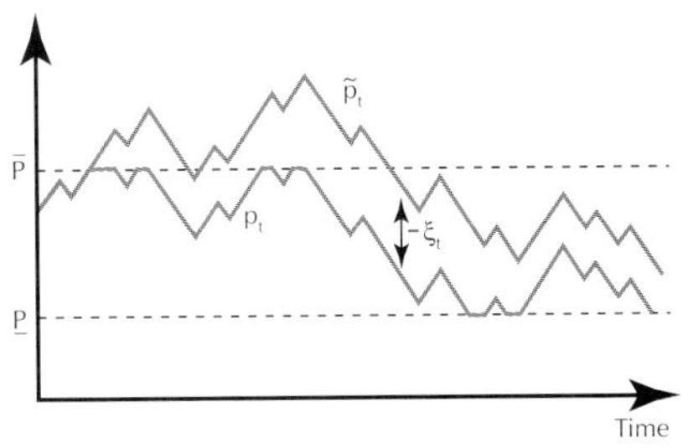

The relationship between the myopic and the competitive price processes is not a simple one. $\tilde{p}$ is potentially unbounded, while p is constrained to lie within the bounds $\overline{P}$ and $\underline{P}$. Assuming that $p_0 = \tilde{p}_0$, the effect of this constraint may be captured by a process ξ_t,

$$p_t = \tilde{p}_t + \xi_t$$

where $\xi_t = \int_0^t \hat{\delta}(p_s, q_s) dq_s$ captures the effect of entry and exit which maintain p_s within $[\underline{P}, \overline{P}]$ for all $s \leq t$. A stylised representation of this relationship is presented in Figure 2.

In spite of the difference between the two processes, Proposition 1 states that the optimal entry and exit policies are identical. In other words, if the mean and variance of prices are invariant to market size, the optimal competitive equilibrium policies need not take into account the effect of entry and exit on the level of prices. Firms may ignore entry and exit altogether and treat the price process as an exogenous diffusion process whose mean and variance are a fixed function of the price level.[12]

Proposition 1. Let $\overline{P}$ and $\underline{P}$ be the optimal entry and exit levels for the exogenous price process defined by

$$d\tilde{p} = \mu(\tilde{p})dt + \hat{\sigma}(\tilde{p})dw \qquad \textbf{(6)}$$

Then $\overline{P}$ and $\underline{P}$ are competitive equilibrium entry and exit triggers for the endogenous price process:

$$dp = \hat{\delta}(p,q)dq + \hat{\mu}(p)dt + \hat{\sigma}(p)dw \qquad \textbf{(5)}$$

Proof. See Appendix.

The proof of Proposition 1 is easy to sketch. Suppose that all firms enter and exit at prices p_u and p_l, respectively, where $p_l < \underline{P} < \overline{P} < p_u$ and $\overline{P}$ and $\underline{P}$ are the myopic triggers. Is this a Nash equilibrium? The answer is no, since a firm does better by following the myopic policies. The reason follows from Bellman's principle: if a firm wants to be active tomorrow at p_u, the optimal policy today is entry at $\overline{P}$, because the myopic price process and the competitive price process are identical until p_u. Taking the limit as $p_u \to \overline{P}$ and $p_l \to \underline{P}$ shows that $\{\underline{P}, \overline{P}\}$ is a Nash equilibrium. Continuity of the optimal policies then shows that this equilibrium exists.

To illustrate this result, I present a simple example in which investment is irreversible.

An example

Suppose that $l = \infty$, so that investment is irreversible. Suppose further that the inverse demand curve takes the form $D(q, x) = x/q$, and that x follows a geometric Brownian motion. It follows from equation (4) that in the absence of investment, the price process also follows a geometric Brownian motion:

$$dp = \mu p dt + \sigma p dw$$

Finally, assume that $c = 0$ to simplify notation.

In this case the problem of the myopic firm is identical to the problem solved by Dixit (1989a). Following Dixit, we can solve for the myopic firm's optimal entry trigger:

$$\bar{P} = \alpha(r - \mu)k/(\alpha - 1)$$

where $\alpha > 1$ is a root of the polynomial $r - (\mu - \frac{1}{2}\sigma^2)\phi - \frac{1}{2}\sigma^2\phi^2$. If σ is equal to zero, then we are back to the certainty case: α equals r/μ, and $\bar{P}$ is equal to rk. As σ increases, α falls toward 1, and $\bar{P}$ rises above rk. The resulting gap between the entry trigger and the annuity value of capital reflects the option value of remaining inactive.

To see that $\bar{P}$ is the competitive entry trigger, let $\Pi(p)$ denote the value of a competitive firm that is active in the market. Treating the firm as an asset, it must be the case that the annuity value of the firm is equal to the sum of the flow return and the expected capital gain:

$$r\Pi(p_t) = p_t + E_t d\,\Pi(p_t)$$

Using Itô's lemma to characterise the expected capital gain, we obtain a differential equation with the following solution:

$$\Pi(p) = p/(r - \mu) + Ap^{\alpha} + Bp^{\beta}$$

where $\alpha > 1$ and $\beta < 0$ are once again roots to $r - (\mu - \frac{1}{2}\sigma^2)\phi - \frac{1}{2}\sigma^2\phi^2$.

It remains to determine A and B. Since Π approaches zero as p approaches zero, B equals zero. A is determined from the following boundary condition[13]:

$$\Pi'(\bar{P}) = 0$$

It is now easily confirmed that $\Pi(\bar{P})$ equals k, and, since firms entering the market at $\bar{P}$ earn zero profits, we have constructed a competitive equilibrium from the myopic policy.

To get some intuition for what is happening, consider $\Pi(p)$ more closely. Let T denote the first time that price reaches $\bar{P}$. Because the present value of profits at T is k,

$$\Pi(p_t) = E_t\int_t^T e^{-rs}p_s ds + e^{-rT}k$$

A little algebraic manipulation yields

$$\Pi(p_t) = \frac{p_t}{r-\mu} - E_t e^{-rT}\left(\frac{p_T}{r-\mu} - k\right)$$

The first term is the present value of profits if there is no further investment in the market. This is the value of an active myopic firm. The second term is the value of waiting until T, investing and receiving an active myopic firm. This is the value of an idle myopic firm. The difference between the value of an active and an inactive competitive firm is therefore the same as that between an active and an inactive myopic firm. Competition therefore does not alter the incentive to trade an idle firm for an active firm.

In the next section we shall see another reason why the timing of the marginal investment decision is the same in both situations: both are reflections of the same maximisation problem. Competitive equilibrium maximises aggregate consumer surplus, whereas the myopic firm maximises its individual contribution to consumer surplus.

A MORE GENERAL CASE

So far, I have abstracted from the effects that entry may have on the infinitesimal mean and variance of the price process. I now show that firms may safely ignore these effects as well. To calculate the competitive equilibrium entry and exit barriers, firms need only consider the current mean and variance of innovations in prices. Changes in these parameters due to future entry and exit will, of course, alter future policies, but they have no influence on current decisions.

In the case in which entry and exit alter the mean and variance of the price process, the competitive entry and exit barriers may be calculated from the following algorithm. Given that the market size is $\bar{q}$ and that prices follow the process identified in equation (4), firms can calculate the policy $\bar{P}(\bar{q})$ and $\underline{P}(\bar{q})$ from a process in which entry and exit have no effect on the mean and variance of prices:

$$d\hat{p} = \hat{\delta}(\hat{p},q)dq + \hat{\mu}(\hat{p}_t,\bar{q})dt + \sigma(\hat{p}_t,\bar{q})dw \qquad (7)$$

Here future entry and exit affect the level of the price process through the first term on the right-hand side of (7), but $\hat{\mu}$ and $\hat{\sigma}$ depend only on the current number of firms in the market $\bar{q}$. This defines a single point of the functions $\bar{P}(q)$ and $\underline{P}(q)$, that associated with $\bar{q}$. The exercise must therefore be repeated each time the size of the market changes.

Proposition 2 formalises this algorithm. Together Propositions 1 and 2 imply that firms may ignore all effects of entry and exit on the price process.

Proposition 2. The competitive entry and exit strategies $\bar{P}(q)$ and $\underline{P}(q)$ can be calculated as follows. Pick an arbitrary $\bar{q}$. Let $\bar{p}$ and $\underline{p}$ be equilibrium entry and exit barriers for the price process $\hat{p}$ defined by

$$d\hat{p} = \hat{\delta}(\hat{p},q)dq + \hat{\mu}(\hat{p}_t,\bar{q})dt + \hat{\sigma}(\hat{p}_t,\bar{q})dw \tag{7}$$

where $\bar{q}$ is fixed at the chosen level. Then, given that the current industry output is $\bar{q}$, $\bar{p}$ and $\underline{p}$ are equilibrium entry and exit barriers for

$$dp = \hat{\delta}(\hat{p},q)dq + \hat{\mu}(p,q)dt + \hat{\sigma}(p,q)dw \tag{8}$$

So $\bar{P}(\bar{q}) = \bar{p}$ and $\underline{P}(\bar{q}) = \underline{p}$.

Proof. See Appendix.

Proposition 2 states that competition insulates a competitive firm from endogenous changes in the mean and variance of the price process. The principle underlying the proposition is related to the idea of subgame replacement in game theory. It makes no difference to the total game if one subgame is replaced by another subgame with identical payoffs. Similarly, entry and exit times can be thought of as times at which the active firms receive payoff k or –l. It makes no difference what the form the price process takes beyond this point, as long as it has value k or –l. In the present model, optimal entry and exit ensure that at any investment time the present value of profits is equal to the cost of adjustment regardless of the future evolution of the shock. Therefore, a competitive firm only needs to worry about the evolution of prices up until the next investment time. This explains why the current investment decision depends only on the current characteristics of the price process.

This discussion ignores one way in which the future might influence the present: future changes may alter the timing of the next investment decision. The proof of Proposition 2 makes use of the continuity of the model which implies that a small change in the capital stock has only a small change on the characteristics of the price process. Future changes in the price process therefore have a negligible impact on the next investor's investment decision, and hence the time at which today's investor receives k or –l.

THE SOCIAL PLANNER'S PROBLEM

Together Propositions 1 and 2 show that firms can ignore other firms when deciding when to invest. To better understand the reason that firms may ignore so much, it is useful to take a look at the problem from another perspective. This will give us some intuition as to the generality of the results, as well as to where and why they might fail.

Recall from the introduction that competitive equilibrium and myopic behaviour are not only directly connected, but are also indirectly connected through the social planner's problem. Since the relationship between competitive equilibrium and the social planner's problem is well-known (see Debreu (1954) and Lucas and Prescott (1971)), I shall take this correspondence as given, and instead concentrate on the relationship between the social planner's problem and that of a myopic firm. The purpose of this discussion is to illustrate the connection between the planner's optimal control problem and the firm's optimal stopping problem. Readers interested in a more rigorous treatment may consult Karatzas and Shreve (1984) or El Karoui and Karatzas (1988). Leahy (1993) provides a discrete time treatment.

Consider a social planner who maximises the present discounted value of social welfare in the form of consumer surplus:

$$\max_{q_t} E\left\{\int_0^\infty \int_0^{q_t} e^{-rt}\left[D(z,x_t) - c\right] dz\, dt - \int_0^\infty e^{-rt} k\, dq_t^+ - \int_0^\infty e^{-rt} l\, dq_t^-\right\} \qquad \textbf{(9)}$$

where q_t^+ and q_t^- represent cumulative increases and decreases in the capital stock, respectively.

At any point in time, the contribution of the $\bar{q}$th unit of capital to social welfare is simply

$$D(\bar{q}, x) - c \qquad (10)$$

if this unit of capital is in use and zero otherwise. The social planner can do no better than to maximise this contribution separately for each $\bar{q}$. By maximising the contribution of each unit, the social planner will maximise their sum. The problem of choosing entry and exit times to maximise (10), however, is precisely the problem that is solved by the myopic firm. The social planner would therefore want to follow the myopic policies. The only remaining issue is whether or not the myopic policies are feasible. Feasibility requires that the myopic policy triggers $\bar{X}(q)$ and $\underline{X}(q)$ be increasing in q.[14] This condition will hold if the return to capital is decreasing in q.

This discussion shows that there is a direct relationship between the problem of the myopic firm and that of the social planner. Either the social planner's problem can be seen as an optimal control problem, the problem of adjusting the industry's capital stock over time so as to maximise equation (9), or it can be seen as a series of optimal stopping problems, the choice of entry and exit triggers that maximise the individual contribution of each separate unit of capital.

This relationship explains what happens to the option value of investment in competitive equilibrium. While it is true that for a competitive firm the value of remaining uncommitted is equal to zero, the option still has value for society as a whole. The issues raised in the literature on irreversible investment do not disappear in competitive equilibrium, they are simply shifted to the aggregate. In essence, endogenising the price process transfers the option from the firm to the social planner.

EXTENSIONS AND LIMITATIONS

I have shown that in a constant-returns-to-scale industry in which uncertainty follows a general diffusion process, risk-neutral investors may ignore the effects of future investment on the level, mean and variance of the price process. The proofs of Propositions 1 and 2 were simplified by several of the conditions in Assumption (*) and Definition (*). Without constant returns, entry and exit would not have occurred at fixed levels, complicating the proof of Proposition 1. Without free entry, the value of an active firm would not have been equal to k at the point of entry, complicating the proof of Proposition 2. Finally, without the continuity of demand, infinitesimal investment and the continuity of the shock, the myopic triggers would not have been continuous in q, complicating both proofs. The purpose of this section is to indicate directions in which these assumptions may be relaxed and ways in which relaxing the assumptions will endanger the myopia result.

This discussion makes two general points. First, if one considers the strategies of a myopic firm, a competitive firm and a social planner, and one notes that one of these strategies is related to the other two, then it must be the case that the other two are related to each other. Hence whenever competitive equilibrium is Pareto optimal, and the planner's optimal control problem may be recast as an optimal stopping problem, myopic policies will be competitive. Conversely, if any two of these policies are unrelated, then it cannot be the case that both of these policies are related to the third. Hence any violation of the Arrow–Debreu assumptions that invalidates the second welfare theorem, and hence the correspondence between competitive equilibrium and the social planner's problem, endangers the optimality of the myopic policies in competitive equilibrium.

Limitations

There are several conditions that are necessary for the myopic policies to be optimal in competitive equilibrium.

First, it is important that investment projects be infinitely divisible. To see this, suppose that all firms were to follow the myopic policies when investment is discrete. If they invest at the myopic entry trigger, then their entry will reduce the price discretely, and all firms will lose money. If they target their postinvestment price to the myopic trigger, then they will have to allow the preinvestment price to rise well above this trigger. These high prices, however, will allow firms already in the market to earn positive profits. Hence the myopic policies cannot be a competitive equilibrium.[15] A similar problem arises if demand is discontinuous. Neither discrete investment nor discontinuous demand, however, destroys the connection between the myopic policies and the social planner's problem.

It is also important that returns to scale are not increasing. With increasing returns it may be profitable to invest large amounts of capital, while incremental projects remain unprofitable. It is, therefore, impossible to completely eliminate the interdependence among investment projects.

Another important limitation of the current chapter is the absence of idiosyncratic shocks. Here it is an open question as to whether a version of the myopia result still holds. In some sense the myopic policies and competitive equilibrium of this chapter represent polar cases; in the former there are only idiosyncratic shocks, whereas in the latter there are only common shocks. This provides some hope for intermediate cases. Of course, much will depend on the specific way in which one introduces idiosyncratic shocks into the model and determines equilibrium.

Possible generalisations

Several of the conditions in Assumption (*) and Definition (*) appear to be unnecessary.

First, it is possible to extend the analysis to incorporate decreasing returns. Decreasing returns do not alter either the link between competitive equilibrium and the social planner's problem or the link between the social planner's problem and the myopic policies.

Second, the free entry is also unnecessary. Since firms can ignore the presence of other entrants, it should not matter how many of these other entrants are actually there. This indicates that the mass of potential entrants can be limited to $\bar{q}$.[16]

It should also be possible to relax the assumption that firms are risk neutral. At first glance, it may seem strange that this issue is irrelevant since risk-averse firms should be especially concerned with future changes in the mean and variance of the price process. Recall, however, the intuition behind Proposition 2 that points of entry are stopping times at which the actual evolution of the price process may be replaced by the cost of entry. In the case of risk aversion, future entry and exit will always occur in such a way as to keep the expected utility of the payoff stream equal to the cost of entry.

A fourth possible extension is to more general stochastic processes for the shock, including discrete time processes, continuous time jump processes, as well as non-Markov processes. I base this conjecture on the connection between the problem of the myopic firm and the social planner. The intuition underlying this connection does not appear to depend on the specification of the shock.

CONCLUSION

This chapter has considered three types of investment strategies: the policies of a competitive firm, the policies of a social planner and the policies of a myopic firm that treats the evolution of the price process as independent of the strategies of other firms. While the connection between the first two types of strategies is generally known, the role of the latter is new. By demonstrating the relationship between the myopic policies and the social planner's policies, the chapter illustrates a general connection between optimal stopping problems and optimal control problems that may be useful in many economic applications. By demonstrating the optimality of the myopic policies in competitive equilibrium, the paper makes possible the direct application to competitive equilibrium of the considerable research on the optimal investment strategy of an individual price-taking firm.

APPENDIX: PROOFS OF THE PROPOSITIONS

A few preliminaries will simplify the exposition of the proofs. The main needs are to set out some notation and to prove that the myopic policies are continuous in the parameters.

Suppose that a firm faces an exogenous price process of the form,

$$dp = \mu(p,q)dt + \sigma(p,q)dw \tag{A1}$$

where q is a fixed parameter and μ and σ are continuous in p and q. Let $V(p,\bar{p},\underline{p},q)$ represent the value of an idle firm when the price process follows (A1), the current price is p, the entry and exit policy is $\{\bar{p},\underline{p}\}$ and q is fixed for all time. Similarly, let $J(p,\bar{p},\underline{p},q)$ denote the value of an active firm under similar circumstances. We assume that $\bar{p}$ and $\underline{p}$ lie in an unspecified compact interval **B** with $\bar{p} > \underline{p}$.[17]

Standard dynamic programming arguments imply that V satisfies the following differential equation:

$$rV - \mu(p,q)V_p - \tfrac{1}{2}\sigma(p,q)^2V_{pp} = 0$$

Similarly, J satisfies

$$rJ - \mu(p,q)J_p - \tfrac{1}{2}\sigma(p,q)^2J_{pp} = p - c$$

These equations may be solved subject to the four boundary conditions[18]:

$$J(p,\bar{p},\underline{p},q) = V(p,\bar{p},\underline{p},q) - k, \quad \forall p \geq \bar{p}$$

$$V(p,\bar{p},\underline{p},q) = J(p,\bar{p},\underline{p},q) - l, \quad \forall p \leq \underline{p}$$

$$\lim_{p \to p_*} V(p,\bar{p},\underline{p},q) = 0$$

$$\lim_{p \to p^*} J(p,\bar{p},\underline{p},q) - E\int_t^{\infty} e^{-rs} p_s ds = 0 \qquad \textbf{(A2)}$$

Here p_* and p^* represent the upper and lower limits of the price space and may depend on the specification of the price process in (A1). The first two conditions in (A2) are value-matching conditions. They state that at the point of entry or exit the difference between an active and inactive firm is equal to the adjustment cost. The third condition says that as the price falls, the prospect that the firm will ever become active diminishes. The last condition says that as price rises the prospect that the firm will ever exit diminishes.

Furthermore, we let

$$V^*(p,q) = \max_{\bar{p},\underline{p}} V(p,\bar{p},\underline{p},q)$$

and

$$J^*(p,q) = \max_{\bar{p},\underline{p}} J(p,\bar{p},\underline{p},q)$$

Finally, let $\bar{p}^*(q)$ and $\underline{p}^*(q)$ denote the policy choices that maximise V and J.

Lemma. V^* and J^* are continuous in p and q. $\bar{p}^*(q)$ and $\underline{p}^*(q)$ are continuous correspondences in q.

Proof. First note that the V and J are continuous in all of their arguments. Continuity in p results from the continuity and infinite variation of the price process. Continuity in $\bar{p}$ and $\underline{p}$ results from the fact that path by path the value of a policy is continuous in the trigger or target unless the trigger or target is a local maximum or minimum for the process, and the latter event is of measure zero. Finally, continuity in q follows from the continuity of $\mu(q)$ and $\sigma(q)$ and standard arguments on the continuity of the solution to a differential equation.[19]

The first statement now follows directly from an application of the Theorem of the Maximum (see Stokey and Lucas (1989), p. 62).

The Theorem of the Maximum also implies that $\bar{p}^*(q)$ and $\underline{p}^*(q)$ are upper hemicontinuous. To get continuity, we prove that the correspondences are convex valued. Suppose that $\{\bar{p}_1, \underline{p}_1\}$ and $\{\bar{p}_2, \underline{p}_2\}$ are both optimal for a given q. Let $p_u = \min\{\bar{p}_1, \bar{p}_2\}$ and $p_l = \max\{\underline{p}_1, \underline{p}_2\}$. Now as both policies are maximal, we have $V(p,\bar{p}_1,\underline{p}_1,q) = V(p,\bar{p}_2,\underline{p}_2,q)$ and $J(p,\bar{p}_1,\underline{p}_1,q) = J(p,\bar{p}_2,\underline{p}_2,q)$ for all $p \in [p_l, p_u]$.

Without loss of generality assume that $\bar{p}_1 < \bar{p}_2$. Define $G(p,\bar{p},\underline{p},q) = J(p,\bar{p}_2,\underline{p}_2,q) - V(p,\bar{p}_2,\underline{p}_2,q)$. Then G must solve

$$rG - \mu(p,q)G_p - \tfrac{1}{2}\sigma(p,q)^2 G_{pp} = p - c$$

subject to the boundary conditions

$$G(p,\bar{p},\underline{p},q) = k, \quad \forall p \geq \bar{p}_2 \text{ and } p = \bar{p}_1$$
$$G(p,\bar{p},\underline{p},q) = -l, \quad \forall p \leq \underline{p}_2$$

Since G is well set and since G equals k at both $\bar{p}_1$ and $\bar{p}_2$, G must equal k at all points in between. But this implies that all $p \in [\bar{p}_1, \bar{p}_2]$ satisfy the boundary conditions (A2) and are points of entry with the same value as $\bar{p}_1$ and $\bar{p}_2$. A similar argument applies to the exit trigger. This shows that the optimal entry and exit triggers form a convex valued correspondence and completes the proof.

With the lemma we may now prove Propositions 1 and 2.

Proposition 1. Let $\bar{P}$ and $\underline{p}$ be the optimal entry and exit levels for the exogenous price process defined by

$$dp = \hat{\mu}(p)dt + \hat{\sigma}(p)dw \tag{A3}$$

Then $\bar{P}$ and $\underline{p}$ are competitive equilibrium entry and exit triggers for the endogenous price process:

$$dp = \hat{\delta}(p,q)dq + \hat{\mu}(p)dt + \hat{\sigma}(p)dw \tag{A4}$$

Proof. As discussed in the text, competitive equilibrium will yield a price process that is regulated at fixed entry and exit barriers. The task is to show that these barriers must be $\bar{P}$ and $\underline{P}$, the optimal entry and exit levels against the unregulated process (A3).

Consider the optimal policy of a single firm given that all other firms enter at p_u and exit at p_l, where $p_u > \bar{P} > \underline{P} > p_l$. The actions of others therefore regulate the price process at p_u and p_l. To find a Nash equilibrium, we must find levels of p_u and p_l such that the single firm finds it optimal to enter and exit with the others. If the optimal policy of the single firm calls for it to be active at p_u and idle at p_l, then it is immediate that $\bar{P}$ and $\underline{P}$ are its optimal entry and exit barriers. This follows from Bellman's optimality principle.[20] Let T denote the first hitting time of either p_u or p_l. Let $Z(\bar{p},\underline{p},p_t)$ denote the flow profit from following an entry-exit strategy of $\bar{p}$ and $\underline{p}$

when the current price is p_t. The optimal policy must then satisfy

$$V^*(p_t) = \max_{\bar{p},\underline{p}} E\left\{\int_0^T Z(\bar{p},\underline{p},p_t)dt + e^{-rT}\tilde{V}(p_T)\right\}$$

where $\tilde{V}(p_T)$ equals $V^*(p_l)$ if $p_T = p_l$ and $J^*(p_u)$ if $p_T = p_u$.[21] As regulated process is identical to the unregulated process until the time T, Z is maximised by $\bar{p} = \bar{P}$ and $\underline{p} = \underline{P}$. We may now take the limit as $p_u \to \bar{P}$ and $p_l \to \underline{P}$ to obtain a Nash equilibrium in which process is regulated at the optimal entry and exit barriers. Thus, the equilibrium policies for (A4) are identical to the optimal policies against (A3).

This leaves open the question of whether it is optimal to be in the market at p_u and idle at p_l. Suppose that for some p_u it was no longer optimal to enter. Then it must be that the value of holding an active firm at p_u is less than k, and since it is not optimal to invest, the value of holding an idle firm is zero:

$$0 = V^*(p_u) \geq J^*(p_u) - k \qquad \textbf{(A5)}$$

Let $\tilde{p}_u$ denote the supremum of the set for which (A5) holds. Applying the lemma, V^* and J^* are continuous functions, so that (A5) holds with equality at $\tilde{p}_u$, and the firm is indifferent between entry and remaining inactive. The previous argument therefore implies that the optimal entry point is $\bar{P} \leq \tilde{p}_u$. Suppose that $\bar{P} < \tilde{p}_u$. As V^* equals zero, there is no value to being an idle firm. J^* is therefore equal to the present discounted value of profits until exit less the present discounted value of the exit cost. This value is strictly monotonic in price, as for each possible path of the shock raising the initial price can only raise future prices.[22] As $\bar{P} < \tilde{p}_u$, this implies that $J(\bar{P}) < k$, which contradicts the assertion that $\bar{P}$ is the optimal entry level. Therefore, it must be the case that $\tilde{p}_u = \bar{P}$ and for all $p_u > \bar{P}$ it is optimal to be active at p_u. A similar argument shows that for all $p_l > \underline{P}$ it is optimal to be inactive at p_l. This completes the proof.

Propositions 2. The competitive entry and exit strategies $\bar{P}(q)$ and $\underline{P}(q)$ can be calculated as follows. Pick an arbitrary $\bar{q}$. Let $\bar{p}$ and $\underline{p}$ be equilibrium entry and exit barriers for the price process defined by

$$dp = \hat{\delta}(p,q)dq + \hat{\mu}(p_t,\bar{q})dt + \hat{\sigma}(p_t,\bar{q})dw \qquad \textbf{(A6)}$$

where $\bar{q}$ is fixed at the chosen level. Then, given that the current industry output is $\bar{q}$, $\bar{p}$ and $\underline{p}$ are equilibrium entry and exit barriers for

$$dp = \hat{\delta}(p,q)dq + \hat{\mu}(p_t,q)dt + \hat{\sigma}(p_t,q)dw \qquad \textbf{(A7)}$$

So $\bar{P}(\bar{q}) = \bar{p}$ and $\underline{P}(\bar{q}) = \underline{p}$.

Proof. Suppose that there are $\bar{q}$ firms in the market, that the entry and exit strategies $\overline{P}(q)$ and $\underline{P}(q)$ have been constructed in accordance with the proposition, and that price follows (A7). We wish to show that the present discounted value of profits is k when the market price equals $\overline{P}(\bar{q})$ and –l when the market price equals $\underline{P}(\bar{q})$.

Given a realisation of the price process, define an excursion as a path segment that begins at an entry or exit barrier, ends at an entry or exit barrier and is interior to the barriers at all times in between. During the excursion the number of firms is constant, so the process may be described by (A6). Since the optimal policy $\{\overline{P}(q), \underline{P}(q)\}$ is continuous by the lemma, each excursion begins and ends at points that the equilibrium entry or exit points for (A6) given the output level over the excursion; that is, if there are $\bar{q}$ firms during the excursion, then the excursion begins at either $\overline{P}(\bar{q})$ or $\underline{P}(\bar{q})$ and ends at either $\overline{P}(\bar{q})$ or $\underline{P}(\bar{q})$.[23] Now as $\overline{P}(\bar{q})$ and $\underline{P}(\bar{q})$ are by construction equilibrium entry and exit barriers for (A6), the expected present discounted value of profits under (A6) is k at $\overline{P}(q)$ and –l at $\underline{P}(q)$. Hence the expected profit during an excursion beginning with entry is $E(k - e^{-rT}\gamma(T))$, where T is the duration of the excursion and $\gamma(T)$ equals k or –l depending on whether the excursion ends with entry or exit. We may repeat this exercise for each excursion of (A7).

The excursions completely determine the present discounted value of profits. This is because the time spent at the barriers is measure zero almost surely. To see this, let S denote the set of entry times, $S = \{t{:}p(q_t, x_t) = \overline{P}(q_t)\}$. S is closed as it is the inverse image of the set {0} under the continuous mapping $p(q_t, x_t) - \overline{P}(q_t)$. Furthermore, S has no interior points since during any arbitrary time interval the price both rises and falls because the shock has infinite variation. As entry can only occur during times in which the shock rises, any arbitrary interval, no matter how small, contains times of inaction.[24] Since the excursions comprise the complement of S, their union is open, and the number of excursions is countable, so we may sum them. Let $T_i(\omega)$ denote the time of the ith excursion for a specific realisation of the shock process $x(\omega)$, where $\omega \in \Omega$ is a particular state of the world. Let $\gamma(\omega,T)$ equal k if the excursion begins with entry and –l if it begins with exit. The present discounted value of profits is therefore

$$V(\omega,T_0) = E\left\{\sum_{i=1}^{\infty}\gamma(\omega,T_i(\omega))e^{-rT_i} - \gamma(\omega,T_{i+1}(\omega))e^{-rT_{i+1}}\right\}$$

Taking the sum path by path, we are left with

$$V(\omega,T_0) = \gamma(\omega,T_0) = \begin{cases} k, & \text{if entry at } T_0 \\ -l, & \text{if exit at } T_0 \end{cases}$$

which confirms that $\overline{P}(q)$ and $\underline{P}(q)$ are in fact equilibrium policies for (A7).

1 See, for example, Bernanke (1983), Brennan and Schwartz (1985), McDonald and Siegel (1986), Pindyck (1988) and Dixit (1989a) for an analysis of the irreversible investment decision from the perspective of an individual price-taking firm, and Pindyck (1988) and Bertola (1989) for an analysis of the optimal policy of a monopolist.

2 The connection between these two problems was first noticed by Karatzas and Shreve (1984), who consider a setting equivalent to an industry with irreversible investment and a linear demand curve. This work has been extended by El Karoui and Karatzas (1988). See Leahy (1993) for an analysis in discrete time.

3 The distinction between the timing of the marginal investment decision and quantities of investment is important. Models of investment by a price-taking firm, such as Abel (1983) or Caballero (1991), will generally not be an accurate guide to the equilibrium quantities of investment in a competitive industry. In treating the price process as exogenous, these models ignore all interaction between industry investment and the price process (Pindyck, 1990).

4 It is important that price be continuous (as a correspondence) in quantity so that entry by an infinitesimal firm does not alter the price process. Otherwise, the price process might depend on whether the marginal firm chose to enter at t_1 or t_2. I shall return to this point in a later section.

5 Many general causes may be reduced to this case. For example, consider a good produced from capital and labour with constant returns to scale. If labour is perfectly flexible, then D may be interpreted as the marginal revenue product of capital after optimising with respect to labour. Often the constant c will be zero in applications.

6 It is trivial to introduce depreciation. It only alters the discount rate.

7 It is assumed that $\mu(x)$ and $\sigma(x)$ satisfy the standard condition for a solution to (2) to exist: for some constant K and $x, y \in \mathbf{R}$,

$$| \mu(x) - \mu(y) | + | \sigma(x) - \sigma(y) | < K | x - y |,$$
$$| \mu(x) |^2 + | \sigma(x) |^2 < K^2(1 + | x |^2)$$

See Karatzas and Shreve (1988), p. 289.

8 There are no second-order terms in q because the output process is variation finite. If this were not the case, the economy would quickly incur infinite costs of adjustment since $k + 1 > 0$.

9 A similar correspondence arises in the exchange rate band literature (see Krugman (1991)). There one can refer to a band either on the exchange rate (prices) or on the fundamentals (shocks).

10 If the entry and exit triggers are constant, x is a Brownian motion, and demand is unit elastic, then price follows a regulated Brownian motion in the sense of Harrison (1985).

11 Lippman and Rumelt (1985) and Edleson and Osband (1989) study special cases in which the mean and variance of the price process are invariant to market size. They find that the optimal policies may be characterised by two stationary price triggers.

12 Dixit (1991a) proves a similar result in the case of price ceilings. Price ceilings differ from regulation, however, in that a price ceiling only bounds the price process over the period for which price would have been excessively high in the absence of the price ceiling. With a price ceiling at p, the price process becomes $\hat{p}_t = \min \{\tilde{p}_t, p\}$. Regulation through entry and exit alters the level of the price process itself as can be seen from Figure 2.

13 This is a smooth contact condition. If it did not hold, the expected capital gain or loss at P would be infinite due to the infinite variation of the price process.

14 If the triggers are not increasing in q, then the application of the myopic policies by the social planner will call for more than q_1 units of capital but less than q_2 units of capital with $q_1 > q_2$. This is not feasible.

15 This is why Dixit (1989b) solves a linked option problem in his model of equilibrium entry and exit. Investment in his model is discrete; therefore, all of his firm's investment decisions are interdependent. The profitability of one depends on the timing of the next.

16 A limit on the number of potential entrants will result in positive profits to all firms. This indicates that it is important that the limit be imposed on the number of potential entrants and not on the size of the market. If there is an entry quota and an unlimited supply of potential entrants, competition will lead firms to enter the market earlier than they would under the myopic policies. The resulting equilibrium is inefficient. See Bartolini (1990).

17 This requires that we let **B** grow until it contains the policies in which we are interested.

18 This characterisation of the optimal policies is adapted from Dixit (1989a).

19 For example an argument may be based on Birkhoff and Rota (1978, Theorem 6.5, Lemma (p. 156), Theorem 10.11).

20 I am indebted to Andrew Caplin for suggesting this argument.

21 q has been dropped from the V^* as in this case the value of a firm is independent of market size.

22 See Caplin and Leahy (1991) for this monotonicity argument.

23 If there were discontinuities, then the level of entry for $P(q_+)$ may differ from that of $P(q)$.

24 In general, it can be shown that the myopic entry and exit barriers are monotonically increasing in q. Hence they are variation finite. As p has quadratic variation, it cannot coincide with a variation finite process over any open interval of time.

BIBLIOGRAPHY

Abel, Andrew B., 1983, "Optimal Investment under Uncertainty", *American Economic Review* 73, pp. 228–33.

Bartolini, Leonardo, 1990, "Exchange Rate Uncertainty and Foreign Investment Quotas", mimeo, Princeton University.

Bernanke, Ben S., 1983, "Irreversibility, Uncertainty, and Cyclical Investment", *Quarterly Journal of Economics* 98, pp. 85–106.

Bertola, Giuseppe, 1989, "Irreversible Investment", Working Paper, Princeton University.

Birkhoff, Gerret and Gian-Carlo Rota, 1978, *Ordinary Differential Equations* (New York: John Wiley & Sons).

Brennan, Michael and Eduardo Schwartz, 1985, "Evaluating Natural Resource Investment", *Journal of Business* 58, pp. 135–57.

Caballero, Ricardo, 1991, "On the Sign of the Investment-Uncertainty Relationship", *American Economic Review* 81, pp. 279–88.

Caplin, Andrew and John Leahy, 1991, "State-Dependent Pricing and the Dynamics of Money and Output", *Quarterly Journal of Economics* 106, pp. 683–708.

Debreu, Gerard, 1954, "Valuation Equilibrium and Pareto Optimum", *Proceedings of the National Academy of Sciences USA* 40, pp. 588–92.

Dixit, Avinash, 1989a, "Entry and Exit Decisions Under Uncertainty", *Journal of Political Economy* 97, pp. 620–38.

Dixit, Avinash, 1989b, "Hysteresis, Import Penetration, and Exchange Rate Pass-Through", *Quarterly Journal of Economics* 104, pp. 205–28.

Dixit, Avinash, 1991a, "Irreversible Investment with Pricing Ceilings", *Journal of Political Economy* 99, pp. 541–57.

Dixit, Avinash, 1991b, "Irreversible Investment and Competition Under Uncertainty", Working Paper, Princeton University.

Edleson, Michael and Kent Osband, 1989, "Competitive Markets with Irreversible Investment", Working Paper.

El Karoui, Nicole and Ioannis Karatzas, 1988, "Integration of the Optimal Risk in a Stopping Problem with Absorption", mimeo, Columbia University.

Harrison, J. Michael, 1985, *Brownian Motion and Stochastic Flow Systems* (New York: John Wiley & Sons).

Karatzas, Ioannis and Steven E. Shreve, 1984, "Connections between Optimal Stopping Problems and Singular Stochastic Control Problems I: Monotone Follower Problems", *SIAM Journal of Control and Optimization* 22, pp. 865–77.

Karatzas, Ioannis and Steven E. Shreve, 1988, *Brownian Motion and Stochastic Calculus* (New York: Springer Verlag).

Krugman, Paul, 1991, "Target Zones and Exchange Rate Dynamics", *Quarterly Journal of Economics* 106, pp. 669–82.

Leahy, John, 1990, "Optimality, Competitive Equilibrium and the Entry and Exit Decisions of Firms", in *Aggregation and Adjustment*, PhD thesis, Princeton University.

Leahy, John, 1993, "On the Connection between Optimal Stopping and Optimal Control Problems in Discrete Time", mimeo, Harvard University.

Lippman, S.A. and R.P. Rumelt, 1985, "Demand Uncertainty and Investment in Industry-Specific Capital", Working Paper, University of California, Los Angeles.

Lucas, Robert E. Jr. and Edward C. Prescott, 1971, "Investment under Uncertainty", *Econometrica* 39, pp. 659–81.

McDonald, Robert and Daniel Siegel, 1986, "The Value of Waiting to Invest", *Quarterly Journal of Economics* 101, pp. 707–27.

Pindyck, Robert S., 1988, "Irreversible Investment, Capacity Choice and the Value of the Firm", *American Economic Review* 78, pp. 969–85.

Pindyck, Robert S., 1990, "A Note on Competitive Investment under Uncertainty", mimeo, Massachusetts Institute of Technology.

Stokey, Nancy L. and Robert E. Lucas, Jr., 1989, *Recursive Methods in Economic Dynamics*, (Cambridge: Harvard University Press).

Part III

Models of Asymmetric Information and Search

6

*Information Revelation Through Option Exercise**

Steven R. Grenadier

Stanford University

In many real-world situations, agents must formulate option exercise strategies with imperfect information. In such a setting, agents may infer the private signals of other agents through their observed exercise strategies. The building of an office building, the drilling of an exploratory oil well and the commitment of a pharmaceutical company toward the research of a new drug all convey private information to other market participants. This chapter develops an equilibrium framework for option exercise games with asymmetric private information. Many interesting aspects of the patterns of equilibrium exercise are analysed. In particular, informational cascades, where agents ignore their private information and jump on the exercise bandwagon, may arise endogenously.

In standard option models the timing of exercise is simultaneous and uninformative. Agents are assumed to be perfectly informed about the parameters of the option, and the optimal exercise policy is a purely deterministic function of the current state variables. However, there are likely to be many realistic situations in which agents are imperfectly and differentially informed. In such cases, agents may impute the private information of others by observing their exercise (or lack of exercise) decisions. The important questions in such a framework are how do agents determine their exercise strategies, and how is information conveyed?[1] Examples abound in both the financial and real option domains. For example, the exercise of

*This chapter was first published in *The Review of Financial Studies* 12(1), 1999, pp. 95–130 and is reprinted with the permission of Oxford University Press. Comments from Raj K. Aggarwal, Kerry Back, Geert Bekaert, Paul Pfleiderer, C.F. Sirmans, Chester Spatt, Jeff Zwiebel, editors David Hirshleifer and Sheridan Titman, two anonymous referees and seminar participants at the University of California, Berkeley, University of Oregon, University of Wisconsin and Yale University are gratefully acknowledged.

an executive stock option may allow the market to impute insider information on the volatility of the company's stock. However, the vast majority of applications are likely to occur in real options markets. In order to aid in the intuition in the model, consider the following two examples of exercise games with private information externalities.

EXAMPLE 1. OIL EXPLORATION

A well-known case of a real option is oil exploration. Paddock, Siegel, and Smith (1988) model the option-like characteristics of offshore oil leases. In addition to possessing the standard features of options (ongoing oil price uncertainty, large lump-sum exercise costs), it also possesses important private information externalities. Due to the leasing practices of the federal government, it is frequently the case that two or more firms lease adjacent tracts of land that may contain an oil deposit (Hendricks and Kovenock (1989)). Prior to obtaining the lease, each firm has obtained private information about the probability of finding a deposit. Each firm obtains a noisy signal of the likelihood of striking oil by combining seismic surveys with in-house expertise. However, the existence and magnitude of the deposit can only be known through drilling. Firms then face a trade-off between the benefits of drilling and potentially obtaining oil earlier and the benefits of waiting for other firms to drill first and reveal information about the size of the deposit. Therefore, in equilibrium, firms must determine their optimal drilling exercise times in a context of private information and ongoing uncertainty.

EXAMPLE 2. REAL ESTATE DEVELOPMENT

Another classic case of a real option is real estate investment. As in Titman (1985), Williams (1993) and Grenadier (1996), the development of real estate is analogous to an American call option on a building, where the exercise price is equal to the construction cost. However, in real estate markets, there is a great deal of private information held by various market participants. Local developers may have particular knowledge of the suitability of particular locations, while large national developers may have greater expertise on marketwide factors and trends. The timing of construction will convey potentially valuable information about the value of a particular developer's private signal. Thus the early entry of a local developer may convey positive information about the potential value of a particular location (just as the failure to enter would convey less positive information).

This chapter develops a model in which private information is conveyed through the revealed exercise strategies of market participants. A set of n agents holds options on a continuous-time stochastic price process, and each must determine the optimal moment at which to exercise. However, agents possess private information about a parameter that affects the option payoff. The private signals are informative, but imperfect. Agents form exercise strategies that are contingent on not only their own signal, but on the revealed signals of other agents through their observed exercise policies. In equilibrium, as the potential payoff from exercise becomes greater, agents will trade off the benefits of early exercise with the benefits of waiting for information to leak out through the actions of others. Equilibrium exercise will be sequential, with the most informed agents allowing the least informed agents to free ride on the information conveyed by their exercise, or failure to exercise.[2]

The model demonstrates that while the observed exercise decisions of other agents may convey valuable private information, there are also times when no information can be inferred. This will be the case when an informational cascade arises. As in the literature on information herding (ie, Banerjee (1992) and Bikhchandani, Hirshleifer and Welch (1992)), an informational cascade occurs when agents ignore their own private information and instead emulate the behaviour of those that preceded them.[3] Specifically, in the present model, an informational cascade can arise endogenously in which all agents, regardless of their private information, exercise immediately. Agents will find their private information overwhelmed by the information conveyed by others and simply jump on the bandwagon.

Such option-exercise cascades may help explain the occurrence of "overbuilding" in real estate markets: pronounced bursts of development in the face of declining rents and property values. Such overbuilding has not been limited to any region, property type or even country. For example, when oil patch cities such as Denver and Houston experienced unprecedented waves of office construction during the early 1980s, local office vacancy rates were already above 27% (Grenadier (1996)). It seems unreasonable to assume that the developers (and their lenders) that had jumped on the bandwagon all had positive private signals about local market conditions. It is possible that a rational informational cascade was formed where agents found their private signals overwhelmed by the signals conveyed by previous developers.

The dynamic equilibrium nature of this model differs significantly from that of the current literature on informational cascades. The results obtained here serve to both generalise and add greater realism to the standard results on informational cascades. In the standard models of informational cascades the ordering of actions is predetermined. That is, the timing of agents' decisions is typically assumed to be exogenous and not determined by any sort of optimisation decision.[4] For example, this preordering assumption would imply that even if an investor had an extremely positive private signal about the prospects of investment, if he was preassigned to be the 1,000th investor in line, he must wait for the first 999 investors to act before he may choose to invest. In this model, however, timing is the essence of the game; agents exercise strategically and the revealed ordering of exercise is entirely endogenous.[5] Thus this model can be applied to real investment decisions where firms are free to choose the timing of their investment outlays. In addition to providing greater realism, the endogenous nature of the timing allows for a greater depth of analysis. Since there is an explicit time dimension underlying the present model, the model can be used to analyse not simply "if" an informational cascade occurs, but also "when". In an investment context, a cascade will develop in both the decision to invest (exercise) and its timing. An additional distinction in the present model is that informational cascades may occur in an environment of ongoing uncertainty. Thus agents not only use the actions of others to inform their behaviour, but also monitor the continuous stochastic movements in an underlying state variable (ie, oil prices, building rents).

An analysis of the equilibrium exercise strategies reveals several interesting features. With diffuse and imperfect information, equilibrium exercise timing will almost always deviate from the full information optimum. Those with the most informative signals will exercise the earliest, and will typically exercise earlier than optimal. Market observers may impute some form of "overbuilding" as is often implied in commercial real estate markets, using *ex post* information. In addition, the patterns of exercise across agents will be seen to differ under markets with varying structures. Markets with large information asymmetries will experience smooth exercise patterns over time, while markets with mild information asymmetries will experience quick bursts of exercise. The patterns of exercise will also vary according to the number of participants. Markets with large numbers of participants will be prone to longer lags between early exercisers but more prone to bursts of exercise for later exercisers.

The model is applied to an analysis of information markets. In markets with diffuse and imperfect information, mechanisms may develop that facilitate the spread of information. Brokers, consultants and newsletters may emerge offering information to market participants. The model is used to analyse the value of such information services, as well as the most likely purchasers of the information.

While in the basic model the exercise of an agent impacts the information set of other agents, there is no direct impact on the actual payoff from exercise. This simplifying assumption facilitates an examination of the information revelation aspects of the option exercise game.[6] However, in many real-world situations, such exercise games will also have direct payoff externalities. In a later section, the model is extended to take into account cases in which there are payoff externalities from exercise. For example, certain technology markets exhibit network externalities in which the benefits from adoption are increasing in the total number of adopters. A classic instance of this was in the battle between VHS and Beta in video cassette recorder technology. It is found that such direct payoff externalities can have a powerful impact on the nature of equilibrium. For example, if there are significant benefits from agglomeration, the likelihood of informational cascades increases substantially. Conversely, if there are negative externalities associated with conformity, the likelihood of informational cascades can fall to zero.

The chapter proceeds as follows. The first section presents the framework and assumptions of the model of option exercise with imperfect and asymmetric information. The next section derives the equilibrium exercise strategies. The third presents a discussion and analysis of exercise cascades, and the fourth analyses the impact of the information environment on the patterns of exercise. The fifth section analyses the sale of information, and the following section generalises the model to the case of payoff externalities. The final section presents the conclusions.

ASSUMPTIONS AND MODEL FORMULATION

Consider a model with n agents, with $n \geq 2$. Each agent holds an identical call option, where agents are free to exercise at any time of their choosing (ie, the option is a perpetual American call option, and exercise decisions are made in continuous time). However, the precise payoff upon exercise is not fully known to any of the agents. In particular, each agent holds an independent signal of the true realised payoff from exercise. Each agent's optimal option exercise strategy will be contingent on not only his own signal, but also on the observed actions of other agents.

In a standard call option selling, exercise yields the difference between the observable value of the underlying asset and the exercise price. Denoting the underlying asset value process by $X(t)$ and the exercise price by I, the payoff from exercise is typically $X(t) - I$. However, in the present model, the payoff from exercise also includes an unobserved random variable, θ, whose realisation directly impacts the option payoff.[7] In particular, let the net payoff from exercise be $\theta \cdot X(t) - I$. For example, $X(t)$ may represent the value of a "typical" office building, θ the uncertain quality of the particular location, and I the cost of construction. Similarly, $X(t)$ may represent the price of oil, θ the grade or quantity of the reserve and I the cost of extraction.

Let the underlying asset value, $X(t)$, evolve as a geometric Brownian motion process:

$$dX = \alpha X dt + \sigma X dz \qquad \textbf{(1)}$$

where α is the instantaneous conditional expected percentage change in X per unit time, σ is the instantaneous conditional standard deviation per unit time, and dz is the increment of a standard Wiener process. α can be decomposed into the difference between the equilibrium expected return on the asset and its dividend rate.

At time 0, the value of θ is unknown. However, each agent holds an independent signal as to its true underlying value. The random variable θ is assumed to equal a constant plus the sum of n independent signals.[8]

Specifically,

$$\theta = \mu + S_1 + S_2 + \cdots + S_n \tag{2}$$

where μ is the expected value of θ and the S_i are independent, mean-zero random variables.[9] Each agent i possesses private information about the true value of θ by observing the true value of signal S_i.

I assume that the information available to an agent is his own private signal, plus the observed actions (exercise decisions) of other agents. The assumption that information transmission is accomplished solely through actions, and not words, is quite common in the literature (ie, Banerjee (1992), Bikhchandani, Hirshleifer and Welch (1992), Welch (1992)). Embodied in this assumption is the belief that only actions are truly believable, and that communication mechanisms such as press releases and information disclosures are not credible. In many situations, agents may have a strong incentive to mislead their opponents, making truthful revelation unlikely. There could be many justifications for the incentive to mislead other agents. Agents may be compensated based on relative performance, they may wish to harm competitors in order to drive out future competition, or they may wish to elicit early exercise by others if there are benefits to agglomeration.

However, this assumption may be too strong in some cases and will be examined at several points in the chapter. If it is possible for agents to collude in their information or write contracts such that information sale is credible, then the information transmission mechanism may be more broad. One possible assumption would be that agents know the signals of all agents that previously exercised, rather than just their actions. The model is easily modified to account for this different assumption, and this is examined in the third section. An outline for the implications on the equilibrium is as follows. At many times under the "only actions revealed" assumption, exercise decisions are fully revealing of agents' signals. During those periods, the equilibrium is identical to the "signals revealed" equilibrium. However, when only actions are observable, there is also the potential for informational cascades to occur, upon which actions become uninformative. If signals were revealed, informational cascades would never occur, and the equilibrium would avoid the information loss due to cascades. With full revelation of signals, equilibrium exercise will always converge to the "correct" decision. While the elimination (or at least mitigation) of informational cascades may be realistic for some markets, there is also considerable anecdotal evidence that agents do sometimes follow the crowd, even when the crowd is incorrect (see Hirshleifer (1995)).

Agents will differ in the quality of their private information, as their signals will differ in the precision with which they forecast the true value of θ. That is, assume that $\text{var}(S_1) > \text{var}(S_2) > \cdots > \text{var}(S_n)$. Therefore, for all $j < k$, agent j has a more precise signal than agent k because he knows the true value of a more dispersed signal. Stated similarly, agent j's conditional expected value of θ has a lower standard error than agent k's.

Assume that the signals, S_i, have simple two point distributions:

$$S_i = \begin{cases} -\epsilon_i & \text{with probability } \frac{1}{2} \\ +\epsilon_i & \text{with probability } \frac{1}{2} \end{cases} \quad \text{for } i = 1, 2, \ldots, n \tag{3}$$

where $\epsilon_1 > \epsilon_2 > \ldots > \epsilon_n$ and $\mu > \sum_{i=1}^{n} \epsilon_i$.[10] At the outset of the game, agents are assumed to know the distributions of competitors' signals, but not their realisations. This common knowledge assumption only requires that agents know one value, ϵ_i, that corresponds to each competitor. This mapping might be based on such observable factors as past reputation, locational proximity or experience within the industry. The assumption that the ϵ_i's are decreasing ensures that the signal variances are decreasing, and the assumption that the sum of the signals is less than μ ensures that even if all of the signals are negative, for high enough value of $X(t)$, the options will eventually be exercised.

In traditional financial option pricing models, the approach to valuation is based on arbitrage arguments, where one uses the fact that one can instantaneously trade the underlying asset and a riskless asset so that the option is precisely replicated. However, such an approach requires assumptions about the liquidity of the underlying asset. Since the majority of applications of this model involve assets such as office buildings, plant and equipment, and technological innovations that are subject to substantial transactions costs, indivisibility, and the inability to be sold short, such arbitrage arguments are particularly questionable. An equilibrium approach relaxes the tradability assumptions needed for arbitrage pricing, although an appropriate equilibrium model must be chosen.[11] For simplicity, I assume risk neutrality so that all assets are priced so as to yield an expected rate of return equal to the risk-free rate, r. This seemingly restrictive assumption can easily be relaxed by adjusting the drift rate to account for a risk premium in the manner of Cox and Ross (1976).

It shall prove useful to briefly consider the optimal exercise strategy in a world of perfect information, where θ is known to all agents. The details of the derivation are provided in the Appendix. In a world of perfect information, all agents will choose to (simultaneously) exercise the first moment that $X(t)$ equals or exceeds an optimal trigger level. Denote the full-information exercise trigger as $X^*(\theta)$, where a closed-form solution appears as equation (A4) in the Appendix. Thus all agents will exercise at $T^*(\theta)$, the first passage time to $X^*(\theta)$, where $T^*(\theta) = \inf[t \geq 0: X(t) \geq X^*(\theta)]$.

In the model, however, agents will often have less than perfect information about the true value of θ. At each point in time, their information set will contain not only their own signal, S_i, but also the information conveyed by the exercise (or lack of exercise) of other agents. Agents will update their information over time by observing the actions of others in an effort to impute the signals of other agents. This attempt to learn from the actions of others will directly shape the equilibrium exercise strategies. In the standard, full-information exercise strategy, the optimal trigger strategy will equate the immediate benefits of exercise with the marginal value of waiting. However, with partial information, agents will also take into account the benefits of waiting for others to reveal information. For example, if you know that if another agent has a high signal he will exercise at a particular stopping time, then it may pay for you to wait for him to reveal his information before you yourself exercise. In the equilibrium solution in the next section, these trade-offs are calculated explicitly.

EQUILIBRIUM WITH INFORMATION REVELATION

In this section I derive an equilibrium for the n-person option exercise game. In this equilibrium, agents will take into account both their own signal as well as infer the signals of other players through their revealed exercise policies. Given all others' strategies, each agent's equilibrium exercise strategy will be optimal.

Before deriving the equilibrium set of exercise strategies, I will provide a bit of intuition. In equilibrium, the most informed investor will reveal his information first through his observed exercise strategy. Thus, at the beginning, all agents will know that if player 1 has good news (ie, $S_1 = +\epsilon_1$), he will exercise early at a relatively low trigger value of X(t). This will be true because S_1 contains the most dramatic information; if the signal is positive, it will pay for agent 1 to exercise even before collecting any information from the other agents' behaviour. Therefore, if investor 1 exercises early, all agents learn that his trigger was positive. If investor 1 doesn't exercise early, then all agents learn that his trigger was negative. Next, agent 2 holds the most dramatic information. Similarly, all will observe his exercise strategy in order to infer his news. Thus information is revealed over time through the observed exercise behaviour of agents 1 to n. Each agent optimally trades off the benefits of immediate exercise with the benefits of waiting to observe the signals revealed by other agents.

I will present the equilibrium in two stages. I will first state the equilibrium for the simplest case where n = 2. I then define the equilibrium for the general case where n ≥ 2.

Equilibrium with two agents

In this equilibrium, agent 1's actions will fully reveal his signal to agent 2. Agent 2 will use this information to inform his exercise strategy. The equilibrium will now be stated and then proven below.

Equilibrium exercise strategies with n = 2

Each agent i's (i = 1,2) exercise strategy will be contingent on the current level of X(t), its signal S_i, and the revealed exercise policy of the other agent. Let $X(0) < X^*(\mu + \varepsilon_1)$.[12] A pair of Bayesian–Nash equilibria exercise strategies can be described as follows:

Agent 1 (the more informed agent): *If* $S_1 = +\epsilon_1$, *then exercise the first moment that* X(t) *equals the trigger* $X^*(\mu + \epsilon_1)$. *If* $S_1 = -\epsilon_1$, *then copy the exercise strategy of agent 2 by exercising immediately after him.*

Agent 2 (the less informed agent): *Wait until the trigger* $X^*(\mu + \epsilon_1)$ *is reached. If agent 1 exercises, then infer that agent 1's signal is* $+\epsilon_1$ *and exercise the first moment that* X(t) *equals or exceeds the trigger* $X^*(\mu + \epsilon_1 + S_2)$. *This will imply immediate exercise if* $S_2 = +\varepsilon_2$. *If instead agent 1 does not exercise, then infer that agent 1's signal is* $-\epsilon_1$ *and exercise the first moment that* X(t) *equals the trigger* $X^*(\mu - \epsilon_1 + S_2)$.[13]

To prove that the above strategies constitute an equilibrium, I will demonstrate that given each player's strategy, the other player's strategy is optimal. First, take agent 1's strategy as given. It is clear that by observing whether agent 1 exercises at the trigger $X^*(\mu + \epsilon_1)$, agent 2 will correctly infer agent 1's signal and will then know the true value of θ. The only scenario in which agent 2 does worse than the full-information optimum is when both agents have positive signals: agent 2 exercises at the trigger $X^*(\mu + \epsilon_1)$, when with full information he would have exercised earlier at $X^*(\mu + \epsilon_1 + \varepsilon_2)$. However, if agent 2 exercises before agent 1, his expectation of θ would be $\mu + \epsilon_2$. With this expectation, the optimal exercise strategy would be to wait until the trigger $X^*(\mu + \epsilon_2)$ is reached. Since this implies waiting beyond the trigger $X^*(\mu + \epsilon_1)$, it will not be optimal to pre-empt the revelation of agent 1's signal.

Now I demonstrate that agent 1's strategy is optimal, taking agent 2's strategy as given. First, note that if $S_1 = -\epsilon_1$, by copying agent 2 he achieves the full-information optimum. Thus he could do no better. Now, if $S_1 = +\epsilon_1$, agent 1 exercises at the trigger $X^*(\mu + \varepsilon_1)$, without knowledge of agent 2's signal, and thus does not achieve the full-information optimum. However, the only possible reason for agent 1 to deviate from this strategy would be to wait for agent 2 to reveal his signal. Thus, suppose agent 1 waits for agent 2 to reveal his signal. This will take place at the trigger $X^*(\mu - \epsilon_1 + \epsilon_2)$: if agent 2 exercises at this trigger his signal is positive, otherwise it is negative. However, regardless of whether or not agent 2 exercises, the full-information strategy will always dictate immediate exercise. This is because for any realised value of S_2, the full-information trigger $X^*(\mu + \varepsilon_1 + S_2)$ is less than the trigger $X^*(\mu - \epsilon_1 + \epsilon_2)$. Since this delaying strategy yields the same expected payoff as the equilibrium strategy, $(\mu + \varepsilon_1)X - I$, but at a suboptimal exercise time, the optimal exercise trigger remains at $X^*(\mu + \epsilon_1)$.

The intuition for the equilibrium is quite simple. If agent 1 has a positive signal, the expected value of exercising will become so great as to overwhelm any benefit from waiting to learn from agent 2's actions. If instead agent 1's signal is low, agent 2 will infer this by noting that agent 1 did not exercise early. Then, once agent 1's information has leaked out, agent 1 knows that agent 2 is fully informed and can therefore copy agent 2's fully informed exercise strategy. In almost all cases, agent 2 achieves the full-information optimum by waiting for agent 1 to reveal his information, and then proceeding to exercise optimally. Only in the case in which both signals are positive does agent 2 regret not exercising earlier. However, agent 2 knows that if agent 1 also has a positive signal, agent 1's optimism will be greater and will always choose to exercise at an earlier moment.

Equilibrium with n ≥ 2 agents

When there are more than two players, much of the intuition from the two-player game continues to hold. Information will stream from the agent with the most informative signal (agent 1) to the agent with the least informative signal (agent n). If agent 1 has a positive signal, he will find it optimal to exercise first, without waiting for more information. If his signal is negative, he will not exercise early, but will simply copy the next agent who exercises. Recognising this, agent 1's signal will be fully revealed by whether or not he exercises early. After agent 1 reveals his signal through his actions, the remaining agents play a game with (n – 1) players, using the information conveyed by agent 1. Now, it is agent 2 who is the most informed agent. The remaining agents will wait for him to reveal his signal before playing an (n – 2) player game. This sequence ends with a two-player game as before.

Therefore, the definition of equilibrium is relatively straightforward. The general n-person game can be defined recursively by the (n – 1)-person game. The only complication in the general n-player game is the potential for informational cascades to occur. Informational cascades will be discussed in greater detail in the following section. In the two-player game, an agent is able to perfectly impute the signal of the other player through his revealed exercise policy. With many players (in fact with n > 3), it may be possible that agents' actions will be utterly uninformative of their private signals. Agents will form a herd, each copying the other by exercising immediately regardless of their private signal.[14]

An informational cascade will occur whenever two consecutive agents reveal positive signals. Suppose agents j and j + 1 each receive positive signals. Agent j will reveal his positive signal by his early exercise. Since agent j + 1 will be even more optimistic (agent j did not know that j + 1 also had a positive signal), agent j + 1 will exercise immediately after agent j. Now, what will agent j + 2 do? If he has a positive signal, he will clearly exercise immediately since he is more optimistic than both previous agents.

But even if he has a negative signal, he will exercise immediately because he is still more optimistic than agent j, who found it optimal to exercise at this moment. This is because when agent j found it optimal to exercise, his expectation was that $E(S_{j+1}) = E(S_{j+2}) = 0$. Even if agent $j + 2$ knows that $S_{j+1} = +\varepsilon_{j+1}$ and $S_{j+2} = -\varepsilon_{j+2}$, he is still more optimistic than agent j because $\varepsilon_{j+1} > \varepsilon_{j+2}$. Thus it will be optimal for agent $j + 2$ to exercise regardless of his private signal. Since agent $j + 3$ will find himself in the same position as agent $j + 2$, he too will find it optimal to exercise immediately regardless of his private signal. This will cascade to all remaining agents; once an informational cascade forms following two consecutive positive signals, each remaining agent will exercise immediately and the game is over.[15]

The discussion above can now be summarised in the statement of equilibrium for the general case of n players.

Equilibrium exercise strategies with $n \geq 2$

Each agent i's ($i = 1, 2, \ldots, n$) exercise strategy will be contingent on the current level of $X(t)$, its signal S_i, and the revealed exercise policies of other agents. Let $X(0) < X^*(\mu + \varepsilon_1)$. Let $\gamma_0 = \mu$, and let $\gamma_j = \gamma_{j-1} + S_j$ for $j = 1, 2, \ldots, n$. If there exists some j such that both $S_j > 0$ and $S_{j+1} > 0$, an informational cascade is said to begin with agent i^c where $i^c = \min[1 \leq j \leq n - 2 : S_j > 0, S_{j+1} > 0]$. Otherwise there will be no informational cascades and we set $i^c = n + 1$. The n agents will follow a set of Bayesian–Nash equilibrium exercise strategies, where the equilibrium strategy for player $i = 1, 2, \ldots, n$ is as follows:

- *If* $i \geq i^c$, *then exercise at the trigger* $X^*(\gamma_{i^c})$.
- *If* $i < i^c$ *and* $S_i > 0$, *then exercise at the trigger* $X^*(\gamma_i)$.
- *If* $i < i^c$, $i < n$ *and* $S_i < 0$, *then copy the next agent that exercises by exercising immediately after him.*
- *If* $i < i^c$, $i = n$ *and* $S_i < 0$, *then exercise the first moment that* $X(t)$ *equals or exceeds the trigger* $X^*(\gamma_n)$.

Figure 1 provides an example of an equilibrium realisation for a 10-player option exercise game. A hypothetical sample path of X(t) is plotted, along with arrows denoting the equilibrium exercise times of the players. In a table beside the graph are the signals of the players. Since player 1 receives a positive signal, he exercises when the trigger $X^*(\mu + S_1) = X^*(5.4)$ is reached at time $t = 0.13$. Agent 1's exercise reveals his signal to the remaining players. Since agent 2 has a negative signal, he delays his exercise until the next player exercises. This inaction reveals his negative signal to the market. Agent 3 has a positive signal and exercises at time $t = 0.195$ when the trigger $X^*(\mu + S_1 + S_2 + S_3) = X^*(5.36)$ is reached. The market infers agent 3's positive signal, and agent 2 immediately copies agent 3. Agent 4 has a negative signal and delays his exercise until the next player exercises. This inaction reveals his negative signal to the market. Since agents 5 and 6 each have positive signals, a cascade forms beginning with agent 5. All

Figure 1 Illustration of equilibrium exercise times

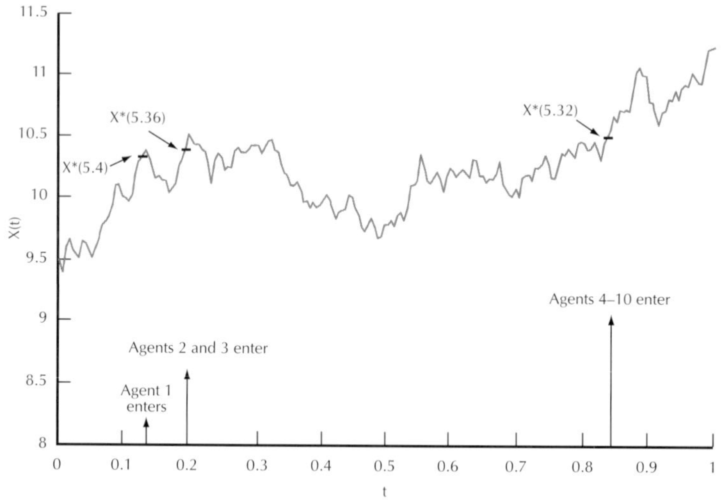

This figure depicts the equilibrium exercise times of 10 agents, for a given sample path of X(t) and a given set of realised signals. The curve depicts a hypothetical sample path of X(t), and the arrows depict the equilibrium exercise times of agents. The exercise strategies follow the definition of equilibrium with $n \geq 2$ agents. Agent 1 has a positive signal and exercises at the trigger $X^*(\mu + s_1)$. The remaining agents update their private information after inferring s_1. Agent 2 has a negative signal and will copy the next agent who exercises. The remaining agents infer s_2 through agent 2's inaction. Agent 3 has a positive signal and exercises at the trigger $X^*(\mu + s_1 + s_2 + s_3)$, revealing s_3 to the market. Agent 2 immediately copies agent 3. Agent 4 has a negative signal and will copy the next agent who exercises, revealing s_4 through his inaction. Since agents 5 and 6 each have positive signals, a cascade forms and all remaining agents exercise at the trigger $X^*(\mu + s_1 + s_2 + s_3 + s_4 + s_5)$. The assumed parameter values are $\alpha = 0.04$, $\sigma = 0.1$, $r = 0.05$, $I = 10$, $X(0) = 9.5$, and $\mu = 5$.

Realised signals	
s_1	+0.40
s_2	–0.36
s_3	+0.32
s_4	–0.28
s_5	+0.24
s_6	+0.20
s_7	–0.16
s_8	–0.12
s_9	+0.08
s_{10}	–0.04

agents that have yet to exercise at time $t = 0.84$ when the trigger $X^*(\mu + S_1 + S_2 + S_3 + S_4 + S_5) = X^*(5.32)$ is reached.

Equilibrium option values

Given the equilibrium exercise strategies of all n players, it is straightforward to value their underlying options. The value of each player's option will simply be the discounted expected value of the payoff, where the payoff is defined by the equilibrium exercise strategies.

The equilibrium defines the exercise trigger for each agent's option, as well as the information set on which it conditions its expected payoff from exercise. In general, the realised exercise trigger for agent j will be a function of all agents' signals. Let $\bar{X}_j(S_1, \ldots, S_n)$ denote agent j's equilibrium exercise trigger, and let $\bar{T}_j(S_1, \ldots, S_n)$ denote the first passage time to the trigger $\bar{X}_j$. For example, for any set of n realised signals, $s_1, \ldots, s_n$, agent j will exercise the first moment X(t) rises to the trigger $\bar{X}_j(s_1, \ldots, s_n)$ at time $\bar{T}_j(s_1, \ldots, s_n)$.

At the beginning of the game, each agent knows only his own signal, S_j. Define the equilibrium value of agent j's option (as of time zero) as $V_j(X; S_j)$. Due to the updating of information, at the moment of exercise each agent forms a conditional expected payoff based on the information learned up to that point. Thus the conditional expectation of θ at the exercise time $\bar{T}_j(s_1, \ldots, s_n)$ will contain all of the information learned by observing the actions of all agents up to time $\bar{T}_j(s_1, \ldots, s_n)$.

The value of the option, $V_j(X; S_j)$, can be derived as follows. Let $G_j(X; s_1, \ldots, s_n)$ denote the "conditional" value of agent j's option, conditional on the realised n signals, $s_1, \ldots, s_n$. The value of the option is then the expected value of G_j, conditional on only the value of agent j's signal.

Just as in the derivation of the full-information option value in the Appendix, the equilibrium conditional option value must satisfy the following differential equation:

$$0 = \frac{1}{2}\sigma^2 X^2 \frac{\partial^2 G_j}{\partial X^2} + \alpha X \frac{\partial G_j}{\partial X} - rG_j \tag{4}$$

$G_j(X; s_1, \ldots, s_n)$ must also satisfy the boundary condition:

$$G_j(\bar{X}_j(s_1, \ldots, s_n); s_1, \ldots, s_n) = \bar{X}_j(s_1, \ldots, s_n) \cdot E\left[\theta \middle| \bar{T}_j(s_1, \ldots, s_n)\right] - I \tag{5}$$

where the expectation implies that when agent j exercises at the equilibrium trigger, he conditions his expectation of θ on all information revealed up to time $\bar{T}_j$. In addition, the boundary condition $G_j(0; s_1, \ldots, s_n) = 0$ must hold.

Finally, to get the *unconditional* value of agent j's option, one must take the expected value of $G_j(X; S_1, \ldots, S_n)$, conditional on agent j's knowledge of S_j. Therefore the equilibrium value of agent j's option, $V_j(X; S_j)$, can be written as

$$V_j(X; S_j) = E[G_j(X; S_1, \ldots, S_n) \mid S_j] \tag{6}$$

In the context of real estate development options, the value of undeveloped land would equal the option value to the owner. Thus the value of agent j's land equals $V_j(X; S_j)$. At all times at which there is no exercise (development), the value of land evolves in a continuous fashion, reflecting only the innovations in the fundamental state variable, X(t). However, at any moment when any other agent begins development (provided that there is no cascade), there is a discontinuous jump in the value of land reflecting the discrete release of private information. Therefore, in the setting of this model, land prices evolve in a continuous fashion corresponding to movements in market fundamentals and jump by discrete amounts following the release of private information.

While closed-form solutions are easily obtained, for large values of n the solutions become quite lengthy. In the section on the sale of information a closed-form solution for $V_1(X; S_1)$ is presented and analysed. In addition, the solution to the system of equations (4), (5) and (6) may be obtained through numerical methods.

ENDOGENOUS EXERCISE CASCADES

An informational cascade is said to occur when agents ignore their own private information and instead emulate the behaviour of previous actors. It is important to note that an informational cascade is not simply the act of simultaneous exercise; for example, simultaneous exercise always occurs in a full-information model. What is distinct about an informational cascade is that agents will ignore their private information and mimic the actions of previous decision makers. As is apparent from the equilibrium in the general model with more than two agents, an informational cascade will occur whenever two consecutive agents reveal positive signals. As agents witness the exercise of better informed agents, they will ignore their own private information and exercise simultaneously. Despite the typically negative connotation associated with herd-like behaviour, this behaviour is entirely rational. While from a social welfare criteria it is suboptimal in that it wastes information, it is individually rational to defer to apparently overwhelmingly positive information.

Many have attributed the pronounced waves of construction in the face of declining rents and property values to irrational overbuilding.[16] In terms of national office construction, Birch (1990) estimates that 43% of all office space ever built in the United States was built during the 1980s. Much of this construction was initiated when the national office vacancy rate was rising rapidly and nearing historically high levels. In local markets, the same phenomenon was occurring. Consider the cases (referred to above) of Denver and Houston. Over the 30-year period from 1960 to 1990, over half of all office construction was completed during the four-year interval 1982–85. During the time of their concentrated burst of construction in 1982–85, each had vacancy rates near 30%.

However, some of the development patterns that emerge in commercial real estate markets may be due to rational, informational cascades. As stated in the introduction, real estate development represents the exercise of an option, where agents possess noisy private signals of building values. Much of the real estate market is private, and such information is regarded as valuable and proprietary. While information is revealed in many ways, observing the actions of developers is clearly an important medium of information conveyance. As in the model, if a few developers have optimistic forecasts of future building values, then it may be entirely rational for other developers to ignore their own information and form a herd. A particular example would be if local developers were optimistic and began development. Then outside developers may also enter the market and begin construction, even if they possess negative signals about the state of the national real estate market.

Although the informational aspects of real estate markets may make them prone to informational cascades, this propensity may be mitigated by the direct payoff externalities associated with development. As discussed in a later section, when there are negative payoff externalities associated with option exercise (ie, the exercise of one agent's option lowers the prospective payoffs on other agents' options), the likelihood of the occurrence of informational cascades can be significantly diminished. Intuitively, if the exercise of an option lowers the payoffs for future option exercise, the remaining agents will delay exercise until building values rise substantially, thus lowering or eliminating the likelihood of a cascade. Even if the private information revealed by previous exercise is significantly positive, agents may still defer exercise because the negative direct payoff externality overwhelms the importance of the private information revelation. In the case of real estate development, there are likely to be first-mover advantages. The first developer to build in a local area can act as a monopolist (perhaps only temporarily) and lock in the most desirable tenants into his building. When future development options are exercised, the rents that can be charged may be lower, as the demand curve for space is likely to be downward-sloping. Therefore the true susceptibility of a real estate market to the occurrence of informational cascades will depend on the interaction between the informational and direct payoff externalities.

As mentioned in the introduction, the informational cascade phenomenon in the present model differs significantly from that in the standard models of information-oriented herding found in the literature (eg, Banerjee (1992) and Bikhchandani, Hirshleifer, and Welch (1992)). First, in this model informational cascades occur in a setting where the ordering of actions (exercise) is determined endogenously. Agents choose the timing of their actions strategically. Second, agents face a continuous rather than binary choice set. That is, they may exercise at any stopping time. Their decision to exercise at a point in time is as informative as their decision not to exercise at a point in time. The timing of decisions is an essential element of the model. Thus herding will develop in both the decision to exercise and the timing of exercise. Finally, agents must make their decision in an environment in which there is ongoing uncertainty, $X(t)$, in addition to the uncertainty revealed through observed actions.

It is straightforward to derive the probability that an informational cascade will occur in a market with n agents. It will be seen that the probability of an informational cascade will approach one as n goes to infinity. Even for moderate sizes of n, the likelihood of an informational cascade will be very high. However, it must be kept in mind that even though there may be a high likelihood of an informational cascade, the timing of such an event could be very far off in the future. This is because even though an informational cascade will occur if two consecutive agents have positive signals, it may take a long time before X(t) rises to the necessary trigger value to entice these agents to exercise and reveal their triggers. I will assume that the stochastic process of X(t) possesses the property that any finite trigger will eventually be reached with probability one. Thus I shall assume $\alpha - 1/2\sigma^2 > 0$.

The probability that an informational cascade forms in a market of size n, p_n, is derived in the Appendix. p_n can be written as

$$p_n = 1 - \frac{4}{\sqrt{5}}\left[\left(\frac{1}{\sqrt{5}-1}\right)^{1+n} + \left(\frac{1}{1+\sqrt{5}}\right)\left(\frac{-1}{1+\sqrt{5}}\right)^{n}\right]$$

$$\forall n \geq 0 \qquad (7)$$

The probability of an informational cascade is increasing and concave. Even for reasonably small values of n, the likelihood of an eventual informational cascade can be very large. With only 10 agents, the probability of an eventual informational cascade is almost 83%. However, as stated previously, the time at which such an event may occur could be far off in the future.

Informational cascades are harmful from a social welfare standpoint because information is wasted. When there are no informational cascades, the observed exercise decisions are fully revealing, as they convey the true private signals of previous agents. Thus if there is no cascade prior to agent j's exercise decision, agent j will know the values of signals 1 to j. However, if there is a cascade beginning with agent $i < j$, agent j will have no information about the values of signals $i + 2$ to $j - 1$. This lost information leads the remaining agents to, on average, make less accurate exercise decisions.

If the assumption that only previous actions are observable is replaced by the assumption that previous signals are observable, this information loss could be avoided. Recall that the only time that actions are not fully revealing is during a cascade. Thus suppose that two consecutive agents receive positive signals and exercise simultaneously. As in the model, the next agent will find it optimal to exercise immediately, regardless of his signal. Now, suppose that this agent also publicises his signal. If this pronouncement is credible, then all remaining agents update their information and an informational cascade will never form. The equilibrium defined earlier for $n \geq 2$ agents can easily be modified to account for this altered assumption. Specifically, the term i^c is always set equal to $n + 1$, and the remaining definition holds.

The assumption that only actions are credible (ie, that actions speak louder than words) is standard in the information herding literature.

Whether this assumption is accurate or not will depend on the type of market structure being considered. There are several factors that can erode the credibility of words. If agents are compensated based on relative performance, rather than absolute performance, then agents will have an incentive to misreport their private signals. This is in keeping with Keynes's observation that "it is better for reputation to fail conventionally than to succeed unconventionally". Another justification is that agents may wish to eliminate future competition. If agents must compete with each other in other markets, this will provide a disincentive from truthfully revealing private information. Similarly, if there are payoff externalities, agents will have an incentive to reveal information in a self-serving fashion.

Under certain conditions it may be possible to overcome the factors making actions the only credible information source. In particular, since the information lost during a cascade is valuable to remaining agents, a market for information sale may develop. However, if it is not possible to write contracts which are *ex post* verifiable, then the information sold will not be credible and will be worthless. Even if the information is credible, a market for information sales may still not develop. As discussed in a later section, if agents can free ride on the purchase of information, then the market price of the information may only represent only a small fraction of the public benefits of information. Thus the market price of the information may not be large enough to compensate for the costs of information acquisition.

AN ANALYSIS OF THE PATTERNS OF EXERCISE

In this section I analyse the patterns that may emerge from equilibrium exercise strategies. For example, in some markets we may observe a period of rapid exercise by all agents. This may not necessarily be caused by informational cascades. Agents may simply find it optimal to act quickly after observing another agent's exercise. Caplin and Leahy (1993) provide an example of a sudden burst of exercise behaviour. Manhattan's Lower Sixth Avenue was experiencing an attenuated retail depression when Bed Bath & Beyond opened the first superstore in the area. Other retailers then quickly exercised their options to locate on Lower Sixth Avenue. Bed Bath & Beyond's entry conveyed powerful information about its confidence that consumers would be willing to travel to such an inconvenient location. In addition, the fact that there was such a long period of inactivity in the area signifies that retailers were also learning from the reticence of others. Conversely, we also find environments in which exercise occurs fairly evenly over time. Agents may only follow up a previous agent's exercise after a cautious delay. Hendricks, Porter and Boudreau (1987) find an even pattern of exercise in oil drilling. They find in a sample of 1,056 wildcat tracts a pattern of drilling that is fairly evenly distributed across the different lease years.

While with perfect information all agents would exercise simultaneously, with imperfect information agents will typically exercise sequentially over time. Agents will find it optimal to balance the benefits of immediate exercise with the benefits of waiting to learn from the actions of others. The agents with the most informative positive signals will exercise early, and those with the least informative signals will exercise later. It will now be demonstrated that a consequence of this trade-off will be that the most informed agents will typically exercise "too early" while the least informed agents will exercise "too late".

Recall that the full-information exercise trigger is $X^*(\theta)$. Thus if agents have perfect information they will exercise the first moment that X(t) rises to the trigger $X^*(\theta)$, denoted as $T^*(\theta)$. Without perfect information, the equilibrium exercise trigger of agent j is $\bar{X}_j(S_1, \ldots, S_n)$. As detailed in the equilibrium, $\bar{X}_j(S_1, \ldots, S_n)$ will in general depend on the signals of all agents. Agent j will exercise the first moment that X(t) rises to the trigger $\bar{X}_j(S_1, \ldots, S_n)$, denoted by $\bar{T}_j(S_1, \ldots, S_n)$.

The following two propositions demonstrate that the most informed agent will typically exercise "too early" and the least informed agent will typically exercise "too late".

Proposition 1. $E[\bar{T}_1(S_1, \ldots, S_n) - T^*(\theta)] < 0$. *Thus, on average, the agent with the most precise signal will exercise earlier than the full-information optimum.*

Proof. See Appendix.

Proposition 1 is simply a result of Jensen's inequality. From the definition of equilibrium for $n \geq 2$ agents, agent 1 exercises at the trigger $X^*[E_1(\theta)]$, where E_1 denotes agent 1's conditional expectation of θ at the moment of exercise. Since $X^*(\theta)$ is convex, a quick application of Jensen's inequality (and the law of iterated expectations) implies that $E[\bar{X}_1(S_1, \ldots, S_n)] < X^*[E^*(\theta)]$. Since agent 1 is expected to exercise at a trigger less than the full-information trigger, he is expected to exercise too early. The intuition for Proposition 1 is as follows. Agent 1 is the most informed agent, and in equilibrium is always the first to reveal his private signal. While other agents face the trade-off between the benefits of immediate exercise and the benefits of waiting for more informed agents to reveal their private signals, agent 1 faces no such trade-off. Agent 1 has no incentive to delay and wait beyond his partial information exercise strategy.

Proposition 2. $E[\bar{T}_n(S_1, \ldots, S_n) - T^*(\theta)] > 0$. *Thus, on average, the agent with the least precise signal will exercise later than the full-information optimum.*

Proof. See Appendix.

The intuition for Proposition 2 is as follows. If no cascade occurs, then agent n always has full information and will never exercise before the full-information trigger. If a cascade occurs, then all agents that participate in the cascade (which always includes agent n) will rush to exercise given the overwhelmingly positive information of those that began the cascade. On average, agents that follow a cascade exercise later than they would have had they possessed full information. Therefore, whether or not a cascade occurs, agent n is expected to exercise at or after the full-information trigger.

Propositions 1 and 2 demonstrate that there will typically be a lag between the exercise of several agents, as agents trade off the benefits of immediate exercise with the benefits of waiting for information to be revealed. I now demonstrate that the degree to which agents wait will depend critically upon the structure of the information environment.

Consider a market with n agents. The expected lag between the exercise of agent j and that of agent j + 1 is $E[T_{j+1}(S_1, \ldots, S_n) - T_j(S_1, \ldots, S_n)]$, denoted as $L_j(S_1, \ldots, S_n)$ for $j = 1, \ldots, n - 1$. This can be expressed as

$$L_j(S_1, \ldots, S_n) = \frac{1}{(\alpha - 1/2\sigma^2)} E\{\ln[X_{j+1}(S_1, \ldots, S_n)/X_j(S_1, \ldots, S_n)]\} \qquad \textbf{(8)}$$

using a slight transformation of a result in Harrison (1985). The expected lag between the exercise of agent j and that of agent j + 1 will be a function of the information environment in the market. This is evident from its dependence on the distribution of the n signals, $S_1, \ldots, S_n$.

I now analyse how different information environments affect the speed of exercise. For example, do agents in markets with significant asymmetries in private information act more cautiously and wait longer between exercise? Are markets with large numbers of players more prone to bursts of exercise? To answer these sorts of questions, I analyse the impact of two important characteristics of the market structure: the distribution of private information and the number of players.

In order to analyse alterations in the structure of the market, I parameterise the information structure in a simple manner. In a market of size n, let $\varepsilon_1, \varepsilon_2, \ldots, \varepsilon_n$ be evenly spaced such that $\varepsilon_n = \delta$, $\varepsilon_{n-1} = 2\delta, \ldots, \varepsilon_1 = n\delta$. Therefore $\varepsilon_j = (n - j + 1) \cdot \delta$, for $j = 1, \ldots, n$. Thus δ represents the magnitude of asymmetries in private information. The greater δ is, the greater the degree to which agents can benefit by learning the signals of others. The effect of changes in δ and n on the clustering of equilibrium exercise are derived through simulation. While the results are presented for a base-case set of parameter values, the qualitative results apply for a wide range of reasonable parameter values.

First, consider the effect of increasing δ, the magnitude of differences in private information. For a fixed number of players, the observed actions of

Table 1 The effect of private information on the time between exercise

	$\delta = 0.01$	$\delta = 0.04$	$\delta = 0.07$
L_1	45.99	186.52	327.77
L_2	21.17	86.51	172.28
L_3	19.71	75.56	160.97
L_4	10.95	52.56	93.81
L_5	8.40	36.50	73.73
L_6	5.48	24.09	52.20
L_7	3.36	14.97	32.49
L_8	1.83	7.67	18.62
L_9	0.67	3.42	6.94

This table presents the expected number of days between the exercise of agents, as the underlying amount of private information varies. This table demonstrates that as the amount of private information increases, the expected lag between the exercise times of any two agents becomes greater. In the model, each of n agents may exercise a call option paying $\theta X(t)$, at any time t, by paying an exercise price of I. X(t) follows a geometric Brownian motion process with drift α and volatility σ. θ is an unknown parameter, whose value is represented by the sum $\mu + S_1 + S_2 + \cdots + S_n$. Agent j knows the realised value of the mean zero signal S_j, whose random value is $+(n - j + 1)\delta$ or $-(n - j + 1)\delta$, with equal probability. As δ increases, the degree of private information that is held by other agents increases. The expected time lag between the exercise of agent j and agent j + 1 is denoted L_j. The expected lag is estimated through performing 10,000 Monte Carlo simulation runs. The assumed parameter values are $\alpha = 0.04, \sigma = 0.1$, r = 0.05, I = 10, X(0) = 9.5, n = 10 and $\mu = 5$.

other players reveal more information about the true state of the market. The key will be to focus on the effect of bad news. This is true because, regardless of the amount of information, there will always be simultaneous exercise between consecutive agents with good news. There will be time between exercise only in the case of good news, followed by bad news, followed by good news. Thus consider the timing of exercise when an agent exercises (due to good news) and the next agent does not exercise immediately (due to bad news). Now, in a market with great distinctions in private information, when an agent reveals (through failing to exercise) that his signal is negative, the next agent to exercise will be more cautious before exercising due to the severity of the revelation. Conversely, if there were only small distinctions in private information, bad news should have little impact on the timing of the next exercise; the next agent with good news will exercise very soon thereafter.

Table 1 depicts the effect of differing levels of private information asymmetry on the clustering of exercise. The table displays the average number of days between exercise in a 10-player game for three levels of δ. There are two important results. First, for any level of δ, the average delay will be greatest for the earliest exercisers. Thus the lag between players 1 and 2 will be the greatest, and the lag between players n – 1 and n will be the smallest. This is because the earliest agents reveal the most informative signals. A negative early signal will inspire the greatest amount of caution. Second, as expected, the greater the magnitude of private information asymmetries, the greater the average lag between exercise. For example, if $\delta = 0.01$, then the expected lag between the exercise of agents 2 and 3 is only about 21 days. However, when $\delta = 0.07$, the expected lag is almost half a year.

Now, consider the effect of increasing the number of players in the market. There will be two effects. First, for any given player, the greater the number of remaining players in a market, the greater the amount of private information that is unknown to the player. This leads to a greater desire to wait for others to reveal their information. In equilibrium, this will lead to a greater lag between the exercise of any given two consecutive players. Second, the greater the number of players, the greater the likelihood of an informational cascade. Once a cascade ensues, the time between exercise is zero for all remaining players. Therefore intuition suggests that the average time between the exercise of later players will fall to zero as the number of agents increases. As the number of agents becomes very large, the equilibrium approaches that of the model of Zhang (1997). In Zhang, all agents form an informational cascade with probability one. In the current model, as the number of agents increases to infinity, all but a finite number of players form a cascade with probability one.

Table 2 The effect of market size on the time between exercise

	n = 5	n = 10	n = 15
L_1	83.37	186.52	321.20
L_2	31.31	86.51	161.33
L_3	21.33	75.55	147.02
L_4	7.72	52.56	110.49
L_5	—	36.50	85.03
L_6	—	24.09	71.46
L_7	—	14.96	51.57
L_8	—	7.66	35.41
L_9	—	3.42	27.74
L_{10}	—	—	19.06
L_{11}	—	—	13.33
L_{12}	—	—	7.51
L_{13}	—	—	4.27
L_{14}	—	—	1.56

This table presents the expected number of days between the exercise of agents, as the number of agents varies. This table demonstrates that as the number of agents increases, the expected lag between the exercise times of earlier agents increases, while the expected lag between the exercise times of later agents decreases. In the model, each of n agents may exercise a call option paying $\theta X(t)$, at any time t, by paying an exercise price of I. X(t) follows a geometric Brownian motion process with drift α and volatility σ. θ is an unknown parameter, whose value is represented by the sum $\mu + S_1 + S_2 + \cdots + S_n$. Agent j knows the realised value of the mean zero signal S_j, whose random value is $+(n - j + 1)\delta$ or $-(n - j + 1)\delta$, with equal probability. The expected time lag between the exercise of agent j and agent j + 1 is denoted L_j. The expected lag is estimated through performing 10,000 Monte Carlo simulation runs. The assumed parameter values are $\alpha = 0.04, \sigma = 0.1$, $r = 0.05$, $I = 10$, $X(0) = 9.5$, $\delta = 0.04$ and $\mu = 5$.

Table 2 depicts the effect of increasing the number of players on the clustering of exercise. The previous intuition is confirmed. First, as the number of players increases, the average lag between any two given players increases. Agents find it in their interest to increase the time between exercise because of the greater amount of private information remaining in markets of greater size. For example, with five players the average lag between the exercise of players 3 and 4 is about 21 days, while with 15 players the average lag is about 147 days. Second, as the size of the market increases, the average lag between the exercise of later players becomes very small, due to the increased likelihood of a cascade. With five players, where there is a 50% probability of a cascade, the average lag between the exercise of the final two players is 7.7 days. However, with 15 players, where there is a 94% probability of a cascade, the average lag between the exercise of the final two players is only about 1.5 days.

The model's empirical implications with respect to the patterns of exercise can be summarised as follows. First, on average, the earliest entrant will exercise earlier than the full-information optimum, while the last entrant will exercise later than the full-information optimum. Second, the average lag between exercise will be greatest for the earliest exercisers. Thus markets will typically grow slowly at first, as the early entrants exercise cautiously. Then markets will pick up steam with a rapid succession of later exercise. Third, markets with the greatest amount of information asymmetries will typically have the greatest lag between exercise. Fourth, markets with the greatest number of participants will have the longest average lags between the agents who exercise first, and the shortest average lags between the agents who exercise last.

THE SALE OF INFORMATION

It is clear that even in an equilibrium in which there is some information revealed through exercise, most agents would be better off having more information. In markets with diffuse and imperfect information, mechanisms may develop that facilitate the spread of information. In such markets, brokers, consultants, newsletters and other services may provide information and advice to those who wish to take advantage of their greater market knowledge.

In the context of the present model, the purchase of additional information by any agent will have a positive externality on all other market participants. The increase in market information will push the industry closer to a fully informed equilibrium. Information will leak out through its use, even to those who did not buy it, as the market will extract the information from the purchaser's observed exercise policy.

Even though the individual agents may not truthfully reveal their information, I assume that firms that sell information will communicate truthfully. They will offer to sell their information, and will charge a price so as to maximise their profits. While many forms of information sale can be envisioned, I shall concentrate on a simple formulation: the sale of perfect information. More detailed treatments of the economics of information sale can be found in Admati and Pfleiderer (1986, 1988).

Suppose an information service offers to sell knowledge about the true value of θ. Any agent who buys this information will then exercise at the full-information optimal trigger, $X^*(\theta)$. What should the information service charge? Who will purchase this information?

If any firm purchases the information, then all other participants will simply emulate the exercise strategy of the perfectly informed agent.[17]

Thus all other agents can free ride off the purchase of information by a single agent. Therefore, at most one agent will purchase the information. The full welfare gains from the provision of information will not be captured by the one agent who pays for the information. If the cost of extracting the information for the information providing firm is larger than the private benefits to the purchaser, the information market will shut down. This may lead to a potential argument for government intervention. Similarly, groups of agents could benefit by forming coalitions that share information. The ability to form such coalitions will depend on such factors as antitrust enforcement and the ability to create mechanisms that ensure truthful information revelation.

The firm that sells the information will sell the information to the agent that will pay the highest price for the information. This will always be firm 1. Firm 1, the firm with the most informative signal, will always pay the most for information, because all other firms have access to his information, but he does not necessarily have access to their information. Agent 1 always reveals his information first and all other agents are at least as well off as agent 1, since they never have less information at their disposal than agent 1.

The value of this information to agent 1 would be the difference between the expected value of exercising his option at the full-information exercise trigger and the value of exercising his option according to the partial-information equilibrium described in the second section. The full-information value of agent 1's option is $W(X; \theta)$, as derived in equation (A3) in the Appendix. Prior to obtaining the information, the expected value of the full-information option to agent 1 is $E[W(X; \theta) \mid S_1]$,where agent 1's expectation of θ is conditioned on knowledge of his own signal S_1. This expectation can be written as

$$E[W(X; \theta) \mid S_1] = \sum_{m_2=1}^{2} \sum_{m_3=1}^{2} \cdots \sum_{m_n=1}^{2} \left(\frac{1}{2}\right)^{n-1} W\left[X; \mu + S_1 + \sum_{j=2}^{n} \varepsilon_j \cdot (-1)^{m_j}\right] \tag{9}$$

The value of agent 1's option without the purchase of information, $V_1(X; S_1)$, is determined by the equilibrium exercise strategy detailed in the second section. This value is conditional on S_1. If $S_1 = +\varepsilon_1$ and agent 1 has the highest signal, he will exercise first at the trigger $X^*(\mu + \epsilon_1)$ and will obtain a value of $W(X; \mu + \epsilon_1)$. If $S_1 = -\varepsilon_1$, then he will exercise when the first agent with a positive signal exercises. For example, agent j will be the first to reveal a positive signal if $S_j = +\epsilon_j$, but $S_i = -\epsilon_i$ for $i < j$. The probability of this occurring is $(\frac{1}{2})^{j-1}$. In that case, agent 1 would exercise at the trigger $X^*(\mu - \epsilon_1 - \cdots - \epsilon_{j-1} + \epsilon_j)$ and will obtain a value of $W(X; \mu - \sum_{k=1}^{j-1}\epsilon_k + \epsilon_j)$. If no agent has a positive signal, all will exercise at the trigger $X^*(\mu - \sum_{k=1}^{n}\epsilon_k)$ and will obtain a value of $W(X; \mu - \sum_{k=1}^{n}\epsilon_k)$. The value of agent j's option,

V_1 (X; S_1), can be written in closed form as

$$V_1(X; S_1) = \begin{cases} W(X; \mu + \epsilon_1) & \text{if} \quad S_1 = +\varepsilon_1 \\ \Sigma_{j=2}^{n} (\tfrac{1}{2})^{j-1} W(X; \mu - \Sigma_{k=1}^{j-1} \epsilon_k + \epsilon_j) & \\ + (\tfrac{1}{2})^{n-1} W(X; \mu - \Sigma_{k=1}^{n} \epsilon_k) & \text{if} \quad S_1 = -\varepsilon_1 \end{cases} \tag{10}$$

Therefore the value of the information to agent 1 will equal the expected net benefit of achieving the full-information option value over the equilibrium partial-information option value. This will be a function of the current level of X and the signal of agent 1, S_1. Denote the value of this information as N(X, S_1). The value of the information can be written as

$$N(X, S_1) = E[W(X; \theta) \mid S_1] - V_1(X; S_1) \tag{11}$$

If the information provider is a monopolist, then it will sell the information for the maximum price, N(X, S_1), provided the cost of extracting the information is less than the expected benefits of selling the information. If the information providing industry is competitive, then free entry will ensure that the information will sell for its production cost, provided the cost is less than the value of the information to agent 1.

EQUILIBRIUM EXERCISE OF OPTIONS WITH PAYOFF EXTERNALITIES

In the basic model, even though information is conveyed through exercise, the actual payoff from exercise is independent of the actions of others. In many real-world situations, exercise produces informational as well as direct payoff externalities. In this section I generalise the model by providing an example of a case in which exercise by one agent directly affects the payoffs on the option of another agent. It will be seen that under certain conditions, positive exercise externalities can increase the likelihood of cascades. Conversely, negative exercise externalities can entirely eliminate the occurrence of cascades. The example is simple, but conveys much of the intuition for more complicated applications. More general discussion of this phenomenon can be found in Dybvig and Spatt (1983) and Katz and Shapiro (1986).

There are numerous examples of payoff externalities from the exercise of options to invest. For example, certain technology markets exhibit network externalities in which the benefits from adoption are increasing in the total number of adopters. A classic instance of this was in the battle between VHS and Beta; as more and more consumers adopted the VHS technology, the benefits for others who adopted VHS became greater. Another example of exercise externalities would be in the case of retail stores deciding whether to enter a retail area. The presence of anchor tenants entices traffic flow into the area. In addition, a diversified array of speciality stores increases traffic flow for the anchors. There are also examples in which the exercise externalities are negative. An example would be firms considering the exercise of an option to begin manufacturing a product with a downward-sloping demand curve. As more firms exercise, the payoffs decline for all.

Consider a modified version of the basic model that illustrates the equilibrium in a game with dependent payoffs from exercise. For simplicity I shall solve for the equilibrium set of exercise strategies for the case in which $n = 2$, where the payoff externality is positive. The nature of the payoff externalities are as follows: the payoff from option exercise when there is only one option exercised is $\theta \cdot X$, as in the basic model. However, when there are two options exercised, each receives an additional payoff of $\kappa \cdot X$, where $\kappa \geq 0$ is a measure of the benefits from agglomeration. Therefore the first agent to exercise receives a payoff of $\theta \cdot X - I$, and will receive an additional payoff of $\kappa \cdot X$ if and when the other agent exercises. The second agent to exercise receives an immediate payoff of $(\theta + \kappa) \cdot X - I$.

The nature of the equilibrium will depend on the magnitude of the agglomeration effect. If κ is small, then the equilibrium exercise strategies are similar to that of the basic model. Agent 1 reveals his signal by his early exercise decision, and agent 2 combines this information with his own private information to condition his exercise strategy. However, if κ is large, then the equilibrium will be quite different. Agent 2 may not always use his private information in forming his exercise strategy. In particular, if agent 1 reveals a positive signal, agent 2 will "herd" and exercise immediately regardless of his private signal. That is, informational cascades can occur even after only one positive signal is revealed.

The equilibrium set of exercise strategies for this example is now summarised.

Equilibrium exercise strategies for a two-player game with positive payoff externalities

In an option exercise game in which there is a positive payoff externality of κ, each agent i's ($i = 1, 2$) exercise strategy will be contingent on the current level of $X(t)$, its signal S_i, and the revealed exercise policy of the other agent. Let $X(0) < X^*[\mu + \epsilon_1 + \max(\kappa/2,\ \kappa - \varepsilon_2)]$. The equilibrium exercise strategies will differ according to whether $\kappa \geq 2\varepsilon_2$. A pair of equilibrium exercise strategies for the two cases is as follows:

Case A. $\kappa \geq 2\varepsilon_2$. Agent 1 (the more informed agent): *If* $S_1 = +\epsilon_1$, *then exercise the first moment that* $X(t)$ *reaches the trigger* $X^*(\mu + \epsilon_1 - \varepsilon_2 + \kappa)$. *If* $S_1 = -\epsilon_1$, *then copy the exercise strategy of agent 2 by exercising immediately after him.*

Agent 2 (the less informed agent): *Wait until the trigger* $X^*(\mu + \epsilon_1 - \varepsilon_2 + \kappa)$ *is reached. If agent 1 exercises, then infer that agent 1's signal is* $+\varepsilon_1$ *and then exercise immediately, regardless of the sign of* S_2. *If instead agent 1 does not exercise,* *then infer that agent 1's signal is* $-\epsilon_1$ *and exercise the first moment that* X(t) *equals or exceeds the trigger* $X^*(\mu - \varepsilon_1 + S_2 + \kappa)$.[18]

Case B. $\kappa < 2\varepsilon_2$. Agent 1 (the more informed agent): *If* $S_1 = +\epsilon_1$, *then exercise the first moment that* X(t) *reaches the trigger* $X^*(\mu + \epsilon_1 + \kappa/2)$. *If* $S_1 = -\epsilon_1$, *then copy the exercise strategy of agent 2 by exercising immediately after him.*

Agent 2 (the less informed agent): *Wait until the trigger*$X^*(\mu + \epsilon_1 + \kappa/2)$ *is reached. If agent 1 exercises, then infer that agent 1's signal is* $+\epsilon_1$ *and exercise the first moment that* X(t) *equals or exceeds the trigger* $X^*(\mu + \varepsilon_1 + S_2 + \kappa)$. *This will imply immediate exercise if* $S_2 = +\varepsilon_2$. *If instead agent 1 does not exercise, then infer that agent 1's signal is* $-\epsilon_1$ *and exercise the first moment that* X(t) *equals or exceeds the trigger* $X^*(\mu - \varepsilon_1 + S_2 + \kappa)$.[19]

The proof that the above strategies constitute an equilibrium appears in the Appendix.

This example illustrates the fact that direct payoff externalities can overwhelm information externalities. Specifically, if the payoff from the benefits of agglomeration are large enough, the information that is revealed by exercise strategies can be totally lost. In Case A, agent 2 will respond to agent 1's early exercise by exercising immediately regardless of his private information. When extended to a market of size $n > 2$, an informational cascade will now take place after the revelation of the first positive signal. Such a market will then be highly informationally inefficient in that many agents will exercise with little or no updating of information. Agents will find the benefits of agglomeration so powerful so as to overwhelm the benefits of learning. Empirically this would imply that markets that display the greatest benefits from agglomeration would be the most prone to cascades. Such environments might include markets for technological innovations and retail development.

In a similar manner to the agglomeration example, it can be shown that for the case of negative exercise externalities, the opposite effect is true. With negative payoffs associated with conformity, agents may never find it optimal to exercise simultaneously. If the externalities are sufficiently negative, agents will always respond to a previous agent's exercise by waiting for X(t) to rise even further as to justify incurring the negative externality. Thus agents' exercise strategies will always be fully revealing, leading to the elimination of informational cascades. Empirically this would imply that markets that display the greatest negative externalities associated with agglomeration would be the least prone to cascades. Such markets might include products with steeply downward-sloping demand curves or with "snob appeal" where consumers value uniqueness over conformity.

CONCLUSION

In a world of perfect information, option exercise can be conducted in isolation. However, in markets with both public and private information, the exercise of options must be determined as part of a strategic equilibrium. This chapter presents a model of equilibrium option exercise policies and information revelation in markets with private signals. The act of exercise provides both a private benefit and a public benefit. The option exerciser receives an expected positive payoff, while other option holders may be able to impute private information. The model provides insights into the patterns of exercise in a variety of realistic economic settings. Cascades and boom-and-bust behaviour may result from rational underlying forces. The model is used to determine the value and nature of market mechanisms that may form to provide private information. In addition, the model is extended to cases in which there exist both information and direct payoff externalities.

APPENDIX

A.1 Derivation of full-information exercise trigger

Let $W(X; \theta)$ denote the value of each agent's option in a world where θ is a known parameter. Using standard arguments (ie, Dixit and Pindyck (1994), Grenadier (1995)), $W(X; \theta)$ must solve the following equilibrium differential equation:

$$0 = \tfrac{1}{2}\sigma^2 X^2 W_{xx} + \alpha X W_x - rW \tag{A1}$$

Equation (A1) must be solved subject to appropriate boundary conditions. These boundary conditions serve to ensure that an optimal exercise strategy is chosen:

$$\begin{aligned} W[X^*(\theta), \theta] &= \theta X^*(\theta) - I \\ W_x[X^*(\theta), \theta] &= \theta \end{aligned} \tag{A2}$$

Here, $X^*(\theta)$ is the value of $X(t)$ that triggers entry. The first boundary condition is the value-matching condition. It simply states that at the moment the option is exercised, the expected payoff is $\theta X - I$. The second boundary condition is the smooth-pasting or high-contact condition.[20] This condition ensures that the exercise trigger is chosen so as to maximise the value of the option.

Closed-form solutions for $W(X, \theta)$ and the full-information trigger $X^*(\theta)$ are easily obtained. The value of the perfect information option strategy can be written as

$$W(X; \theta) = \begin{cases} \left(\dfrac{I}{\gamma - 1}\right)^{1-\gamma} \left(\dfrac{\theta}{\gamma}\right)^{\gamma} X^{\gamma} & \text{for} \quad X < X^*(\theta) \\ \theta X - I & \text{for} \quad X \geq X^*(\theta) \end{cases} \tag{A3}$$

where

$$X^*(\theta) = \frac{\gamma}{\gamma - 1} \cdot \frac{I}{\theta}$$

$$\gamma = \frac{-(\alpha - \sigma^2/2) + \sqrt{(\alpha - \sigma^2/2)^2 + 2r\sigma^2}}{\sigma^2} > 1 \qquad \textbf{(A4)}$$

and where $\alpha < r$ to ensure convergence. Note that $X^{*\prime}(\theta) < 0$, $X^{*\prime\prime}(\theta) > 0$.

A.2 Proofs of Propositions 1 and 2

Proposition 1. $E[T_1(S_1, \ldots, S_n) - T^*(\theta)] < 0$. *Thus, on average, the agent with the most precise signal will exercise earlier than the full-information optimum.*

Proof. Define I_1 as the information available to agent 1 at the moment he exercises. I_1 will include the realisation of S_1, plus any information (a subset of $(S_2, \ldots, S_n)$) revealed by other agents. The equilibrium describes the exercise strategy of agent 1 as follows. If agent 1 has a positive signal, he exercises at the trigger $X^*(\mu + \varepsilon_1)$. If agent 1 has a negative signal, and agent j is the first agent to reveal a positive signal through exercise, then he exercises at the trigger $X^*(\mu - \sum_{k=1}^{j-1} \varepsilon_k + \varepsilon_j)$. If no agent has a positive signal, then he exercises at the trigger $X^*(\mu - \sum_{k=1}^{n} \varepsilon_k)$. From this specification of agent 1's equilibrium exercise strategy, it follows that $\bar{X}_1(S_1, S_2, \ldots, S_n) = X^*[E_1(\theta)]$, where E_1 denotes agent 1's conditional expectation of θ at the moment of exercise.

Since $X^*(\theta)$ is convex, Jensen's inequality implies that $X^*[E_1(\theta)] < E_1[X^*(\theta)]$. Using the law of iterated expectations, $E[\bar{X}_1] = E\{X^*[E_1(\theta)]\} < E\{E_1[X^*(\theta)]\} = E[X^*(\theta)]$. Therefore $E[\bar{X}_1 - X^*(\theta)] < 0$. Thus, on average, agent 1 exercises earlier than the full-information exercise time.[21]

Proposition 2. $E[T_n(S_1, \ldots, S_n) - T^*(\theta)] > 0$. *Thus, on average, the agent with the least precise signal will exercise later than the full-information optimum.*

Proof. The proof will proceed in two parts. The first part will demonstrate that when no cascades occur, agent n always exercises at or after the full-information trigger. The second part will demonstrate that if a cascade occurs, agent n is expected to exercise after the full-information trigger.

First, assume that no cascades occur. From the definition of equilibrium for $n \geq 2$ agents, agent n will know the true value of θ when he exercises. He will never exercise prior to the full-information trigger.

Second, assume that a cascade occurs, beginning with agent $i \leq n - 2$ (noting that the latest point at which a cascade can begin is with agent $n - 2$). According to the equilibrium, $i^c = i$, agents $i + 1$ to n will exercise simultaneously at the trigger $X^*[E_i(\theta)]$, where $E_i(\theta)$ denotes agent i's conditional expectation of θ at the moment of his exercise. Thus, conditional on a cascade beginning with agent i, $X_n = X^*[E_i(\theta)]$. Given that a cascade begins with agent i, we know that agent i + 1 exercises immediately and reveals $S_{i+1} = +\varepsilon_{i+1}$. Therefore the distribution of θ, conditional on agent i + 1's information, is $E_i(\theta) + \varepsilon_{i+1} + \Sigma_{j=i+2}^{n} S_j$.

I shall prove that $X^*[E_i(\theta)] > E_{i+1}\{X^*[E_i(\theta) + \varepsilon_{i+1} + \Sigma_{j=i+2}^{n} S_j]\}$ by induction. First, let i = n – 2. Conditional on a cascade beginning with agent n – 2,

$$\begin{aligned} X^*[E_{n-2}(\theta)] &> X^*[E_{n-2}(\theta) + \varepsilon_{n-1} - \varepsilon_n] \\ X^*[E_{n-2}(\theta)] &> \tfrac{1}{2}\{X^*[E_{n-2}(\theta) + \varepsilon_{n-1} + \varepsilon_n] + X^*[E_{n-2}(\theta) + \varepsilon_{n-1} - \varepsilon_n]\} \\ X^*[E_{n-2}(\theta)] &= E_{n-1}\{X^*[E_{n-2}(\theta) + \varepsilon_{n-1} + S_n]\} \end{aligned} \tag{A5}$$

where the first two lines follow from the fact that $X^*(\theta)$ is decreasing and that $\varepsilon_{n-1} > \varepsilon_n$, and the third line follows from the definition of agent n – 1's conditional expectation.

Next, assume that $X^*[E_{i+1}(\theta)] > E_{i+2}\{X^*[E_{i+1}(\theta) + \varepsilon_{i+2} + \Sigma_{j=i+3}^{n} S_j]\}$. I now demonstrate that $X^*[E_i(\theta)] > E_{i+1}\{X^*[E_i(\theta) + \varepsilon_{i+1} + \Sigma_{j=i+2}^{n} S_j]\}$. Conditional on a cascade beginning with agent i,

$$\begin{aligned} X^*[E_i(\theta)] &= X^*[E_{i+1}(\theta) - \varepsilon_{i+1}] \\ &> E_{i+2}\left\{X^*\left[E_{i+1}(\theta) - \varepsilon_{i+1} + \varepsilon_{i+2} + \sum_{j=i+3}^{n} S_j\right]\right\} \\ &= E_{i+2}\left\{X^*\left[E_i(\theta) + \varepsilon_{i+2} + \sum_{j=i+3}^{n} S_j\right]\right\} \\ &> E_{i+2}\left\{X^*\left[E_i(\theta) + \varepsilon_{i+1} - \varepsilon_{i+2} + \sum_{j=i+3}^{n} S_j\right]\right\} \\ &= E_{i+1}\left\{X^*\left[E_i(\theta) + \varepsilon_{i+1} - \varepsilon_{i+2} + \sum_{j=i+3}^{n} S_j\right]\right\} \\ &> E_{i+1}\left\{X^*\left[E_i(\theta) + \varepsilon_{i+1} + \sum_{j=i+2}^{n} S_j\right]\right\} \end{aligned} \tag{A6}$$

where the first and second equalities follow because when a cascade begins with agent i, $E_{i+1}(\theta) = E_i(\theta) + \varepsilon_{i+1}$, where the third equality follows because $E_{i+2}[f(\Sigma_{j=i+3}^{n} S_j] = E_{i+1}[f(\Sigma_{j=i+3}^{n} S_j]$, for any function f, where the first inequality follows by assumption (in our induction proof), and where the second and third inequalities follow from the fact that $X^*(\theta)$ is decreasing.

Since we have proved that conditional on $i^c = i$, the inequality in Equation (A6) holds, applying the law of conditional iterated expectations yields $E[X_n | i^c = i] > E[X^*(\theta) | i^c = i]$. Therefore we have proved by induction that, conditional on the occurrence of a cascade beginning with any agent i, the exercise of agent n is expected to be at a trigger greater than the full-information trigger.

A.3 Derivation of the probability of informational cascades

Define the probability that a cascade forms in a market of size n as p_n. p_n can be derived by recursion. The probability that a cascade occurs in a market of size n is equal to the probability that a cascade occurs among the first n – 1 agents plus the probability that a cascade does not occur among the first n – 1 agents, but does occur with the final agent. In order for there to be no cascade among the first n – 1 agents, but for a cascade to occur with n agents, it must be the case that there is no cascade for the first n – 3 agents, a negative signal for agent n – 3, and positive signals for agents n – 2 and n – 1. This can then be written recursively as

$$p_n = p_{n-1} + (1 - p_{n-3}) \cdot (\tfrac{1}{2})^3, \quad \text{for } n > 3 \tag{A7}$$

and with $p_1 = 0$, $p_2 = 0$, $p_3 = 1/4$.

This third-order difference equation has the solution

$$p_n = 1 - \frac{4}{\sqrt{5}}\left[\left(\frac{1}{\sqrt{5}-1}\right)^{1+n} + \left(\frac{1}{1+\sqrt{5}}\right)\left(\frac{-1}{1+\sqrt{5}}\right)^{n}\right] \quad \forall n \geq 0 \tag{A8}$$

A.4 Proof of equilibrium exercise strategies for example with payoff externalities

Case A. $\kappa \geq 2\varepsilon_2$

To prove that the strategies constitute an equilibrium, I will demonstrate that given each player's strategy, the other player's strategy is optimal. First, I demonstrate that agent 2's strategy is optimal, taking agent 1's strategy as given. Note that agent 2 achieves the full-information optimal exercise strategy (and therefore cannot be improved), with the sole exception of when $S_1 = +\varepsilon_1$ and $S_2 = +\varepsilon_2$. When both signals are positive, agent 1 might prefer to exercise earlier than the equilibrium trigger. This would mean that agent 2 would forgo the benefits of learning agent 1's private signal. However, if agent 2 pre-empts agent 1, what will agent 1 do? If $S_1 < 0$, then agent 1 will not exercise immediately, but will wait until the trigger $X^*(\mu - \varepsilon_1 + \varepsilon_2 + \kappa)$ is reached. Thus, at best (if agent 2 exercises very early, agent 1 might still defer exercising), agent 1 will exercise immediately along with agent 2 only if $S_1 > 0$. Therefore agent 2's expected payoff from early exercise would be $(\mu + \varepsilon_2 + \kappa/2) \cdot X - I$ plus an expected payoff of $\kappa/2$ in the future at the trigger $X^*(\mu - \varepsilon_1 + \varepsilon_2 + \kappa)$. Since the future payoff is fixed, the optimal moment to receive this expected payoff is at $X^*(\mu + \varepsilon_2 + \kappa/2)$. Since $\kappa \geq 2\varepsilon_2$, $\mu + \varepsilon_2 + \kappa/2 < \mu + \varepsilon_1 - \varepsilon_2 + \kappa$, and thus $X^*(\mu + \varepsilon_2 + \kappa/2) > X^*(\mu + \varepsilon_1 -$

$\varepsilon_2 + \kappa$), which implies that agent 2 is always better off waiting until the trigger $X^*(\mu + \varepsilon_1 - \varepsilon_2 + \kappa)$ before exercising, even if his signal is positive.

Now, taking agent 2's strategy as given, I demonstrate the optimality of agent 1's strategy. If agent 1 has a negative signal, then his equilibrium strategy achieves the full-information optimum. Thus I focus on the case in which $S_1 = +\varepsilon_1$, where the equilibrium strategy is to exercise at the trigger $X^*(\mu + \varepsilon_1 - \varepsilon_2 + \kappa)$. First, I show that it never pays to exercise earlier. If agent 1 exercises earlier, then (at best) agent 2 will exercise immediately only if $S_2 > 0$. This is because if agent 2 has a negative signal, his optimal strategy is to wait until the trigger $X^*(\mu + \varepsilon_1 - \varepsilon_2 + \kappa)$ is reached. Thus an earlier than equilibrium exercise by agent 1 will provide an expected immediate payoff of $(\mu + \varepsilon_1 + \kappa/2) \cdot X - I$ plus an expected payoff of $\kappa/2$ in the future (when agent 2 exercises). Since the future payoff is fixed, the optimal moment to receive this payoff is at $X^*(\mu + \varepsilon_1 + \kappa/2)$. Since $\kappa \geq 2\varepsilon_2$, $\mu + \varepsilon_1 + \kappa/2 < \mu + \varepsilon_1 - \varepsilon_2 + \kappa$, which implies that it is better for agent 1 to always wait at least until the trigger $X^*(\mu + \varepsilon_1 - \varepsilon_2 + \kappa)$ is reached before exercising. Second, I show that it never pays to exercise later. The only reason for agent 1 (with $S_1 > 0$) to wait would be to "pretend" to have a negative signal and learn agent 2's private signal when the trigger $X^*(\mu - \varepsilon_1 + \varepsilon_2 + \kappa)$ is reached. However, this information will prove useless because agent 1 will exercise immediately regardless of the sign of S_2. This is true because even if $S_2 = -\varepsilon_2$, the optimal exercise trigger is $X^*(\mu + \varepsilon_1 - \varepsilon_2 + \kappa)$, which implies immediate exercise. Since the "pretend" strategy implies waiting until $X^*(\mu + \varepsilon_1 - \varepsilon_2 + \kappa)$ and then always exercising immediately, this implies an expected payoff of $(\mu + \varepsilon_1 + \kappa) \cdot X - I$. The value of receiving this payoff is decreasing at any point above the trigger $X^*(\mu + \varepsilon_1 + \kappa)$, which implies it is not optimal to wait beyond the equilibrium trigger $X^*(\mu + \varepsilon_1 - \varepsilon_2 + \kappa)$.

Case B. $\kappa < 2\varepsilon_2$

First, I demonstrate that agent 1's strategy is optimal, taking agent 2's strategy as given. If agent 1 has a negative signal, then his equilibrium strategy achieves the full-information optimum. Thus I focus on the case in which $S_1 = +\varepsilon_1$, where the equilibrium strategy is to exercise at the trigger $X^*(\mu + \varepsilon_1 + \kappa/2)$. I begin by showing that agent 1 will never enter earlier than the trigger $X^*(\mu + \varepsilon_1 + \kappa/2)$. Agent 1 knows that if agent 2 exercises immediately after him if $S_2 = +\varepsilon_2$, and at the trigger $X^*(\mu + \varepsilon_1 - \varepsilon_2 + \kappa)$ if $S_2 = -\varepsilon_2$, then his expected payoff from being the first to exercise is $(\mu + \varepsilon_1 + \kappa/2) \cdot X - I$ paid immediately, plus $\kappa/2 \cdot X$ in the future when the trigger $X^*(\mu + \varepsilon_1 - \varepsilon_2 + \kappa)$ is hit. Since the future payoff is fixed, the optimal moment to receive this payoff is at $X^*(\mu + \varepsilon_1 + \kappa/2)$. Thus there is no incentive to exercise earlier. The only potential reason to wait longer would be to acquire information on agent 2's signal. In order to do so, agent 1 would have to wait until the trigger $X^*(\mu - \varepsilon_1 + \varepsilon_2 + \kappa)$ for knowledge of S_2. However, it will always prove optimal for agent 1 to exercise immediately at the trigger

$X^*(\mu - \varepsilon_1 + \varepsilon_2 + \kappa)$, regardless of information about the sign of S_2. This is because the immediate payoff would be $(\mu + \varepsilon_1 + S_2 + \kappa) \cdot X - I$, and the optimal exercise strategy for such a payoff is to exercise the first moment X(t) equals or exceeds the trigger $X^*(\mu + \varepsilon_1 + S_2 + \kappa)$, which is always less than $X^*(\mu - \varepsilon_1 + \varepsilon_2 + \kappa)$. Since this strategy of waiting for more information provides precisely the same expected payoff as the equilibrium strategy, but at a stopping time greater than the time at which the value maximising trigger $X^*(\mu + \varepsilon_1 + \kappa/2)$ is reached, it never pays to exercise later.

Now, taking agent 1's strategy as given, I demonstrate the optimality of agent 2's strategy. It is clear that by observing whether agent 1 exercises the first moment that X(t) equals or exceeds $X^*(\mu + \epsilon_1 + \kappa/2)$, agent 2 will correctly infer agent 1's signal and will then know the true value of θ. It is never optimal to wait longer than the full-information trigger. It would also never be optimal for agent 2 to take the lead and exercise earlier. Even if agent 2 is optimistic ($S_2 = +\varepsilon_2$), prior to learning agent 1's signal his optimal exercise strategy would be to wait until the trigger $X^*(\mu + \varepsilon_2 + \kappa/2)$, as was shown above in demonstrating agent 1's optimal strategy. Since this trigger is greater than the trigger $X^*(\mu + \epsilon_1 + \kappa/2)$, it will prove optimal to wait for S_1 to be revealed.

1 In a full information setting, the strategic exercise of options has been the subject of several articles. Emanuel (1983), Constantinides (1984) and Spatt and Sterbenz (1988) analyse the strategic exercise of warrants and convertible securities. Grenadier (1996) provides a game-theoretic equilibrium to a strategic option exercise game and applies the model toward analysing real estate developer decisions.

2 Back (1993) provides an extension of the Kyle (1985) model of continuous insider trading that includes asymmetric information and options. While there is no strategic exercise feature in the model, information is conveyed by trading in the options. This information transmission eliminates the typical redundancy of the option. Childs, Ott and Riddiough (1996) analyse option exercise in a context in which the underlying asset value is imperfectly observable.

3 Herding behaviour may also result from agency-related reputational problems. See Scharfstein and Stein (1990) and Zwiebel (1995).

4 Bikhchandani, Hirshleifer and Welch (1992) conjectured that with endogenous timing those agents with the most precise information will move first.

5 In Zhang (1997), agents also choose when to act. In his model, however, informational cascades always occur at the moment the first agent acts, after an initial lag. Chamley and Gale (1994) consider a discrete-time investment model under incomplete information where agents choose when to invest. In their model, agents' signals are of equal precision.

6 Grenadier (1996) models an option exercise game where exercise results in an increase in the number of firms in an industry, and thus lowers the cashflows available to competing firms.

7 It is assumed that exercise does not reveal the true value of θ to those that have not yet exercised. As long as θ remains even partially unobservable, the model's results are unaffected. This assumption is more likely to hold in private, segmented markets such as real estate development where institutional frictions such as construction lags and slow information leakage make observability difficult. Relaxation of this assumption is simple in this model's equilibrium framework. The general n-person equilibrium is transformed into the two-person equilibrium (defined in the second section) upon the moment of first exercise. The remaining agents act simultaneously as the "less informed agent" in the two-person equilibrium.

8 The assumption of independent signals is made to simplify the discussion. In essence, agents' differing signals represent idiosyncratic components of the true state variable. The model could be modified to allow for correlated signals by adding a random common component to θ, and making the signals equal to the sum of the common components plus noise.

9 If one wishes to visualise the problem as an option on a random number of shares of X, then $\mu = 1$ would signify that the expected number of underlying shares is one.

10 The assumption of simple, discrete signal distributions is common in the literature on investment cascades. Allowing for continuous signal distributions would greatly complicate the equilibrium derivation, as well as imply very different asset price dynamics. With continuous signal distributions, information would evolve continuously, as agents refine their posterior distribution of θ through observing the actions of other agents. This implies that asset prices will evolve continuously, as both public information on X(t) and private information on θ evolve continuously over time. In the current setup with discrete signals, asset prices evolve continuously during periods in which no private information is revealed, but jump by discrete increments when private information is revealed. Such mixed asset price dynamics may be particularly descriptive of real-options investments discussed in this article.

11 In addition, asymmetric information can itself lead to the failure to price options by arbitrage. As demonstrated by Back (1993) in a model of continuous insider trading, trading in the option affects the flow of information, making a seemingly redundant asset not capable of being priced by arbitrage. As such, an equilibrium pricing approach must be used.

12 This assumption, while not essential, ensures that equilibrium exercise does not occur in a discrete jump at an arbitrary initial starting point of the game.

13 Off-equilibrium actions play a very limited role in the basic model. Since an agent's exercise payoff does not depend on what later agents do, there is no incentive to make an out-of-equilibrium move to try to influence a later player. Thus there is a multiplicity of off-equilibrium conjectures associated with this equilibrium.

14 As discussed in note (7), if the true value of θ is revealed when the first agent exercises, the equilibrium is "truncated". That is, once the option is first exercised (by the agent with the greatest signal), all of the remaining agents are fully informed and exercise the first moment X(t) equals or exceeds the trigger $X^*(\theta)$.

15 It is interesting to briefly consider the impact of allowing agents to possess options with differing payoff functions. For example, agents may have different costs of exercise. While much of the spirit of the model will remain (agents trade off the benefits of exercise with the benefits of waiting to learn from the actions of others), there will be one important difference. When agents have different payoff functions, it will not necessarily be the case that information is transmitted from the most informed agents to the least informed agents. If an agent has a low exercise price, he may exercise earlier than another agent with a more precise signal, as his lower exercise price overcomes his information disadvantage. This may lead to a lessening of the likelihood of informational cascades, as the information inferred from prior exercise may no longer overwhelm the remaining private information.

16 Grenadier (1996) provides an alternative rational explanation for such seemingly irrational developer behaviour.

17 I am assuming that the identity of any agent that purchases information becomes common knowledge. It may be possible in certain industry environments for this information to be kept secret. However, the purchaser might still be identified by his exercise strategy.

18 I make the following off-equilibrium assumptions. If agent 1 exercises at any time prior to the trigger $X^*(\mu + \varepsilon_1 - \varepsilon_2 + \kappa)$, then agent 2 infers that agent 1 has a positive signal. Agent 2 will then exercise according to the trigger $X^*(\mu + \varepsilon_1 + S_2 + \kappa)$. Similarly, if agent 2 exercises before the trigger $X^*(\mu + \varepsilon_1 - \varepsilon_2 + \kappa)$, then agent 1 infers that agent 2 has a positive signal and exercises the first moment that X(t) equals or exceeds the trigger $X^*(\mu + S_1 + \varepsilon_2 + \kappa)$.

19 I make the following off-equilibrium assumptions. If agent 1 exercises prior to the trigger $X^*(\mu + \varepsilon_1 + \kappa/2)$, then agent 2 infers that agent 1 has a positive signal and then exercises

according to the trigger $X^*(\mu + \varepsilon_1 + S_2 + \kappa)$. Similarly, if agent 2 exercises prior to the trigger $X^*(\mu + \varepsilon_1 + \kappa/2)$, then agent 1 infers that agent 2 has a positive signal and then exercises according to the trigger $X^*(\mu + S_1 + \varepsilon_2 + \kappa)$.

20 See Merton (1973) for a derivation of the high-contact condition.

21 Recall that we assume that $\alpha > \frac{1}{2}\sigma^2$ in order for the expectation with respect to first passage times to exist.

BIBLIOGRAPHY

Admati, A. and P. Pfleiderer, 1986, "A Monopolistic Market for Information", *Journal of Economic Theory* 39, 400–38.

Admati, A. and P. Pfleiderer, 1988, "Selling and Trading on Information in Financial Markets", *American Economic Review* 78, pp. 96–103.

Back, K., 1993, "Asymmetric Infommation and Options", *Review of Financial Studies* 6, pp. 435–72.

Banerjee, A.V., 1992, "A Simple Model of Herd Behavior", *Quarterly Journal of Economics* 107, pp. 797–817.

Bikhchandani, S., D. Hirshleifer and I. Welch, 1992, "A Theory of Fads, Fashion, Custom, and Cultural Change as Informational Cascades", *Journal of Political Economy* 100, pp. 992–1026.

Birch, D., 1990, *America's Future Office Space Needs*, National Association of Industrial and Office Parks, Washington, DC.

Caplin, A. and J. Leahy, 1993, "Miracle on Sixth Avenue: Information Externalities and Search", Discussion Paper 1665, Harvard Institute of Economic Research, Harvard University.

Chamley, C. and D. Gale, 1994, "Information Revelation and Strategic Delay in a Model of Investment", *Econometrica* 62, pp. 1065–85.

Childs, P., S. Ott and T. Riddiough, 1996, "Incomplete Information, Exercise Policy, and the Valuation of Claims on Noisy Real Assets", Working Paper, University of Kentucky.

Constantinides, G., 1984, "Warrant Exercise and Bond Conversion in Competitive Markets", *Journal of Financial Economics* 13, pp. 371–97.

Cox, J.C. and S.A. Ross, 1976, "The Valuation of Options for Alternative Stochastic Processes", *Journal of Financial Economics* 3, pp. 145–66.

Dixit, A.K. and R.S. Pindyck, 1994, *Investment Under Uncertainty*, Princeton University Press.

Dybvig, P.H. and C.S. Spatt, 1983, "Adoption Externalities as Public Goods", *Journal of Public Economics* 20, pp. 231–47.

Emanuel, D.C., 1983, "Warrant Valuation and Exercise Strategy", *Journal of Financial Economics* 12, pp. 211–35.

Grenadier, S.R., 1995, "Valuing Lease Contracts: A Real Options Approach", *Journal of Financial Economics* 38, pp. 297–331.

Grenadier, S.R., 1996, "The Strategic Exercise of Options: Development Cascades and Overbuilding in Real Estate Markets", *Journal of Finance* 51, pp. 1653–79.

Harrison, J.M., 1985, *Brownian Motion and Stochastic Flow Systems*, John Wiley & Sons, New York.

Hendricks, K. and D. Kovenock, 1989, "Asymmetric Information, Information Externalities, and Efficiency: The Case of Oil Exploration", *RAND Journal of Economics* 20, pp. 164–82.

Hendricks, K., R.H. Porter and B. Boudreau, 1987, "Information and Returns in OCS Auctions: 1954-1969", *Journal of Industrial Economics* 35, pp. 517–42.

Hirshleifer, D., 1995, "The Blind Leading the Blind: Social Influence, Fads and Informational Cascades", in K. Ierulli and M. Tommasi (eds.), *The New Economics of Human Behaviour*, Cambridge University Press.

Katz, M.L. and C. Shapiro, 1986, "Technology Adoption in the Presence of Network Externalities", *Journal of Political Economy* 94, pp. 822–41.

Kyle, A.S., 1985, "Continuous Auctions and Insider Trading", *Econometrica* 53, pp. 1315–35.

Merton, R.C., 1973, "Theory of Rational Option Pricing", *Bell Journal of Economics and Management Science* 4, pp. 141–83.

Paddock, J.L., D.R. Siegel and J.L. Smith, 1988 "Option Valuation of Claims on Real Assets: The Case of Offshore Petroleum Leases", *Quarterly Journal of Economics* 103, pp. 479–508.

Scharfstein, D. and J. Stein, 1990, "Herd Behavior and Investment", *American Economic Review* 80, pp. 465–79.

Spatt, C.S. and F.P. Sterbenz, 1988, "Warrant Exercise, Dividends, and Reinvestment Policy", *Journal of Finance* 43, pp. 493–506.

Titman, S., 1985, "Urban Land Prices Under Uncertainty", *American Economic Review* 75, pp. 505–14.

Welch, I., 1992, "Sequential Sales, Learning, and Cascades", *Journal of Finance* 47, pp. 695–732.

Williams, J.T., 1993, "Equilibrium and Options on Real Assets", *Review of Financial Studies* 6, pp. 825–50.

Zhang, J., 1997, "Strategic Delay and the Onset of Investment Cascades", *RAND Journal of Economics* 28, pp. 188–205.

Zwiebel, J., 1995, "Corporate Conservatism and Relative Compensation", *Journal of Political Economy* 103, pp. 1–25.

7

*Pricing Real Assets with Costly Search**

Joseph T. Williams

University of British Columbia

Markets for many real assets are characterised by sequential search followed by bilateral bargaining between matched buyers and sellers. For a category of real assets, the joint, intertemporal valuation problems of buyers, owners, and sellers, and the associated Nash pricing function are solved explicitly. In equilibrium, the average transaction price is a noisy, proportional random walk, and the liquidity premium is positive for matched owners. Depending on the values of the parameters, the liquidity premium can be substantial. In a related problem of optimal development with costly search, the optimal exercise point, cost of development, and value of the undeveloped asset are calculated analytically. With search, development can occur sooner and undeveloped assets have lower market values than the standard solution without search.

Markets for many real assets are decentralised. Common examples are categories of real estate within a metropolitan area, including single and multifamily, housing, commercial and industrial properties, and retail centres. Other examples are types of consumer and producer durables, small businesses, and firms acquired through mergers and takeovers.[1] In all of these markets, buyers must search for sellers and/or sellers must search for buyers. Once a potential buyer and seller meet, they must bargain to determine a price. If the parties cannot agree on a price, then they must continue to search. Search is costly because buyers and sellers are mismatched to their assets while searching. The bargaining solution must reflect the options of both the buyer and seller to continue searching, the benefit to both of a match, and the buyer's subsequent costly search as a seller once

*This chapter was first published in *Review of Financial Studies* 8(1), 1995, pp. 55–90 and is reprinted with the permission of Oxford University Press. The author would like to thank A. Bick, R. Uppal, C. Spatt (editor), the referee, and participants in seminars at Arizona, ASU, Berkeley, SFU, SMU, UBC, and the AFA/AREUEA. Research support was provided by the Bureau of Asset Management at UBC.

his match fails. In turn, the benefit from a match must reflect the systematic or common component of the rent or cashflow from the asset. For example, with single-family houses in a metropolitan area the common component could be the local rent on apartments or the implicit rent from owner-occupied houses. A realistic solution for housing must allow the common component of rent to change stochastically through time as conditions in the local rental market change.

Costly search with pairwise bargaining has been studied extensively in steady state.[2] In steady state all factors affecting the matching of potential buyers and sellers, the bargaining between matched agents, and thereby the pricing of real assets are constant through time. This analytically convenient assumption allows careful, detailed study of the relationship between matching, bargaining and prices. However, it also requires that prices change through time only in response to the changing idiosyncratic component of matches between different buyers and sellers. Most importantly, it precludes stochastic changes through time in aggregate state variables or common factors that affect the cashflows from assets. With housing this precludes unpredictable changes in local employment or income that could affect the common component of the implicit rent on housing services. In option pricing theory such common factors are critical to the pricing of assets.

In this chapter the pricing process is derived in continuous time for a category of real assets with costly search and bilateral bargaining. Under the assumptions of the model, the expected price or current market value of a representative asset is a continuous, proportional random walk with a constant mean and variance per unit of time, while the average transaction price is a proportional random walk with noise. In equilibrium the liquidity premium on an asset is positive and possibly substantially so, depending on the values of the parameters.[3] Also, optimal development of a real asset, conditional on a match, is accelerated by costly search. Simultaneously, the resulting value of the option to develop or, equivalently, the undeveloped asset is depressed relative to the standard solution without search. Finally, some of the comparative statics are surprising. For example, if buyers search more intensively and thereby arrive on average more frequently to inspect the assets of sellers, then these assets become more liquid. In equilibrium these more liquid assets have a higher expected price but the same liquidity premium and the same distribution of selling times. In this model the higher reservation values for more liquid assets and the resulting higher prices offset in equilibrium all effects of more intensive search on expected yields and selling times.

Related work on real assets is limited.[4] In Baldwin and Meyer (1979), firms must search for investments. Opportunities to invest arrive in a Poisson process, and past investments expire in another Poisson process. Because the representative firm has the capacity to invest in only one project at a time, its required rate of return or reservation value for new projects increases in the average arrival rate and decreases in the average departure rate. The pricing problem in continuous time is solved explicitly for a risk-neutral firm with a discount rate equal to zero. In Wheaton (1990) a simple but innovative model of matching in steady state is applied to owner-occupied housing. There, families can switch between two possible types and three possible states. The types can be interpreted as small versus large families, whereas the states are matched homeowners, unmatched homeowners searching for another house, and matched homeowners seeking to sell their previous house. Comparative statics are calculated for the price in steady state, the fraction of buyers, the expected time to sale, and the effort expended on search. Both articles differ in focus from this chapter and hence differ in critical details. The first model effectively has both buyers and owners of real assets but no sellers. Without sellers, risky real assets cannot be priced in equilibrium. The second model is effectively static. Without a stochastically evolving state variable, like the common component of rent or implicit rent on real assets, developed and undeveloped property cannot be priced like options in equilibrium and liquidity premiums cannot be calculated.

The model is presented in the first section. By assumption, the cash inflow on each asset is the product of two components. The common component is realised on all assets and observed by all agents, while the residual is specific to the match and observed by the buyer and seller. The common component is a geometric Wiener variate, and the residual is an independent Pareto variate with an independent Poisson jump to zero when the match between an owner and his asset is broken. Also, buyers search independently in the same Poisson process, and all agents agree on the Nash bargaining solution. Under these conditions the unique equilibrium is identified in the second section. The pricing function is linear homogeneous in the common component. The price per unit of the common component is a constant plus an independent Pareto variate, and the expected price or, equivalently, the current market value of the asset is a geometric Wiener variate. With costly search the expected liquidity premium on each asset is a positive constant. Finally, for each asset the time to sale and fraction of assets for sale have exponential and truncated β-distributions, respectively.

This solution is then extended in several directions. In the third section it is applied to undeveloped assets. With costly search, development can occur at any value of the common component of cash inflows at or above the optimal value at which the owner first accepts an offer to develop his property. Both this optimal point and the resulting value of an undeveloped asset are below the corresponding values without costly search. The cost of development, as determined by the Nash bargaining solution, is increasing and concave in the common component and asymptotic to a linear upper bound. In the fourth section the solution is used to identify necessary and sufficient conditions for the employment of brokers by buyers and sellers. Throughout, the comparative statics are calculated analytically, if possible, and numerically otherwise for plausible values of the parameters. From the numerical computations, for example, brokers are found to be more valuable in markets with persistently low demand than in markets with persistently high demand. In turn, these markets are characterised by higher liquidity premiums, higher expected times to sale, and larger fractions of assets for sale. Other extensions are also discussed in this section, and the results are summarised in the last section. All technical details appear in the Appendix.

MODEL

Buyers inspect the assets of sellers, and sellers offer their assets for sale. At time t the market for the representative category of real assets has B_t buyers and S_t sellers, each with one asset for sale. Each buyer inspects the assets of sellers in the same process. During the period from time t to time $t + \Delta t$, the representative buyer b arrives to inspect the asset of some seller with the probability, $\alpha\Delta t + o(\Delta t)$. As usual, the residual $o(\Delta t)$ represents terms of smaller order than Δt. During the same period buyer b sees no asset with probability $1 - \alpha\Delta t + o(\Delta t)$ and multiple assets with probability $o(\Delta t)$. The numbers of inspections per buyer per period are stationary and independent across both buyers and periods. In this case, buyers inspect assets for sale in independent, identical Poisson processes. The constant Poisson arrival rate per unit of time, $\alpha > 0$, reflects the efficiency of search, as described in the fourth section. For analytical convenience, all sampling of specific assets for sale is purely random, with replacement period by period and independent across buyers. Sampling with replacement is an increasingly accurate approximation to the more realistic assumption of sampling without replacement as progressively more assets are offered for sale: $S_t \to \infty$. With these assumptions the representative buyer b visits the representative seller s over the period of length Δt beginning at time t, with the probability: $(\alpha/S_t)\Delta t + o(\Delta t)$. Ignoring terms of smaller order than Δt, seller s then receives during this period offers from $n_{t+\Delta t}$ buyers with the binomial probabilities:

$$\Pr\{n_{t+\Delta t} = n\} = \binom{B_t}{n}\left(\frac{\alpha\Delta t}{S_t}\right)^n\left(1 - \frac{\alpha\Delta t}{S_t}\right)^{B_t - n}$$

for $n = 0, \ldots, B_t$.

Markets for real assets can have many buyers and sellers. Primary examples are single-family residences, including detached houses and condominiums, and consumer durables, like automobiles – especially in major metropolitan areas. Also, some categories of commercial real estate, small businesses, and producer durables can have reasonably large numbers of buyers and sellers. Accordingly, let the number of buyers and sellers, B_t and S_t, increase without bound, such that the ratio of buyers and seller, B_t/S_t, converges to a positive constant ρ: $B_t, S_t \to \infty$ with $B_t/S_t \to \rho > 0$. The constant ρ can be interpreted as the ratio of buyers to sellers in a quasi-steady state. This steady state excludes short-term, random variations in the ratio of buyers to sellers, B_t/S_t, around its mean ρ but includes long-term trends that may force the mean ρ away from 1. For example, a metropolitan area may lose jobs and thereby residents for decades, so that ρ remains below 1 for residential real estate. Another metropolitan area may grow so rapidly that public controls on development constrain the construction of new housing, so that ρ stays above 1. Under these limiting conditions, the above binomial density converges to a Poisson density with the mean $\alpha\rho\Delta t$. With progressively shorter and shorter periods, $\Delta t \to 0$, this Poisson density converges to

$$\Pr\{n_t = 0\} = 1 - \alpha\rho\Delta t + o(\Delta t), \quad \Pr\{n_t = 1\} = \alpha\rho\Delta t + o(\Delta t) \tag{1}$$

Under the above assumptions, seller s receives, over a short interval of length Δt beginning at time t, offers from n_t buyers with the Poisson probabilities (1).

When a buyer acquires an asset, he immediately begins to receive rent or services from the asset. This rent or implicit rent is realised continuously through time until the match between the owner and his asset is broken. If buyer b purchases the asset of seller s, then he receives from time $t + \Delta t$ the rent: $x_t y_{bs} \Delta t$. The component x_t is common to all matches, dependent on time t, and known to all agents. The residual y_{bs} is idiosyncratic to the match, constant through time, and observed only by the buyer b and seller s, conditional on a match. The common component x_t evolves continuously though time in a proportional random walk with a constant mean and variance per unit of time. Thus, x_t follows a geometric Wiener process,

$$\frac{dx}{x} = \mu dt + \sigma dz \tag{2}$$

with the standard Wiener variate z_t and the constant mean μ and variance σ^2 per unit of time. The residual rents y_{bs} are independently and identically distributed for all possible pairs or matches bs of buyer b to the asset of seller s. The common distribution for the residual rents can reflect the restricted search by the representative category of buyers to its preferred category of assets. By assumption, this distribution is Pareto with a finite mean,

$$\Pr\{y_{bs} \leq y\} = 1 - \left(\frac{\eta}{y}\right)^{\theta} \tag{3}$$

for all $y \geq \eta > 0$ with $\theta > 1$. The Pareto distribution has a finite mth moment if and only if $m < \theta$. This distribution has been used empirically to measure the size of cities, the returns on securities, and the distribution of income. With (3) the residual rent y_{bs} has the conditional mean: $E[y_{bs} \mid y_{bs} \geq y] = \theta y/(\theta - 1)$ for all $y \geq \eta$. The unconditional mean, $\eta\theta/(\theta - 1)$, is 1 if $\eta = 1 - 1/\theta$.

The match between an owner and his asset is broken by a random event. With residential real estate the event could be a change in the size of the owner's family, his income from employment or the location of his employment. Examples are marriages, births, deaths, divorces, departures to college, retirement and new jobs at distant locations. When such an event occurs, the match bs disappears and the residual rent y_{bs} drops to zero. Without delay the owner then offers his asset for sale. This event – the termination of a match followed immediately by an offer to sell – occurs during any period of length Δt with the probability, $\beta\Delta t + o(\Delta t)$, satisfying $\beta > 0$. By assumption, the event is independent across both matches and periods. In this case, the length or duration of a match has an exponential distribution with the mean, $1/\beta$. This simple specification is selected mainly for its analytical convenience.[5] In fact, illiquid assets, like single-family houses, are sometimes offered for sale to rebalance portfolios or to buy bigger properties when constraints on downpayments are met.[6] Also, the exponential distribution has no memory, so that the probability of a move does not depend on the duration of the stay.[7]

In each period all buyers are identical before visiting an asset for sale, while all sellers are identical at all times. All agents are risk-neutral with a common discount rate. Without further loss of generality, this discount rate can be interpreted as the riskless rate of interest.[8] Measured per unit of time, the riskless rate is the constant ι. The subsequent problem has a finite solution if and only if this constant satisfies $\iota > \mu$. When buyer b inspects the asset of seller s, both the buyer and seller observe the residual rent y_{bs}. Next, the buyer and seller negotiate. If there exists a price that both prefer to no transaction, then they agree on the Nash bargaining solution and trade; otherwise they disagree and do not trade. Thereby, buyer b purchases the asset of seller s if his rent $x_t y_{bs}$ exceeds or equals a lower bound $V(x_t)$. This lower bound is subsequently shown to be the optimal stopping point or reservation value of both the buyer and seller. It is determined endogenously and calculated explicitly in the subsequent solution under the restriction that V is continuous on $x_t > 0$. The associated ratio, $v_t = V(x_t)/x_t$, is called the unit reservation value. Finally, if buyer b purchases the asset of seller s, then he pays the price from the subsequent Nash bargaining solution: $P_{bs}(x_t) \equiv P(x_t, y_{bs})$. Without further loss of generality, the seller then pays the transaction cost: $\gamma P_{bs}(x_t)$. The constant, $0 \leq \gamma < 1$, is the proportional transaction cost. After a transaction the seller leaves the market.

Such strong assumptions require some motivation. If agents are risk averse and the capital market is complete, then the subsequent results are changed only slightly. Everywhere in the subsequent solution the expected growth rate μ of the state variable x_t in (2) is replaced by a constant, risk-adjusted, expected growth rate if the following conditions are satisfied. Instantaneous changes in the common component x_t are perfectly locally correlated with concurrent changes in the price of a portfolio of securities that can be traded continuously without transaction costs in a perfectly competitive capital market.[9] Through time the price of this portfolio evolves continuously in a proportional random walk with a constant percentage mean and variance per unit of time. In this case, the portfolio of substitute securities has per unit of time a constant excess expected rate of return per unit of standard deviation κ. Conditional on this constant, the risk-adjusted expected rate of growth of the state x_t is the constant $\mu - \kappa\sigma$. Again, this risk-adjusted mean would replace the mean μ everywhere in the subsequent analysis. Finally, because matches are either made or broken in completely independent events, the resulting changes in the expected utilities of both buyers and sellers are fully diversifiable. These Poisson jumps can then be valued by the argument in Naik (1991).

Several other assumptions must also be motivated. Risk-neutral buyers and sellers could have different discount rates, reflecting their relative impatience for transactions. These discount rates affect the bargaining powers of buyers and sellers and thereby the allocation of surplus in the Nash bargaining solution but otherwise do not alter the subsequent solution. As previously described, the Nash bargaining solution is supported in models of costly search by non-cooperative bargaining under symmetric information. By contrast, symmetric information within the bargaining problem between a matched buyer and seller is more problematic and is motivated mainly by its analytical convenience. Typically, a buyer has private information about his personal preferences for the asset and his urgency to buy, while the seller has private information about the quality of his asset and his urgency to sell. Clearly, a rational agent reveals truthfully his private information in a negotiation only when it improves this personal welfare. However, buyers are typically interviewed by both sellers or their brokers and potential lenders. Brokers of both the buyer and seller act legally as agents of the seller in most jurisdictions, and lenders typically demand detailed information. Although buyers sometimes approach lenders only after negotiating with sellers, sales contracts commonly have financing as a contingency. To protect the buyer, sales contracts commonly have physical inspection of both the asset and the seller's records as another contingency.

Buyers, sellers and owners have no limit on the length of time during which they can search for a match or hold an asset. In this case, before buyer b can meet a seller and inspect his asset, he has the expected utility

$U_b(x_t)$, conditional only on the current common component of rent x_t. The buyer's current expected utility or indirect value is

$$U_b(x_t) = e^{-\iota\Delta t}\left\{\alpha\Delta t\, E_x\left[\left(\frac{\eta}{v_{t+\Delta t}}\right)^{\theta} E_{x+\Delta x}\left[U_{bs}(x_{t+\Delta t}) - P_{bs}(x_{t+\Delta t})\middle|\, y_{bs} \geq v_{t+\Delta t}\right]\right]\right.$$

$$\left. + E_x\left[\left(1 - \alpha\Delta t\left(\frac{\eta}{v_{t+\Delta t}}\right)^{\theta}\right)U_b(x_{t+\Delta t})\right]\right\} + o(\Delta t) \quad \textbf{(4)}$$

In equation (4) the expectation E_x is calculated conditional on the current state, $x_t = x$, whereas the expectation, $E_{x+\Delta x}[\cdot \mid y_{bs} \geq v_{t+\Delta t}]$, is calculated conditional on the subsequent state, $x_{t+\Delta t} = x + \Delta x$, and the event of a transaction, $y_{bs} \geq v_{t+\Delta t} = V(x_{t+\Delta t})/x_{t+\Delta t} = V(x + \Delta x)/(x + \Delta x)$. Also, the residual terms of order Δt reflect the probability of a buyer's arrival, $\alpha\Delta t + o(\Delta t)$. If buyer b is matched to the asset of seller s at time t, then his expected utility is $U_{bs}(x_t)$. This expected utility satisfies the second recursion:

$$U_{bs}(x_t) = E_x\left[\int_t^{t+\Delta t} e^{-\iota(t'-t)} x_{t'} y_{bs}\, dt'\right] + e^{-\iota\Delta t}\{\beta\Delta t\, E_x\left[U_s(x_{t+\Delta t})\right]$$

$$+ (1 - \beta\Delta t)\, E_x[U_{bs}(x_{t+\Delta t})]\} + o(\Delta t) \quad \textbf{(5)}$$

Finally, for seller s the same asset has the expected utility $U_s(x_t)$:

$$U_s(x_t) = e^{-\iota\Delta t}\left\{\alpha\rho\Delta t\, E_x\left[\left(\frac{\eta}{v_{t+\Delta t}}\right)^{\theta}(1 - \gamma)E_{x+\Delta x}\left[P_{bs}(x_{t+\Delta t})\middle|\, y_{bs} \geq v_{t+\Delta t}\right]\right]\right.$$

$$\left. + E_x\left[\left(1 - \alpha\Delta t\left(\frac{\eta}{v_{t+\Delta t}}\right)^{\theta}\right)U_s(x_{t+\Delta t})\right]\right\} + o(\Delta t) \quad \textbf{(6)}$$

These recursions can be interpreted as follows. In (4) and (6) each agent's current valuation equals his discounted expected value from the next period. Agent b buys: in (4) with the approximate probability $\alpha\Delta t(\eta/v_{t+\Delta t})^{\theta}$, while agent s sells in (6) with the approximate probability $\alpha\rho\Delta t(\eta/v_{t+\Delta t})^{\theta}$. Each probability equals the approximate probability of a match – $\alpha\Delta t$ for a buyer and $\alpha\rho\Delta t$ for a seller – times the probability of a transaction conditional on a match. If a transaction occurs, then buyer b becomes an owner and realises the net gain: $U_{bs}(x_{t+\Delta t}) - P_{bs}(x_{t+\Delta t})$. Similarly, seller s receives after a transaction the net price: $(1 - \gamma)P_{bs}(x_{t+\Delta t})$. Immediately thereafter the seller leaves the market. In (5) a matched owner receives over an interval of time Δt the rents indicated in the integral and loses his match with the approximate probability $\beta\Delta t$. Without a match he immediately offers his asset for sale and then values it as a seller. With a match his status stays the same and his valuation changes only with the evolution of the state x_t.

These valuation functions must satisfy two sets of boundary conditions. With the geometric Wiener process (2), once the rent $x_t y_{bs}$ reaches zero, it remains there forever. Thus, all utilities must be zero at $x_t = 0$:

$$U_b(0) = 0, \qquad U_{bs}(0) = 0, \qquad U_s(0) = 0 \tag{7}$$

Also, to prevent explosive growth and collapse in prices; the utilities per dollar of implicit rent must be bounded above. In this case, there must exist two positive constants, ζ_b and ζ_s, and a positive, increasing function ζ with the values, $\zeta_{bs} = \zeta(y_{bs})$, such that the utilities satisfy

$$U_b(x_t) \le \zeta_b x_t, \qquad U_{bs}(x_t) \le \zeta_{bs} x_t, \qquad U_s(x_t) \le \zeta_s x_t \tag{8}$$

as $x_t \to \infty$. These boundary conditions are familiar from many valuation models of contingent claims.

The Nash bargaining solution has the following properties. Because both buyer b and seller s have the same information, they agree to trade if and only if both benefit from the transaction:

$$P_{bs} \le U_{bs} - U_b, \qquad U_s \le (1-\gamma)P_{bs} \tag{9}$$

If a transaction occurs in state x_t then buyer b pays the price $P_{bs}(x_t)$, becomes an owner, and immediately realises the gain, $U_{bs}(x_t) - U_b(x_t) - P_{bs}(x_t)$. Simultaneously, seller s receives after transaction costs the net price, $(1-\gamma)P_{bs}(x_t)$, and transfers to the buyer an asset with the value $U_s(x_t)$. With a transaction the total gain is allocated in the following fractions: $1/(1+\lambda)$ to the buyer and $\lambda/(1+\lambda)$ to the seller. The constant, $\lambda > 0$, is the seller's relative share of the surplus. This constant can be specified exogenously as a Nash, cooperative bargaining solution. Alternatively, it can be determined as a subgame perfect solution to a non-cooperative bargaining problem with complete information, as described in the next section. In either case, the seller's gain $(1-\lambda)P_{bs} - U_s$, must equal $\lambda/(1+\lambda)$ times the total gain, $U_{bs} - U_b - U_s - \gamma P_{bs}$. Equivalently, the seller's gain equals λ times the buyer's gain:

$$(1-\gamma)P_{bs} - U_s = \lambda(U_{bs} - U_b - P_{bs}) \tag{10}$$

This allocation determines the price $P_{bs}(x_t)$ in equilibrium. Given this price, the unit reservation value v_t then follows from either inequality in (9). At this optimal stopping value, the buyer's residual rent y_{bs} just satisfies the inequalities in (9) with the Nash bargaining price in (10).

EQUILIBRIUM

In this section the previous problem is solved. The solution includes the utility functions of the representative buyer U_b, matched owner U_{bs}, and seller, U_s, as well as the pricing function P_{bs} and the unit reservation value v_t. A non-cooperative bargaining solution to the seller's relative share of the surplus λ is identified. Finally, the expected time to sale of an advertised asset and the expected fraction of assets for sale are determined in equilibrium. Throughout this section and the next, all subscripts are dropped from the expectation E_x, the unit reservation value v_t, and the common component of rent x_t, and the current values are identified simply as E, v, and x.

The expected utilities of the representative buyer, matched owner and seller must satisfy three differential equations. These equations are derived as follows. In (4) to (6), divide all terms by $\Delta t > 0$; let $\Delta t \to 0$; and exploit the previously assumed continuity of the unit reservation value v. For buyer b the expected utility U_b, from (4), must satisfy the differential equation:

$$0 = \frac{1}{2}\sigma^2 x^2 U_b'' + \mu x U_b' - \iota U_b + \alpha\left(\frac{\eta}{v}\right)^{\theta} E\left[U_{bs} - U_b - P_{bs} \middle| y_{bs} \geq v\right] \quad \mathbf{(11)}$$

for all $x \geq 0$. Again, the expectation E is calculated conditional on the current state x. Similarly, once buyer b is matched to the asset of seller s, his expected utility U_{bs} from (5) is determined by the valuation equation:

$$0 = \tfrac{1}{2}\sigma^2 x^2 U_{bs}'' + \mu x U_{bs}' - \iota U_{bs} + \beta(U_s - U_{bs}) x y_{bs} \quad \mathbf{(12)}$$

for $x \geq 0$. Finally, for seller s the expected utility U_s from (6) satisfies the equation

$$0 = \frac{1}{2}\sigma^2 x^2 U_s'' + \mu x U_s' - \iota U_s + \alpha\rho\left(\frac{\eta}{v}\right)^{\theta} E\left[(1 - \gamma)P_{bs} - U_s \middle| y_{bs} \geq v\right] \quad \mathbf{(13)}$$

for $x \geq 0$.

These are familiar valuation equations. Only one term in each equation is new. In (11) the conditional expectation is the expected gain of switching from a buyer to an owner immediately after a purchase. If a transaction occurs, then buyer b acquires the asset of seller s and realises the net gain $U_{bs}(x) - U_b(x) - P_{bs}(x)$. Conditional on a match, a transaction occurs if and only if the residual rent y_{bs} exceeds or equals the unit reservation value v. This event occurs with the probability $(\eta/v)^{\theta}$, and a match occurs with the approximate probability α per unit of time. In (13) the final terms are similar. If a transaction occurs, then seller s leaves the market with the net gain: $(1 - \gamma)P_{bs}(x) - U_s(x)$. Also, a match occurs with the approximate probability $\alpha\rho$ per unit of time, reflecting the ratio of buyers to sellers ρ. In (12) a matched owner receives per unit of time the approximate rents xy_{bs} and loses his match with the approximate probability β per unit of time. Without a match he immediately offers his asset for sale and realises the loss of utility: $U_b(x) - U_{bs}(x)$.

The valuation functions of the representative buyer, owner and seller, as well as the pricing function and the seller's reservation value are determined by (7) to (13). As shown in the Appendix, this problem has a unique solution. The solution is linear homogeneous in the common component of the return on the asset:

$$P_{bs}(x) = p_{bs}x, \qquad V(x) = vx,$$
$$U_b(z) = u_b x, \qquad U_{bs}(x) = u_{bs}x, \qquad U_s(x) = u_s x$$

for $x \geq 0$. Here, the coefficients are $p_{bs} \equiv P_{bs}(1)$, $v \equiv V(1)$, $u_b \equiv U_b(1)$, $u_{bs} \equiv U_{bs}(1)$, and $u_s \equiv U_s(1)$. Henceforth, these constants are called unit values. The linear homogeneity in (14) follows from the proportionality in the geometric Wiener process (2), the independence between the stochastic increment in (2) and the random variable in (3), the linear homogeneity of the owner's utility U_{bs} in the rent xy_{bs}, and the boundary conditions, (7) and (8). Specifically, the linear homogeneous solution (14) satisfies the system of differential equations, (11) to (13); the boundary conditions, (7) and (8); and the Nash bargaining conditions, (9) and (10). The remaining, non-linear solutions to the differential equations, (11) to (13), are precluded by the boundary conditions, (7) and (8).

The constants in this solution are identified below, starting with the unit reservation value. As shown in the Appendix, the unit reservation value in (14) is the constant:

$$v = \eta[\max(1,\phi)]^{1/\theta} \qquad \textbf{(15)}$$

In (15) the new constant ϕ is

$$\phi \equiv \frac{\alpha}{(1-\gamma+\lambda)(\theta-1)(\iota-\mu)}\left[(1-\gamma+\lambda\rho) - \frac{\beta\lambda\rho(1-\gamma)}{\beta+\iota-\mu}\right]$$

As before, a matched buyer and seller agree to trade if and only if the residual rent exceeds or equals the unit reservation value: $y_{bs} \geq v$. Thus, a transaction can fail to occur if and only if $v > \eta$ or, equivalently, $\phi > 1$. With $\phi > 1$, the unit reservation value (15) simplifies to $v = \eta\phi^{1/\theta}$. For this case, comparative statics can be calculated analytically when the seller's relative share λ is a constant, defined independently of all other parameters. This occurs with either the Nash cooperative bargaining solution or the non-cooperative bargaining solution with equal numbers of buyers and sellers, $\rho = 1$. If the relative share λ is constant, then the unit reservation value v is increasing in both the average arrival rate of buyers α and the proportional transaction cost paid by sellers γ. Also, it is decreasing in both the average rate at which matches are broken β and the difference between the riskless rate of interest and the expected growth rate of the common component, $\iota - \mu$. Finally, with the Nash cooperative solution the reservation value v is increasing in the ratio of buyers to sellers ρ.

Using this reservation value, the utilities per unit in (14) can be calculated as shown in the Appendix. These values, u_b, u_{bs} and u_s, are the constants:

$$u_b = \frac{1 - \gamma}{1 - \gamma + \lambda} \frac{\alpha\eta^\theta v^{1-\theta}}{(\theta - 1)(\iota - \mu)(\beta + \iota - \mu)} \quad \textbf{(16)}$$

$$u_{bs} = \frac{\beta u_s + y_{bs}}{\beta + \iota - \mu} \quad \textbf{(17)}$$

and

$$u_s = \lambda\rho u_b \quad \textbf{(18)}$$

The constants, (16) to (18), apply respectively to buyer b, and owner with the match bs and seller s. With (14) to (18), the solution is effectively separated into two components: a common, intertemporal component x in (14) and an idiosyncratic, constant component in (15) to (18).

The constants in (15) to (18) are effectively bargaining solutions in steady state. In this steady state the common component of rent x is constant and conveniently normalised at $x = 1$. In (16) the utility of buyer b is the expected unit payoff per period from becoming an owner, $\alpha(\eta/v)^\theta \, E[u_{bs} - u_b - p_{bs} \mid y_{bs} \geq v]$, discounted as an annuity at the rate, $\iota - \mu$. This result follows from (A15) in the Appendix. The complicated expression in (16) also reflects the solution for the unit reservation value v, the owner's utility per unit u_{bs} and the price per unit p_{bs}. In (17) the utility of an owner with the match bs is the unit payoff per period from ownership y_{bs} plus the expected unit payoff from becoming a seller βu_s, discounted as an annuity at the rate, $\beta + \iota - \mu$. This discount rate reflects the average arrival of broken matches β. In (18) the utilities of buyers and sellers are proportional, with a constant of proportionality equal to the ratio of buyers to sellers ρ times the seller's relative share of the surplus λ. When buyers and sellers are equal in number, the two utilities are equal: $\rho = 1 = \lambda$. Finally, the unit reservation value (14) follows from the utilities of the representative buyer and seller, as shown in (14), (16) and (17).

The remaining constant to be determined in equilibrium is the price per dollar of the common component. This unit price p_{bs} is calculated in the Appendix from the Nash bargaining price in (10) and the utilities per unit,

(16) to (18). It is

$$p_{bs} = \frac{\psi v + \lambda y_{bs}}{(1 - \gamma + \lambda)(\beta + \iota - \mu)} \quad \textbf{(19)}$$

In (19) the new constant ψ is

$$\psi \equiv \lambda(1 - \gamma)\min(1,\phi)\frac{\rho(\beta + \iota - \mu) - [\iota + \beta(1 - \lambda\rho) - \mu]}{\lambda\rho(\beta + \iota - \mu) + (1 - \gamma)[\iota + \beta(1 - \lambda\rho) - \mu]}$$

For some values of the parameters, a poorly matched buyer and seller prefer not to trade. If the unit value (15) satisfies $v > \eta$ and thereby $\phi > 1$, then both buyer b and seller s reject all matches bs with $\eta \leq y_{bs} \leq v$. Otherwise, they reject no matches: $v = \eta$. The former case is not only more plausible than the latter case but also less complicated expositionally than the general case, $v \geq \eta$. For this reason the parameters in (15) are assumed henceforth to satisfy $\phi > 1$. The subsequent results hold for the more general case, $\phi \geq 1$, with the constant ϕ everywhere replaced by the constant $\max(1, \phi)$.

Typically, average prices are reported in public statistics. If the average residual rent of all successful buyers is $\bar{y}$, then from (19) the average price is $\bar{p}x$, with $\bar{p} \equiv (\psi v + \lambda \bar{y})/[(1 - \gamma + \lambda)(\beta + \iota - \mu)]$. As the number of buyers and sellers becomes large, the average residual rent of successful buyers $\bar{y}$ converges almost surely to the conditional mean, $E(y_{bs} \mid y_{bs} \geq v) = \theta v/(\theta - 1)$. In turn, the average unit price $\bar{p}$ converges almost surely to the expected unit price:

$$\bar{p} \to E(p_{bs} | y_{bs} \geq v) = \frac{\psi + \dfrac{\lambda\theta}{\theta - 1}}{(\beta + \iota - \mu)(1 - \gamma + \lambda)} v \quad \textbf{(20)}$$

The seller's relative share λ is a constant with either the Nash cooperative bargaining solution or the subsequent non-cooperative bargaining solution. In either case, the expected unit price on the right side of (20) is a positive constant. In this model, all buyers and sellers are identical and all potential matches are identical *ex ante*. Conditional on a sale, the representative asset then has the expected price $\bar{p}x$ from (20). For an illiquid asset this expectation is a measure of its current market value to risk-neutral investors. With (9) and (14), it is greater than the seller's valuation $u_s x$ in (18) and less than owners' average valuation $E(u_{bs} \mid y_{bs} \geq v)x$ from (17). Moreover, this measure is consistent with common practice. When searching for a listing agent, homeowners commonly ask prospective brokers: "At what price can I expect to sell my house?" or "What is my house worth?" An honest answer would be an expected price, like (20).

The pricing functions, (19) and (20), have several implications. The average price $\overline{px}$ implied by (19) is a noisy, proportional random walk. The noise arises from the variation through time in the average residual rent of successful buyers $\overline{y}$. It disappears asymptotically in large markets with many concurrent trades of comparable assets: $\overline{y} \to \theta v/(\theta - 1)$. In the limit the expected price from (20) follows the same geometric Wiener process (2) as the common component of rent x. Also, illiquid assets have a positive liquidity premium. From (2) and (20), the value of the representative asset grows at the expected rate μ. If the utilities, (16) to (18), and thereby the unit price from (19) are to be finite, then this mean μ must be less than the riskless rate of interest ι. An unmatched seller expects the rate of return μ but receives no rent, $y_{bs} = 0$, and hence earns the negative premium, $\mu - \iota < 0$. By contrast, a matched owner expects on his asset the total yield: $\mu + E[y_{bs} \mid y_{bs} \geq v]/E[p_{bs} \mid y_{bs} \geq v]$. This yield is calculated using the expected unit price in (20) as the measure of current market value. This produces prior to a purchase the positive liquidity premium:

$$\mu - \iota + \frac{E[y_{bs} | y_{bs} \geq v]}{E[p_{bs} | y_{bs} \geq v]} = \mu - \iota + \frac{\theta(\beta + \iota - \mu)(1 - \gamma + \lambda)}{(\theta - 1)b + \lambda\theta} > 0 \quad \textbf{(21)}$$

This premium must be positive because the matched owner must be compensated for his expected loss after his match is broken. The magnitude of the premium must compensate owners for both the time spent searching for matches and the proportional cost of transactions.

The expected time to sale for the representative asset is determined in equilibrium. For the asset of seller s, the remaining time on the market t_s^m is distributed exponentially under the assumptions above (1). With this distribution the expected time to sale is the inverse of the Poisson arrival rate calculated from (1), (3), and (15):

$$E(t_s^m) = \frac{\phi}{\alpha\rho} \quad \textbf{(22)}$$

As the number of assets offered for sale increases, $S_t \to \infty$, the average time to sale converges almost surely to (22). If $\rho = 1$, so that $\lambda = 1$, then the expected time to sale (22) is independent of the average arrival rate of buyers α, increasing in the proportional transaction cost γ, and decreasing in both the average arrival rate of broken matches β and the excess rate of return $\iota - \mu$.

The expected fraction of assets on the market is also determined in equilibrium. Under the above assumptions, the expected fraction of all assets that are offered for sale at any one time is equal to the expected fraction of time that any one asset is offered for sale. Let t_s^n be the remaining time that asset s is off the market or, equivalently, the interval to its next time on the market. From the assumptions in the first section, the time not on the market t_s^n has an exponential distribution with the mean, $1/\beta$. This distribution is independent of the above exponential distribution for the time on the market t_s^m. As shown in the Appendix, the share of time that an asset spends on the market, $m_s \equiv t_s^m/(t_s^m + t_s^n)$, has a truncated β-distribution of the second kind with the mean[10]:

$$E(m_s) = \frac{\beta\phi}{\alpha\rho - \beta\phi}\left[\frac{\alpha\rho}{\alpha\rho - \beta\phi}\ln\left(\frac{\alpha\rho}{\beta\phi}\right) - 1\right] \tag{23}$$

This mean is the expected fraction of time spent on the market by any one asset. Under the above assumptions, it equals the expected fraction of all assets on the market at any one time. In larger markets with more buyers and sellers, the average time of each asset on the market and thereby the average fraction of all assets on the market converges almost surely to (23).

The seller's share of the surplus from a match depends on the Nash bargaining solution in (10). As previously mentioned, the Nash bargaining solution can be viewed as either axiomatic or the equilibrium of a non-cooperative bargaining game with symmetric information. This equilibrium can be derived as the limit in continuous time of a sequence of unique, perfect equilibriums to bargaining games in discrete time with costly delay. Given the linear homogeneity in (14), the surplus per dollar of the state x can be divided as in the steady-state solution of Rubinstein and Wolinksy (1985) and Binmore and Herrero (1988a, 1988b).[11] In this non-cooperative bargaining solution, the seller's relative share of the surplus λ is

$$\lambda = \frac{\alpha\rho + \iota\phi}{\alpha + \iota\phi} \tag{24}$$

In (24) the constant riskless rate of interest per unit of time ι is the common discount rate of buyers and sellers, and the constants, α/ϕ and $\alpha\rho/\phi$, are the Poisson arrival rates in equilibrium of transactions for buyers and sellers, respectively. Given equal numbers of buyers and sellers, $\rho = 1$, a matched buyer and seller share the surplus equally, $\lambda = 1$. With larger ratios of buyers to sellers ρ, transactions occur more rapidly on average for sellers relative to buyers, so that sellers can bargain successfully for larger relative shares λ in (24).

With the non-cooperative bargaining solution (24), the comparative statics can be complicated, especially for the expected pricing coefficient (19). Consequently, all comparative statics are computed numerically for plausible values of the parameters. In Table 1 buyers and sellers are equal in number, $\rho = 1$, so that the surplus is always shared equally, $\lambda = 1$. In Table 2 demand is either low, $\rho = 0.8$, or high, $\rho = 1.2$, so that the seller's relative share λ is affected by the values of the parameters in (24). The parameters

Table 1 Expected price per dollar of common rent, expected liquidity premium, expected time to sale and expected fraction of assets on the market.

Parameter values for base case: medium demand

$\alpha = 10000$ $\beta = 0.10$ $\gamma = 0.10$ $\eta = 0.95$

$\theta = 20$ $\iota = 0.08$ $\mu = 0.02$ $\rho = 1.0$

α average arrival rate of buyers			β average rate of broken matches		
5000	**10000**	**15000**	**0.05**	**0.10**	**0.15**
6.880	7.123	7.269	9.383	7.123	5.734
0.1572	0.1572	0.1572	0.1058	0.1572	0.2085
0.6175	0.6175	0.6175	0.6883	0.6175	0.5804
0.1295	0.1295	0.1295	0.0887	0.1295	0.1596

γ proportional transaction cost			η minimum residual rent		
0.05	**0.10**	**0.15**	**0.90**	**0.95**	**1.0**
7.0169	7.123	7.232	6.748	7.123	7.498
0.1604	0.1572	0.1541	0.1572	0.1572	0.1572
0.6101	0.6175	0.6253	0.6175	0.6175	0.6175
0.1285	0.1295	0.1305	0.1295	0.1295	0.1295

θ exponent of matching probability			ι riskless rate of interest		
10	**20**	**30**	**0.04**	**0.08**	**0.12**
12.35	7.123	5.997	11.52	7.123	5.160
0.1605	0.1572	0.1561	0.1208	0.1572	0.1935
1.304	0.6175	0.4046	1.593	0.6175	0.0983
0.2013	0.1295	0.0988	0.2245	0.1295	0.4017

μ expected growth rate of rent			ρ ratio of buyers to sellers		
0.00	**0.02**	**0.04**	**0.8**	**1.0**	**1.2**
5.986	7.123	8.792	5.269	7.123	8.867
0.1754	0.1572	0.1390	0.2343	0.1572	0.1142
0.4848	0.6175	0.8706	0.8044	0.6175	0.4968
0.1111	0.1295	0.1596	0.1523	0.1295	0.1129

In each vertical quadruplet, the top number is the expected price per dollar of the common component of rent (20). The second number is the expected liquidity premium (21). The third number is the expected time to sale (22). The bottom number is the expected fraction of assets on the market (23). The column heading is the value of the indicated parameter, and the values of all other parameters are indicated in the base case. The base case with medium demand has an equal number of buyers and sellers: $\rho = 1$. In the base case the residual rent has a mean of 1. The unit of time is one year. The parameters are defined in more detail below Table 2.

in the base case are intended to represent owner-occupied residences. For residences the large average arrival rates α in the table include searches by buyers through multiple listing services without subsequent site visits. Most of the comparative statics are consistent with conventional wisdom. For example, more turnover of owner-occupied housing, as reflected in a higher average rate of broken matches β, reduces the expected transaction price of owner-occupied housing (20), raises the required liquidity pre-

Table 2 Expected price per dollar of common rent, expected liquidity premium, expected time to sale, expected fraction of assets on the market and seller's fraction of the surplus

Parameter values for base case: low demand

$\alpha = 10000$ $\beta = 0.10$ $\gamma = 0.10$ $\eta = 0.95$
$\theta = 20$ $\iota = 0.08$ $\mu = 0.02$ $\rho = 0.8$

α average arrival rate of buyers			**β average rate of broken matches**		
5000	**10000**	**15000**	**0.05**	**0.10**	**0.15**
5.089	5.269	5.377	7.135	5.269	4.178
0.2343	0.2343	0.2343	0.1583	0.2343	0.3101
0.8044	0.8044	0.8044	0.8839	0.8044	0.7627
0.1523	0.1523	0.1523	0.1047	0.1523	0.1871
0.4475	0.4475	0.4475	0.4477	0.4477	0.4473
γ proportional transaction cost			**η minimum residual rent**		
0.05	**0.10**	**0.15**	**0.90**	**0.95**	**1.0**
5.136	5.269	5.409	4.991	5.269	5.546
0.2417	0.2343	0.2268	0.2343	0.2343	0.2343
0.7969	0.8044	0.8123	0.8044	0.8044	0.8044
0.1514	0.1523	0.1523	0.1523	0.1523	0.1523
0.4474	0.4475	0.4475	0.4475	0.4475	0.4475
θ exponent of matching probability			**ι riskless rate of interest**		
10	**20**	**30**	**0.04**	**0.08**	**0.12**
9.308	5.269	4.408	8.204	5.269	3.893
0.2337	0.2343	0.2343	0.1783	0.2343	0.2895
1.694	0.8044	0.5274	2.119	0.8044	0.5177
0.2321	0.1523	0.1173	0.2605	0.1523	0.1159
0.4505	0.4475	0.4465	0.4483	0.4475	0.4474
μ expected growth rate of rent			**ρ ratio of buyers to sellers**		
0.00	**0.02**	**0.04**	**1.0**	**1.2**	**1.4**
4.471	5.269	6.421	3.431	5.269	7.123
0.2625	0.2343	0.2057	0.3930	0.2343	0.1572
0.6279	0.8044	1.143	1.126	0.8044	0.6175
0.1309	0.1523	0.1869	0.1854	0.1523	0.1295
0.4468	0.4475	0.4486	0.3829	0.4475	0.5000

In each vertical quintuplet, the top four numbers match the entries in the previous table. The bottom number is the seller's fraction of the surplus, $\lambda/(1+\lambda)$, calculated from (24). This base case with low demand has fewer buyers than sellers $\rho = 0.8$. The subsequent base case with high demand has more buyers than sellers: $\rho = 1.2$. In all other details Tables 1 and 2 are identical.

mium (21), reduces the expected time to sale (22), and raises the expected fraction of houses on the market (23). Similarly, higher demand for housing, as reflected in a higher ratio of buyers to sellers ρ, raises the expected price (20) but reduces the liquidity premium (21), the expected time to sale (22) and the expected fraction of assets for sale.

Table 2 (continued)

Parameter values for base case: high demand

$\alpha = 10000$ $\beta = 0.10$ $\gamma = 0.10$ $\eta = 0.95$
$\theta = 20$ $\iota = 0.08$ $\mu = 0.02$ $\rho = 1.2$

α average arrival rate of buyers			**β average rate of broken matches**		
5000	**10000**	**15000**	**0.05**	**0.10**	**0.15**
8.565	8.867	9.048	11.39	8.867	7.266
0.1142	0.1142	0.1142	0.0764	0.1142	0.1518
0.4968	0.4968	0.4968	0.5607	0.4968	0.4633
0.1129	0.1129	0.1129	0.7722	0.1129	0.1393
0.5436	0.5436	0.5436	0.5433	0.5436	0.5437

γ proportional transaction cost			**η minimum residual rent**		
0.05	**0.10**	**0.15**	**0.90**	**0.95**	**1.0**
8.825	8.867	8.910	8.400	8.867	9.334
0.1149	0.1142	0.1135	0.1142	0.1142	0.1142
0.4896	0.4968	0.5044	0.4968	0.4968	0.4968
0.1118	0.1129	0.1140	0.1129	0.1129	0.1129
0.5436	0.5436	0.5435	0.5436	0.5436	0.5436

θ exponent of matching probability			**ι riskless rate of interest**		
10	**20**	**30**	**0.04**	**0.08**	**0.12**
15.20	8.867	7.494	14.92	8.867	6.304
0.1185	0.1142	0.1128	0.1972	0.1142	0.1399
1.050	0.4968	0.3253	1.257	0.4968	0.0856
0.1782	0.1129	0.0854	0.1972	0.1129	0.3262
0.5416	0.5436	0.5442	0.5431	0.5436	0.5436

μ expected growth rate of rent			**ρ ratio of buyers to sellers**		
0.00	**0.02**	**0.04**	**1.0**	**1.2**	**1.4**
7.378	8.867	11.10	7.123	8.867	13.24
0.1269	0.1142	0.1015	0.1572	0.1142	0.0784
0.3920	0.4968	0.1394	0.6175	0.4968	0.5746
0.0967	0.1129	0.6955	0.1295	0.1129	0.1238
0.5440	0.5436	0.5429	0.5000	0.5436	0.5791

α – average arrival rate of buyers, equal to the probability per unit of time that the representative buyer inspects the asset of some seller.

β – average rate of broken matches, equal to the probability per unit of time that the match between the representative owner and his asset terminates.

γ – proportional transaction cost, equal to the total transaction cost divided by the sales price.

η – minimum rent, equal to the lower support of the Pareto distribution (3) for the residual rent.

θ – exponent of matching probability, equal to the exponent in the Pareto distribution (3).

ι – riskless rate of interest equal to the discount rate per unit of time for each buyer and seller.

μ – expected growth rate of rent, equal to the proportional drift in the diffusion process (2).

ρ – ratio of buyers to sellers, equal to the current number of buyers divided by the current number of sellers.

Some of the comparative statics are surprising. All of these surprising results reflect the subtle interaction in equilibrium between parameters and the expected price (20). For example, assets with higher average arrival rates of potential buyers α are more liquid. Surprisingly, these more liquid assets do not have lower liquidity premia (21). Instead, these assets have higher unit reservation values (15) and thereby higher expected unit prices (20). Moreover, the higher prices exactly offset the more rapid arrival of buyers, so that the extra liquidity alters in equilibrium neither the liquidity premium (21), the expected time to sale (22), nor the expected fraction of assets for sale (23). For the same reason, assets with lower proportional transaction costs λ have higher, not lower liquidity premiums. Again, the unit reservation values (15) are sufficiently higher in equilibrium, so that higher transaction costs are reflected in higher prices (20), not higher liquidity premiums (21). Also, increments in the proportional transaction cost γ decrease the net expected price $(1 - \gamma)\pi x$ more with higher ratios of buyers to sellers ρ. This is consistent with the result in the fourth section that brokers are more valuable in markets with persistently low demand, $\rho < 1$, than in markets with persistently high demand, $\rho > 1$.

DEVELOPMENT

Developers and owners of undeveloped assets must also search for matches. Given the previous results, this search can be modelled very simply, as follows. Potential developers appear at the property of the representative owner with the Poisson arrival rate δ. The constant, $\delta > 0$, replaces the product $\alpha\rho$ in the Poisson density (1). If a developer arrives at time t, then he proposes to transform the undeveloped asset with the current market value V(x) into a developed asset with the current market price πx. For the undeveloped asset the market value V(x) in this section is distinct from and should not be confused with the reservation value from the previous sections. For the developed asset the constant, $\pi > 0$, is the current market price per dollar of the common component x. This pricing function requires a perfectly competitive market for developers. Each developer must regard both the cost of development and the expected price of a developed asset as independent of his decision to develop.[12] Without further loss of generality, all development can then be assumed to be instantaneous, once it is started.[13] Also, all development is assumed to occur at a fixed density, such as the legal maximum.[14] Under these assumptions, if the developed asset is immediately offered for sale, then its market price per dollar of common rent is $\pi = u_s$, the seller's unit utility in (18). Given a fixed density, the cost of development to the developer is specified exogenously as the constant, $\kappa > 0$.[15] If the owner of the undeveloped asset receives an offer to develop at time t, then he can immediately accept the offer with any probability q satisfying $0 \leq q \leq 1$. Otherwise, the owner rejects the offer.

Once an owner and developer meet, they must bargain to determine the price of development. The price or cost of development $C(x)$ determines the allocation between the owner and developer of the total profit or capital gain from development, $\pi x - \kappa - V(x)$. With the cost $C(x)$ the owner has the capital gain, $\pi x - C(x) - V(x)$, and the developer earns the profit, $C(x) - \kappa$. Again, the Nash cooperative bargaining solution is used, and the owner is assumed to receive the fraction ω of the total gain, while the developer is allocated the residual fraction, $1 - \omega$. The constant, $0 < \omega < 1$, replaces the seller's share, $\lambda/(1 + \lambda)$, in the previous analysis. This Nash bargaining solution determines the cost of development;

$$C(x) = \omega\kappa + (1 - \omega)[\pi x - V(x)] \tag{25}$$

for all $x \geq 0$.

This simple specification of the cost of development is consistent with equilibrium under the following conditions. Suppose that each owner of an undeveloped asset currently accepts a developer's offer with the optimal probability $Q(x)$. Also, represent by $R(x)$ the current ratio of developers' aggregate capacity to units of undeveloped assets. Given the Poisson matching of developers and owners, each developer currently earns per unit of time on each unit of capacity the expected operating revenue: $\delta Q(x)[C(x) - \kappa]/R(x)$. Finally, assume that to maintain his capacity each developer incurs per unit of time the constant cost $K(x)$. If developers can adjust capacity either up or down without additional cost, then the relative number of developers $R(x)$ must satisfy in equilibrium the relationship: $R(x) = \delta Q(x)[C(x) - \kappa]/K(x)$. Otherwise, developers would adjust their collective capacity. In this case, the constant arrival rate of developers δ is consistent with competitive entry and exit.

With these simplifying assumptions, the value of the undeveloped asset must satisfy a familiar differential equation:

$$0 = \max_{0 \leq q \leq 1} \left\{ \frac{1}{2}\sigma^2 x^2 V'' + \mu x V' - \iota V + \delta\omega q(\pi x - \kappa - V) \right\} \tag{26}$$

for all $x \geq 0$. For some finite constant, $\zeta > 0$, this differential equation must satisfy the usual boundary conditions:

$$V(0) = 0, \qquad V(x) \leq \zeta x \tag{27}$$

as $x \to \infty$. This problem differs in only one detail from a standard option pricing problem without costly search. In (26) the last term is the expected benefit from immediate development. It equals the probability per unit of time of a developer's offer δ times the owner's probability of acceptance q times the expected benefit conditional on immediate development. With the Nash bargaining solution (25), this conditional benefit is the expected value of the developed property πx minus the cost of development κ minus the foregone value of the undeveloped property V(x).

As shown in the Appendix, this problem has a unique solution. The owner of the undeveloped asset optimally accepts an offer to develop with the probability;

$$Q(x) = \begin{cases} 0 & 0 \leq x < x^* \\ 1 & x^* \leq x < \infty \end{cases} \tag{28}$$

At all values of the common component of rent on the developed asset x below the optimal switching point x^*, the owner rejects the offer to develop. Otherwise, he always accepts the offer. The optimal switching point in (28) is the constant:

$$x^* \equiv \frac{\kappa}{\pi} \frac{1 + \dfrac{\xi_1}{\xi_0 - \xi_1} \dfrac{\delta\omega}{\delta\omega + \iota}}{1 - \dfrac{1 - \xi_1}{\xi_0 - \xi_1} \dfrac{\delta\omega}{\delta\omega + \iota - \mu}} \tag{29}$$

The constant (29) depends on two new parameters:

$$\xi_0 \equiv \left(\frac{1}{2} - \frac{\mu}{\sigma^2}\right) + \sqrt{\left(\frac{1}{2} + \frac{\mu}{\sigma^2}\right)^2 + 2\frac{\iota - \mu}{\sigma^2}}$$

and

$$\xi_1 \equiv \left(\frac{1}{2} - \frac{\mu}{\sigma^2}\right) - \sqrt{\left(\frac{1}{2} - \frac{\mu}{\sigma^2}\right)^2 + 2\frac{\iota + \delta\omega}{\sigma^2}}$$

These constants satisfy the inequalities: $\xi_0 > 1$ and $\xi_1 < 0$. Given the optimal development policy in (28) and (29), the undeveloped asset has the value:

$$V(x) = \begin{cases} (\pi x^* - \kappa)\left(\dfrac{x}{x^*}\right)^{\xi_0} & 0 \leq x < x^* \\ \delta\omega\left(\dfrac{\pi}{\delta\omega + \iota - \mu} x - \dfrac{\kappa}{\delta\omega + \iota}\right) & \\ + \left(\dfrac{\iota - \mu}{\delta\omega + \iota - \mu} \pi x^* - \dfrac{\iota\kappa}{\delta\omega + \iota}\right)\left(\dfrac{x}{x^*}\right)^{\xi_1} & x^* < x < \infty \end{cases} \tag{30}$$

This solution is shown in Figure 1. As illustrated, the valuation function V is everywhere increasing and strictly convex in the common component

Figure 1 Cost of development and value of an undeveloped asset

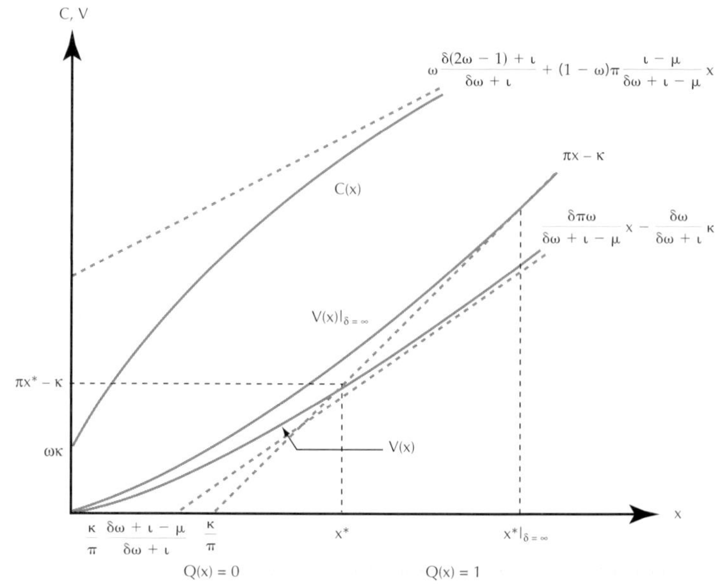

The cost of development $C(x)$ and the value of an undeveloped asset $V(x)$ depend on the current common component of rent x from a developed asset. Developers arrive at the Poisson rate δ and offer to transform the undeveloped asset into a developed asset at the price or cost $C(x)$. For a developer the cost is κ. The owner always accepts the current offer if and only if the expected current price of the developed asset minus the cost of development exceeds or equals the current value of the undeveloped asset: $Q(x) = 1$ if and only if $x \geq V(x)$. This occurs if and only if the common component exceeds or equals the optimal switching point: $x \geq x^*$. Otherwise, the owner always rejects the offer, $Q(x) = 0$. As developers arrive on average more frequently, $\delta \to \infty$, the solution converges to the standard option price without costly search: $V(x) \to V(x)|_{\delta = \infty}$ and $x^* \to x^*|_{\delta = \infty}$.

of rent x. It crosses the exercise value, max $[0, \pi x - \kappa]$, at the optimal switching point (29). That is, from the maximand in (26) switching optimally occurs when the value of the undeveloped asset equals the value of the developed asset minus the cost of development, $V(x^*) = \pi x^* - \kappa$. At all lower values of the common component, $0 \leq x < x^*$, the value of the undeveloped asset exceeds the value of the developed asset minus the cost of development, and the owner optimally rejects in (28) all offers of development. At higher values of x, the reverse relationship holds and the owner accepts all offers to develop. Also, as $x \to \infty$, the valuation function V approaches asymptotically the unexercised value, $\max\{0, [\delta\pi\omega/(\delta\omega + \iota - \mu)]x - [\delta\kappa\omega/\delta\omega + \iota]\}$. Both the switching point x^* and the valuation function V are increasing in the arrival rate of developers δ, the owner's share of the surplus from development ω, and the variance σ^2 of the growth rate in the common component of rent x. In turn, these properties of the valuation function (30) determine through the Nash bargaining solution the properties of the development cost (25). As illustrated in Figure 1, the cost function C is increasing, strictly concave, and asymptotic to its linear upper bound. It is also decreasing in the above parameters, δ, ω and σ^2.

The solution is related to the standard option pricing result without costly search. As potential developers arrive increasingly more frequently, $\delta \to \infty$, the optimal switching point (29) converges to the optimal exercise point without costly search:

$$x^* \to \frac{\kappa}{\pi}\frac{\xi_0}{\xi_0 - 1} \tag{31}$$

Simultaneously, the value of the undeveloped property (30) converges to the corresponding value without costly search:

$$V(x) \to (\pi x^* - \kappa)\left(\frac{x}{x^*}\right)^{\xi_0} \tag{32}$$

for $0 \le x \le x^*$. Alternatively, as developers arrive less frequently, $\delta \to 0$, the undeveloped asset becomes worthless and switching occurs as soon as the option to develop comes into the money: $V(x) \to 0$ and $x^* \to \kappa/\pi$. The optimal switching point x^* in (29) is increasing in the Poisson arrival rate of developers δ, bounded below by κ/π and bounded above by (32).

EXTENSIONS

Brokerage

Buyers and sellers now have a choice. Each buyer b can search for an asset either with or without a broker. Similarly, each seller s can offer his asset for sale with a broker and then pay the proportional transaction cost γ upon a sale or, alternatively, sell without a broker and thereby pay only a minimal closing cost. For notional convenience, the latter cost is set equal to zero. For both buyers and sellers, brokers can improve the efficiency of search.[16] Buyers arrive more slowly without brokers α_0 than with brokers $\alpha_1 : 0 < \alpha_0 < \alpha_1 < 1$. The resulting expected utilities per unit from the previous section are u_{bi} and u_{si} for buyer b and seller s. The index i identifies the presence of a broker; the representative buyer or seller has no broker if $i = 0$ and a broker if $i = 1$. In the previous equilibrium buyer b searches with a broker if and only if $u_{b0} \le u_{b1}$. Similarly, seller s searches with a broker if and only if $u_{s0} \le u_{s1}$. Because buyers and sellers have proportional utilities in (18), buyer b optimally uses a broker if and only if seller s optimally employs a broker. This result also holds if brokers improve the efficiency of search only when both the buyer and seller have a broker. The latter assumption may be more realistic for one important category of assets. With single-family houses, only brokers have access to the local multiple listing service. In this case, a broker for a buyer (seller) may search more efficiently than a non-broker only for a seller (buyer) with a broker.

Conditions under which brokers are used in equilibrium can now be identified. Let λ_i represent the seller's relative share (24) with $i = 0, 1$. Again, select values of the parameters such that some matches do not produce transactions: $v > \eta$. In this case, both buyers and sellers use brokers if and only if the ratio of arrival rates, α_1/α_0, satisfies the inequality:

$$\frac{\alpha_1}{\alpha_0} \geq \frac{1 - \gamma + \lambda_1}{(1-\gamma)^\theta (1 + \lambda_0)} \left[1 - \gamma \frac{\beta(1 - \lambda_1 \rho) + \iota - \mu}{\beta + (\iota - \mu)(1 + \lambda_0 \rho)} \right]^{\theta - 1} \tag{33}$$

calculated from (15) and (16). This inequality is necessary and sufficient for brokerage in the previous equilibrium. In other words, buyers and sellers employ brokers if and only if brokers improve sufficiently the efficiency of search. If $\rho = 1$ so that $\lambda_0 = \lambda_1 = 1$, then the lower bound on the right side of (33) increases in both the arrival rate of broken matches β and the proportional transaction cost γ and decreases in the excess return $\iota - \mu$. Also, with the parameter values in the lower right corner of Table 1, the lower bound in (33) increases in the ratio of buyers to sellers ρ. In this sense, brokers are more valuable, other things being equal, in markets with longer matches, lower transaction costs, higher interest rates, lower growth rates of rent and persistently lower ratios of buyers to sellers.

Alternatively, brokers can improve the efficiency of search, not by increasing the frequency of inspections, but by decreasing the probability of a mismatch, conditional on an inspection. Typically, brokers screen assets for their clients. With single-family houses, for example, brokers for buyers can select sites to be visited based on characteristics of design, construction, maintenance and neighbourhoods. This observable heterogeneity can be introduced very simply into the previous model, as follows. Suppose that brokers increase for the buyers the lower bound in the matching distribution (2) from η_0 to η_1. In this second case, both buyers and sellers use brokers if and only if the ratio of lower bounds, η_1/η_0, satisfies the inequality:

$$\frac{\eta_1}{\eta_0} \geq \left[\frac{1 - \gamma + \lambda_1}{(1 - \gamma)(1 + \lambda_0)} \right]^{1/\theta} \left[\frac{\frac{1 - \gamma + \lambda_1 \rho}{1 - \gamma}(\beta + \iota - \mu) - \beta \rho \lambda_1}{(1 + \lambda_0 \rho)(\beta + \iota - \mu) - \beta \lambda_0 \rho} \right]^{1 - 1/\theta} \tag{34}$$

also calculated from (15) and (16). Again, this inequality is necessary and sufficient for brokerage in the previous equilibrium. If $\rho = 1$ so that $\lambda_0 = \lambda_1 = 1$, then the lower bound on the right side of (34) also increases in both β and γ and decreases in $\iota - \mu$. Again, with the parameter values in the lower right corner of Table 1, the lower bound in (34) increases in the ratio ρ. Thus, the comparative statics from (33) and (34) are identical.

Heterogeneous assets

Previously, the representative category of assets was homogeneous in the following sense: before an inspection the residual rent of buyers y_{bs} was independently and identically distributed across all assets within the category. In fact, buyers or their brokers can observe some attributes of assets before an inspection. Again, with single-family houses these observable attributes can include characteristics of design, construction, maintenance and neighbourhoods. This observable heterogeneity can be introduced into the previous model as follows. Suppose that all observable attributes of the representative class of assets can be clustered into J categories: $j = 1, \ldots, J$. Also, suppose that each buyer restricts his search to one category of assets: in other words, each buyer initially chooses his preferred category of assets over which to search, based on some unmodelled heterogeneity across buyers. Categories are distinguished only by the lower bound in the Pareto distribution (3) for residual rent. For all assets in category j, this lower bound is η_j. Without further loss of generality, these lower bounds can be arranged in increasing order: $\eta_1 < \eta_2 < \cdots < \eta_J$. For example, poorly designed, constructed and maintained houses in bad neighbourhoods can be clustered in lower categories j with smaller lower bounds η_j. Otherwise, all assets in the representative class are indistinguishable by potential buyers before an inspection.

With this modification the results in the previous section are changed only slightly. For category j, the constant η is replaced everywhere by the constant η_j. This generates the utilities, u_{bj}, u_{bsj}, u_{sj}, the reservation value v_j and the price p_{bsj}. In (15) the reservation value v_j is linear in η_j, and from (19) the average price $\bar{p}_j$ is linear in v_j. As a result, the lower bound η can be interpreted as the average minimum residual rent across all assets, and v can be interpreted as the resulting average reservation value from (15). If assets for sale are drawn randomly from the population of all assets, then the average unit price, $\bar{p} \equiv \Sigma_j \bar{p}_j$, is again approximately (20) in large samples. This single change in the lower bound η does not affect the expected liquidity premium (21), the expected time to sale (22) or the expected fraction of assets on the market (23).

Hot and cold markets

In the first section the ratio of buyers to sellers was assumed to be constant through time. The constant ratio, $r_t \equiv B_t/S_t = \rho$, greatly simplified the analysis. With this strong assumption all functions are linear homogeneous in (14). In this case, the problem effectively separates into two, analytically tractable parts: an intertemporal pricing problem, as reflected in the linear functions (14), and a steady-state bargaining problem, as reflected in the constant coefficients, (15) to (19). The non-cooperative bargaining solution to the seller's relative share of the unit surplus λ can then be identified in (24). More realistically, the ratio of buyers to sellers r_t

changes stochastically over time t as conditions change in the market. Moreover, movements in the ratio r_t may be correlated with changes in the current common component of rent x_t. In at least one special case, the pricing problem can be solved explicitly. Suppose that the ratio r_t depends only on the current common component of rent, $r_t = R(x_t)$. Assume that the rental function R has only two steps, $R(x_t) = \rho_o$ for $0 \leq x_t < \xi$ and $R(x_t) = \rho_1$ for $\xi \leq x_t < \infty$. The constants in this function satisfy $0 < \xi < \infty$ and $0 < \rho_1 < 1 < \rho_1 < \infty$. In other words, the market can be either cold or hot. Demand is low, $r_t = p_0 < 1$, in a cold market, $x_t < \xi$, and high, $r_t = \rho_1 > 1$, in a hot market, $x_t \geq \xi$. In this case, the previous problem can be solved explicitly if the seller's relative share of the surplus λ is constant, as in the cooperative Nash solution.

Under these assumptions the pricing function is no longer linear homogeneous. To see this, consider the expected pricing function E(P) that replaces the expected pricing function from (20). Also, consider two other expected pricing functions: $E(P_m)$ for m = 0, 1. The first function holds in a perpetually cold market, m = 0, with low demand, $r_t = \rho_0$, whereas the second holds in a perpetually hot market, m = 1, with high demand, $r_t = \rho_1$. By the argument above (19), these two expected pricing functions are linear homogeneous in the common component x_t. The expected pricing function E(P) is then increasing everywhere, strictly convex in a cold market and strictly concave in a hot market. It is bounded below by the function in a cold market $E(P_0)$ and above by the function in a hot market $E(P_1)$.

Finally, the function E(P) is tangent to $E(P_0)$ at $x_t = 0$ and asymptotic to $E(P_1)$ as $x_t \to \infty$. As a result, both the expected growth rate of prices and the variance of the growth rate of prices are highest during intermediate states x_t. These occur during transitions between hot and cold markets.

SUMMARY

Many real assets are traded in decentralised markets. In these markets buyers, sellers or their brokers must search for matches. Once a potential buyer and seller meet, they must bargain to determine a price. This produces three related problems of valuation, one each for the representative, buyer, owner and seller. The three differential equations must be solved simultaneously with the Nash pricing function. In this chapter the problem is solved analytically under the following conditions. The rent realised on each asset is the product of a common component and a residual that is specific to the match between the buyer and seller. The common component of rent is assumed to follow a geometric Wiener process whereas the residual rent is an independent Pareto variate with an independent Poisson jump to zero when the match between the owner and his asset is broken. All buyers search independently in the same Poisson process, and all matched buyers and sellers agree on the Nash bargaining solution. Finally, all agents are risk neutral with identical discount rates, or the capital market is complete. In equilibrium the unique pricing function is shown to be linear homogeneous in the common component of cashflows, and both the pricing coefficient and the liquidity premium are identified. The expected time to sale and expected fraction of assets for sale are also derived in equilibrium. Next, a problem of optimal development with costly search is solved. Both the optimal decision to develop and the resulting value of an undeveloped asset are calculated explicitly. Several extensions are also discussed.

This chapter has three main results. First, under a reasonably plausible set of conditions, the average transaction price for a category of illiquid assets is a noisy, proportional random walk. The noise decreases with increasingly larger markets, and the process approaches a proportional random walk. Second, under the same conditions the expected liquidity premium can be calculated explicitly. The liquidity premium is positive, and its comparative statics are consistent with conventional wisdom – with some surprises. The premium is positive because a matched owner must be compensated for the time spent searching for a match and the transaction cost incurred conditional on a match. Finally, if the owner of an undeveloped asset must search for a developer, then he alters his optimal exercise policy for his option to develop. On average he develops sooner as developers arrive on average less frequently. As the average arrival rate of developers decreases, so does the market value of his undeveloped property or, equivalently, his option to develop.

Many extensions of this model are possible. In this chapter the liquidity premium can be calculated explicitly only because the portfolio problem of the representative investor is not considered. Instead, the owner is assumed to offer his illiquid asset for sale when his match is broken by an exogenous event, like death, divorce, departure of children to school or a new job in a different location. In fact, an owner may sell his illiquid asset to rebalance his portfolio. Consider Grossman and Laroque (1990). There, the representative individual can trade both risky and riskless securities without transaction costs and a consumer durable (eg, his house) subject to proportional transaction costs. The securities are perfectly divisible, but the durable is completely indivisible. The prices of the risky securities are geometric Brownian motions, and the flow of services from the real asset is proportional to its price. Also, the utility of consumption is isoelastic, and no non-durables can be consumed. Under these assumptions the optimal trading points for the durable are characterised analytically and computed numerically. Numerical solutions may also be possible with non-durable consumption, a risky durable good, and potentially binding constraints on down payments, as in Stein (1993). Similarly, it should be possible to add costly search. This would endogenise the times at which owners offer their assets for sale. Also, it would introduce issues of portfolio balance with financial constraints.

Models of intertemporal pricing with costly search may also have interesting empirical implications. Markets for many real assets, including single-family houses, have two puzzling properties: rates of return are positively correlated with previous trading volumes, and price/earnings multiples are procyclical. Neither empirical observation is consistent with the equilibriums of standard financial models. Indeed, if operating returns are procyclical and business cycles are common knowledge, then price/earnings multiples should be countercyclical, rather than procyclical. Imagine that investors first search across local markets or categories of assets, acquire and process local data, and only then buy assets. In equilibrium investors may than have poorly diversified portfolios with binding financial constraints. Also, information about markets may spread only slowly through time, and past prices may affect current demands.

APPENDIX

Characterisation of the pricing function and reservation value

From (10) the pricing function P_{bs} must satisfy

$$P_{bs} = \frac{U_s + \lambda(U_{bs} - U_b)}{1 - \gamma + \lambda} \tag{A1}$$

for $x \geq 0$. Also, with (10) the two inequalities in (9) are equivalent. The pricing function (A1) can then be eliminated from these inequalities to yield the equivalent condition:

$$U_{bs} - U_b - \frac{1}{1 - \gamma} U_s \geq 0 \tag{A2}$$

for $x \geq 0$. Given the Nash pricing function (A1), this inequality is necessary and sufficient for a transaction to occur. The reservation pricing rule – trade if and only if $y_{bs} \geq v$ – is an optimal stopping rule if and only if the left side of (A2) is decreasing in y_{bs}. In this case, the reservation value v is the smallest value y_{bs} satisfying (A2):

$$0 = \left[U_{bs} - U_b - \frac{U_s}{1 - \gamma} \right] \Bigg|_{y_{bs} = v} \tag{A3}$$

for $x \geq 0$. Subsequently, problem (4) to (10) is solved, ignoring the required monotonicity in (A2). The resulting solution is then shown to have this monotonicity.

Derivation of (14) to (19)

The system of differential equations (11) to (13) is solved, subject to the boundary conditions, (7) and (8), in several steps, as follows. First, define the function $F \equiv \lambda\rho U_b - U_s$. Equations (11) and (13) have the weighted difference

$$0 = \tfrac{1}{2}\sigma^2 x^2 F'' + \mu x F' - \iota F \qquad \textbf{(A4)}$$

Also, the boundary conditions, (7) and (8), have the weighted differences:

$$F(0) = 0, \qquad F(x) \le (\lambda\rho\zeta_b - \zeta_s)x \qquad \textbf{(A5)}$$

The differential equation (A4) has two real roots. One root is defined below (29) as the constant, $\xi_0 > 1$. The other root is identical to the second constant below (29), $\xi_1 < 0$, with one modification: $\delta\omega$ is replaced by $\alpha\lambda\rho/(1+\lambda)$. Given the boundary conditions in (A5), the coefficients associated with these roots must be zero. Thus, (A4) and (A5) have the unique solution: $F(x) = 0$. In other words, the system, (7), (8), (11) and (13), must uniquely satisfy the condition:

$$U_s = \lambda\rho U_b \qquad \textbf{(A6)}$$

With (A6) the system of differential equations simplifies to (11) and (12). The homogeneous equation associated with (12) has $xy_{bs} = 0$. As a result, any solution to the homogeneous system associated with (11) and (12) does not depend on y_{bs}. Using (A1) and (A6), this homogeneous system can be written as

$$0 = \tfrac{1}{2}\sigma^2 x^2 U_b'' + \mu x U_b' - \iota U_b + \frac{\alpha}{1-\gamma+\lambda}\left(\frac{\eta}{\nu}\right)^{\theta}\left[(1-\gamma)U_{bs} - (1-\gamma+\lambda\rho)U_b\right] \qquad \textbf{(A7)}$$

$$0 = \tfrac{1}{2}\sigma^2 x^2 U_{bs}'' + \mu x U_{bs}' - \iota U_{bs} + \beta(\lambda\rho U_b - U_{bs}) \qquad \textbf{(A8)}$$

Now, define the function $G \equiv U_b - kU_{bs}$. From (A7) and (A8), calculate the weighted difference:

$$0 = \frac{1}{2}\sigma^2 x^2 G'' + \mu x G' - \left[\iota + \beta\lambda\rho k + \alpha\frac{1-\gamma+\lambda\rho}{1-\gamma+\lambda}\left(\frac{\eta}{\nu}\right)^{\theta}\right]G + \left\{\alpha\frac{1-\gamma}{1-\gamma+\lambda}\left(\frac{\eta}{\nu}\right)^{\theta} + \left[\beta - \alpha\frac{1-\gamma+\lambda\rho}{1-\gamma+\lambda}\left(\frac{\eta}{\nu}\right)^{\theta}\right]k - \beta\lambda\rho k^2\right\}U_{bs} \qquad \textbf{(A9)}$$

Also, the boundary conditions from (A5) are

$$G(0) = 0, \qquad G(x) \le (\zeta_b - k\zeta_{bs})x \qquad \textbf{(A10)}$$

The quadratic in (A9) has one positive root,

$$k = \frac{1}{2\lambda\rho}\sqrt{\left[1 - \frac{\alpha}{\beta}\frac{1-\gamma+\lambda\rho}{1-\gamma+\lambda}\left(\frac{\eta}{\nu}\right)^{\theta}\right]^2 + 4\frac{\alpha}{\beta}\frac{1-\gamma}{1-\gamma+\lambda}\left(\frac{\eta}{\nu}\right)^{\theta}}$$

$$+ \frac{1}{2\lambda\rho}\left[1 - \frac{\alpha}{\beta}\frac{1-\gamma+\rho\lambda}{1-\gamma+\lambda}\left(\frac{\eta}{\nu}\right)^{\theta}\right] \quad \textbf{(A11)}$$

With (A11) the argument above (A6) produces for (A9) and (A10) the unique solution: $G(x) = 0$. In other words, with the boundary conditions, (7) and (8), and the constant (A11), the homogeneous system, (A7) and (A8), must uniquely satisfy

$$U_b = kU_{bs} \quad \textbf{(A12)}$$

The unique homogeneous solution to (11) to (13) can now be identified. Inserting (A12) into the remaining homogeneous equation (A8) yields

$$0 = \tfrac{1}{2}\sigma^2 x^2 U''_{bs} + \mu x U'_{bs} - [\iota + \beta(1 - \lambda\rho k)]U_{bs} \quad \textbf{(A13)}$$

The parameter (A11) satisfies $k < 1/\lambda\rho$. In this case, the argument above (A6) produces for (A13) and the associated boundary conditions in (7) and (8) the unique solution: $U_{bs}(x) = 0$. Combining this homogeneous solution for (13) with the previous results, (A6) and (A12), yields for the system, (7), (8), and (11) to (13), the unique homogeneous solution:

$$U_b = U_{bs} = U_s = 0 \quad \textbf{(A14)}$$

To complete the argument, it is sufficient to show that (14) is a particular solution to the system, (7), (8) and (11) to (13). Clearly, (14) satisfies both boundary conditions. Inserting (14) into (11), (12) and (A6) then produces the unit values:

$$u_b = \frac{\alpha}{\iota - \mu}\left(\frac{\eta}{\nu}\right)^{\theta} E[u_{bs} - u_b - p_{bs} | y_{bs} \geq \nu] \quad \textbf{(A15)}$$

(17) and (18). The associated unit price from (A1) is

$$p_{bs} = \frac{u_s + \lambda(u_{bs} - u_b)}{1 - \gamma + \lambda} \quad \textbf{(A16)}$$

Also, with (14), (17) and (18), the unit value v in (13) satisfies

$$\nu = \left[\frac{1 - \gamma + \lambda\rho}{1 - \gamma}(\beta + \iota - \mu) - \beta\lambda\rho\right]u_b \quad \textbf{(A17)}$$

In turn, (17), (18), (A16) and (A17) imply that

$$E[u_{bs} - u_b - p_{bs} \mid y_{bs} \geq \nu]$$

$$= \frac{1-\gamma}{1-\gamma+\lambda} E\left[\frac{\beta\lambda\rho u_b + y_{bs}}{\beta+\iota-\mu} - \frac{1-\gamma+\lambda\rho}{1-\gamma} u_b \middle| y_{bs} \geq \nu\right]$$

$$= \frac{1-\gamma}{1-\gamma+\lambda} \frac{\nu}{(\theta-1)(\beta+\iota-\mu)} \quad \text{(A18)}$$

To derive (16), insert (A18) into (A15) and rearrange terms. For (15) insert (16) into (A17). Finally, (19) follows from (16) to (18) and (A16). This solution satisfies the monotonicity condition in (A2).

Derivation of the mean (23)

With $m_s = t_s^m/(t_s^m + t_s^n)$, the time from purchase to the next initial advertisement t_s^n satisfies $t_s^n = t_s^m(1 - m_s)/m_s$. In this case, the conditional density of the fraction m_s is $\Pr\{m_s = m | t_s^m = t\} = \Pr\{t_s^m = [(1 - m)t/m]\}(t/m^2) = \beta(t/m^2) \exp[-\beta(1 - m)t/m]$. Thus, its density is the mixture:

$$\Pr\{m_s = m\} = \frac{\alpha\beta\rho}{am^2} \int_0^\infty t \exp\left[-\left(\beta\frac{1-m}{m} + \frac{\alpha\rho}{\phi}\right)t\right] dt$$

$$= \frac{\alpha\beta\rho}{[\alpha\rho m + \beta\phi(1-m)]^2} \quad \text{(A19)}$$

for $0 \leq m \leq 1$.

Derivation of (28) to (30)

Because the maximand in (26) is linear in q, the optimal control Q(x) is bang-bang in (28). With the probability, $q = Q(x)$, the differential equation (26) has the general solution:

$$V(x) = \begin{cases} v_0 x^{\xi_0} & 0 \leq x < x^* \\ v_1 x^{\xi_1} + \delta\omega\left(\frac{\pi}{\delta\omega+\iota-\mu}x - \frac{\kappa}{\delta\omega+\iota}\right) & x^* \leq x < \infty \end{cases} \quad \text{(A20)}$$

The remaining roots are eliminated by the boundary conditions (27). In (A20) the constants, v_0 and v_1, are uniquely determined by the continuous differentiability of V at the switching point, $x = x^*$. These constants are

$$\nu_1 = \delta\omega\left(\frac{1-\xi_1}{\xi_0-\xi_1}\frac{\pi x^*}{\delta\omega+\iota-\mu} + \frac{\xi_1}{\xi_0-\xi_1}\frac{\kappa}{\delta\omega+\iota}\right)x^{*-\xi_0} \quad \text{(A21)}$$

and

$$\nu_1 = -{}_{\delta}\omega\left(\frac{1-\xi_0}{\xi_1-\xi_0}\frac{\pi x^*}{\delta\omega+\iota-\mu}+\frac{\xi_0}{\xi_1-\xi_1}\frac{\kappa}{\delta\omega+\iota}\right)x^{*-\xi_1} \tag{A22}$$

Finally, the switching point (29) uniquely satisfies the condition:

$$V(x^*) = \pi x^* - \kappa \tag{A23}$$

Collectively, (A20) to (A23) determine the switching point (29). This switching point, combined with (A20) to (A22), yields the valuation function (30).

1 Some decentralised markets have huge trading volumes. For example, from 1989 to 1991 the annual average trading volume for single-family residences in the United States was US$395 billion for existing units and US$84 billion for new units, as reported in the *Federal Reserve Bulletin*, October 1992. This total was about one-third of the concurrent volume of round lots on the NYSE and about one-fifth of the concurrent volume on all stock exchanges in the United States. Measured in terms of the stock of assets, decentralised markets for residential real estate are even larger relative to organised stock exchanges.

2 Initially, the trading price between matched agents was specified as a cooperative Nash bargaining solution. Important examples are Diamond (1982) and Mortensen (1982). Later, the price was determined in the equilibrium of a non-cooperative bargaining game with complete information. Important papers include Rubinstein and Wolinsky (1985), Gale (1986, 1987), and Binmore and Herrero (1988a, 1988b). Incomplete information was introduced more recently in Wolinsky (1990) and Samuelson (1992).

3 Liquidity premiums are derived theoretically and documented empirically in Amihud and Mendelson (1986, 1989, 1991). Other theoretical derivatives appear in Constantinides (1986) and Vayanos and Vila (1992). These models have costly transactions but no costly search. The choice between searching for transactions and trading in a central market is explored in Pagano (1989).

4 Without exception the literature on real options has ignored costly search. Applications of option pricing theory to real assets include Brennan and Schwartz (1985), Titman (1985), McDonald and Siegel (1985, 1986), Majd and Pindyck (1987), Paddock, Siegel, and Smith (1988), Clarke and Reed (1988), Dixit (1989), Sick (1989), Capozza and Sick (1991), Williams (1991, 1993), and Bar-Ilan, Sulem and Zanello (1992).

5 Exogenous sales avoid a difficult stopping problem: the optimal values of the continuous state variable x_t at which to offer an illiquid, indivisible asset for sale, subject to proportional transaction costs. To solve this problem, Grossman and Laroque (1990) must make strong assumptions: the homeowner consumes no non-durables and effectively regards his house as a consumer durable and not as a risky asset in his portfolio.

6 In Stein (1993) the latter point is discussed, but the difficult stopping problem with an endogenous pricing process for properties is not solved.

7 Prepayment rates on residential mortgages show a strong season effect during the first few years. A recent example is Schwartz and Torous (1989).

8 The discount rate and the riskless rate do not appear separately in the subsequent valuation equations, (11) to (13). In these equations the constant ι can be interpreted as either the common discount rate or the riskless rate of interest.

9 It may be possible to generalise this assumption of spanning as follows. The second term in (2) can be decomposed into a systematic component and a residual. The systematic component must be spanned by the stochastic returns on traded securities, whereas the residual is completely diversifiable by construction. In this case, the Brownian residual can be written as the limit of a sequence of Poisson variates, and the argument in Naik (1991) can be applied to the Poisson variates.

10 A β-distribution of the second kind is related to the F-distribution. See Kendall and Stuart (1977) p. 163.

11 For example, see Corollary 1 in Binmore and Herrero (1988a).

12 This can occur with atomistic developers and atomistic owners of undeveloped real estate at the expanding boundary of a monocentric city, as in Capozza and Sick (1991). Downward sloping demand curves for developed assets appear in Pindyck (1988), Dixit (1989), Bar-Ilam, Sulem and Zanello (1992), and Williams (1993). Developers can be imperfectly competitive in Williams.

13 Because all agents are price takers, time to build is easily introduced. The price πx must be replaced by the present value of the expected future price upon completion of construction. This avoids the complication in Bar-Ilan, Sulem and Zanello (1992).

14 The optimal density of development is explored in Clarke and Reed (1988) and Williams (1991).

15 The implications of capacity constraints among developers are studied in Williams (1993).

16 Additional properties of a steady state are developed in Yinger (1981).

BIBLIOGRAPHY

Amihud, Y. and H. Mendelson, 1986, "Asset Pricing and the Bid-Ask Spread", *Journal of Financial Economics* 17, pp. 223–49.

Amihud, Y. and H. Mendelson, 1989, "The Effects of Beta, Bid-Ask Spread, Residual Risk, and Size on Stock Returns", *Journal of Finance* 44, pp. 479–86.

Amihud, Y. and H. Mendelson, 1991, "Liquidity, Maturity, and the Yields on US Treasury Securities", *Journal of Finance* 46, pp. 1411–25.

Baldwin, C. and R. Meyer, 1979, "Liquidity Preference under Uncertainty: A Model of Dynamic Investment in Illiquid Opportunities", *Journal of Financial Economics* 7, pp. 347–74.

Bar-Ilan, A., A. Sulem and A. Zanello, 1992, "Time to Build and Capacity Choice", Working Paper, Department of Economics, Dartmouth College and Tel-Aviv University.

Binmore, K. and M. Herrero, 1988a, "Matching and Bargaining in Dynamic Markets", *Review of Economic Studies* 55, pp. 17–32.

Binmore, K. and M. Herrero, 1988b, "Security Equilibrium", *Review of Economic Studies* 55, pp. 33–48.

Brennan, M. and E. Schwartz, 1985, "Evaluating Natural Resource Investments under Uncertainty", *Journal of Business* 58, pp. 135–57.

Capozza, D. and G. Sick, 1991, "Pricing Risky Land: The Option to Develop", Working Paper, Graduate School of Business Administration, University of Michigan.

Clarke, H. and W. Reed, 1988, "A Stochastic Analysis of Land Development Timing and Property Valuation", *Regional Science and Urban Economics* 18, pp. 367–82.

Constantinides, G., 1986, "Capital Market Equilibrium with Transactions Costs", *Journal of Political Economy* 94, pp. 842–62.

Diamond, P., 1982, "Wage Determination and Efficiency of Search Equilibrium", *Review of Economic Studies* 49, pp. 217–27.

Dixit, A., 1989, "Entry and Exit Decisions under Uncertainty", *Journal of Political Economy* 97, pp. 620–38.

Gale, D., 1986, "Bargaining and Completion: Characterisation (Part I) and Existence (Part II)", *Econometrica* 51, pp. 785–818.

Gale, D., 1987, "Limit Theorems for Markets with Sequential Bargaining", *Journal of Economic Theory* 43, pp. 20–54.

Grossman, S. and G. Laroque, 1990, "Asset Pricing and Optimal Portfolio Choice in the Presence of Illiquid Durable Consumption Goods", *Econometrica* 58, pp. 25–51.

Kendall, M. and A. Stuart, 1977, *The Advanced Theory of Statistics* (vol. 1, Fourth Edition), Charles Griffin, London.

Majd, S. and R. Pindyck, 1987, "Time to Build, Option Value, and Investment Decisions", *Journal of Financial Economy* 18, pp. 7–27.

McDonald, R. and D. Siegel, 1985, "Investment and the Valuation of Firms When There is an Option to Shut Down", *International Economic Review* 26, pp. 331–49.

McDonald, R. and D. Siegel, 1986, "The Value of Waiting to Invest", *Quarterly Journal of Economics* 101, pp. 707–27.

Mortensen, D., 1982, "Property Rights and Efficiency in Mating, Racing, and Related Games", *American Economic Review* 72, pp. 968–79.

Naik, V., 1991, "Option Trading Strategies for Hedging Jumps in Asset Values and Return Volatilities", Working Paper, Faculty of Commerce, University of British Columbia.

Paddock, J., D. Siegel and J. Smith, 1988, "Option Valuation Claims on Real Assets: The Case of Offshore Petroleum Leases", *Quarterly Journal of Economics* 103, pp. 479–508.

Pagano, M., 1989, "Trading Volume and Asset Liquidity", *Quarterly Journal of Economics* 104, pp. 255–74.

Pindyck, R., 1988, "Irreversible Investment, Capacity Choice, and the Value of the Firm", *American Economic Review* 78, pp. 969–85.

Rubinstein, A. and A. Wolinsky, 1985, "Equilibrium in a Market with Sequential Bargaining", *Econometrica* 53, pp. 1133–50.

Samuelson, L., 1992, "Disagreement in Markets with Matching and Bargaining", *Review of Economic Studies* 59, pp. 177–85.

Schwartz, E. and W. Torous, 1989, "Prepayment and the Valuation of Mortgage-Backed Securities", *Journal of Finance* 44, pp. 375–92.

Sick, G., 1989, *Capital Budgeting with Real Options*, Monograph 1989–93, Salomon Brothers Center, Stern School of Business, New York University.

Stein, J., 1993, "Prices and Trading Volume in the Housing Market: A Model with Downpayment Effects", Working Paper, Sloan School of Management, MIT.

Titman, S., 1985, "Urban Land Prices under Uncertainty", *American Economic Review* 75, pp. 505–14.

Vayanos, D. and J. Vila, 1992, "Equilibrium Interest Rate and Liquidity Premium under Proportional Transaction Costs", Working Paper, Sloan School of Management, MIT.

Wheaton, W., 1990, "Vacancy, Search, and Prices in a Housing Market Matching Model", *Journal of Political Economy* 61, pp. 1270–92.

Williams, J., 1991, "Real Estate Development as an Option", *Journal of Real Estate Finance and Economics* 4, pp. 191–208.

Williams, J., 1993, "Equilibrium and Options on Real Assets", *Review of Financial Studies* 6, pp. 825–50.

Wolinsky, A., 1990, "Information Revelation in a Market with Pairwise Meetings", *Econometrica* 58, pp. 1–24.

Yinger, J., 1981, "A Search Model of Real Estate Broker Behavior", *American Economic Review* 71, pp. 591–605.

8

*Bidding for the Antamina Mine: Incentives in a Real Options Context**

Alberto Moel[†] and Peter Tufano

Harvard Business School; [†]AAA Asset Management

This chapter studies the bidding for a copper mine that was offered for sale by the Peruvian government as part of the country's privatisation programme. The mine itself had a valuable real option component, in the form of the right to develop the mine after completing exploration, which we analyse using Monte Carlo methods. The novel aspect of the transaction was the type of bid requested by the Peruvian government, which essentially requested bidders to state both the *premium* that they would pay and *exercise price* (pledged development expenditure) they would set for this real option. This structure gave rise to incentives which affected the amount that firms would offer, their preferences between bidding premium and exercise price, the identity of bidders, the likelihood of ultimate development, and the likelihood of *ex post* renegotiation of the contract.

Since the publication of the seminal work on real options, it has been acknowledged that real option analysis is a valuable analytical tool that allows managers to quantify the value of flexibility. While the rate of adoption of real option analysis has been uneven across industries, it has been embraced most enthusiastically in the natural resource industries, where academic research first proposed the application of these techniques. This chapter details the bidding for a classic real option in a natural resource industry: the right to develop a copper deposit.

*This chapter was first published in *Project Flexibility, Agency and Competition*, and is reprinted with the permission of Oxford University Press. It is based on the Harvard Business School teaching case, "Bidding for Antamina" (297–054) and the associated classroom note, "Copper and Zinc Markets 1996" (297–055). The authors would like to thank executives of RTZ-CRA Limited, Lenos Trigeorgis, Eduardo Schwartz, Antonio Tarnawiecki, Martha Amram, and attendees of the real options workshop hosted by Analysis Group Economics and the Columbia University/Ernest & Young/UCLA Real Options conference for their comments on the original paper. Finally, they thank the MBA and Executive Education students at Harvard Business School for their classroom comments on the case. Funding for this project was provided by the Harvard Business School Division of Research as part of the Global Financial Systems project.

In 1996, the Peruvian government privatised a number of state-owned assets. One of the first was Antamina, a polymetallic ore deposit about 500 kilometres north of Lima, Peru. The winning bidder would have two years to explore the property before deciding whether or not to develop the site. This ability to wait-and-see gave rise to a classic development option, which bidders would need to incorporate into their valuation. The first section of the chapter discusses the practical issues involved in carrying out a real options analysis, in the tradition of the work of Paddock, Siegel and Smith (1988) and others.

What is unique about bidding for Antamina is not the valuation of the project, but rather the real option *embedded in the bidding rules* set up by the Peruvian government, and the resultant implications for behaviour of the bidders and the ultimate developer of the deposit. In effect the auction rules had firms submit both the premium and the exercise price of the real option to develop the property. By allowing bidders to set both elements of the bid, the government may have substantially reduced the problem of winners' curse and motivated bidders to propose very high investment levels in the plant. However, it simultaneously increased the likelihood that the ultimate winner would walk away from the project. Furthermore, the rules set up an important time-inconsistency problem, giving incentives for renegotiation after initial exploration of the property was completed.

The first section of this chapter provides background data on the Antamina mine and the privatisation process set up by the Peruvian government. The second section analyses the value of the mine, using now-standard real option valuation techniques. The third section of the chapter analyses the bidding rules put in place by the Peruvian government and the incentives brought about by these rules.

BACKGROUND ON THE ANTAMINA MINE[1]

Beginning in the early 1990s, the Peruvian government sought to return many of its state-owned firms to private ownership. The government planned privatisations that would raise US$5.75 billion in cash and investment commitments in 1996–99. One of the larger entities to be sold was Centromin, Peru's largest state-owned mining company, which controlled seven mines, a metallurgical complex, four hydroelectric plants, a railway system, port facilities and numerous undeveloped natural resource deposits. The firm had been owned by US-based Cerro de Pasco Corporation until 1974, when it was nationalised by the Peruvian government. After two decades of state-ownership, the government attempted to sell the entire company in 1994 and 1995, but failed. In 1996, the plan was to sell Centromin's eleven properties piecemeal, with the two first parcels being a gold mine and the Antamina mine.

The Antamina mine, located 482 kilometres north of Lima, Peru, was a rich polymetallic copper and zinc deposit. Based on engineering reports made public by the government as part of the auction, the property had proven and probable ore reserves estimated at 127 million metric tons of ore containing 1.7% copper and 0.8% zinc. Centromin's management had publicly stated that these figures vastly underestimated the likely reserves of the property. While part of the ore body had been well-studied by geologists, there had been virtually no geologic study of a large portion of the deposit. As a result, there was large uncertainty about the size of the reserves.

From a practical perspective, none of the potential bidders could carry out anything but limited geological work on the property before the bids were due, and therefore all of the bidders had to rely on their analysis of surveys produced by Cerro de Pasco and Centromin. In addition, the government was not willing to represent or guarantee the amount of reserves. However, based on previous experience, some industry experts estimated that the actual reserves might range from 100 million to 175 million metric tons.

A feasibility study, which would largely consist of additional geological exploration through drilling, would more precisely establish the amount and quality of ore in the Antamina property. Such exploration, which would cost approximately US$24 million, was expected to take about two years, and would be completed before mine construction began. Once that additional geological work had been completed, the developer of the property would be "more confident" of the expected reserves, ie, within ±20%.

Were the subsequent geological work to suggest the mine was economically exploitable, the property would have to be developed in order to be able to extract the copper and other metals. Antamina was located in a remote region in the mountains of Peru at 4000 metres above sea level, and 200 kilometres from the ocean. Developing this site would include the construction of roads, mining rigs, crushing plants and other ancillary facilities, and the purchase and transport of heavy mining equipment, and was expected to last three years. Based on the available information, the capital expenditures to develop the mine (not including the feasibility study) ranged from US$581 million to US$622 million (in 1996 US dollars) and were a function of the amount of ore found.

Table 1 shows estimates of mine life, yearly production, capital expenditures and operating costs for three development scenarios corresponding to three potential outcomes of the quantity of ore in the deposit. These estimates were taken from discussions with industry experts. The expected scenario can be thought to represent the mean or median of the distribution of the possible ore reserves, the low scenario to represent 1 to 1.5 standard deviations below the mean (corresponding to the 100 million metric tons estimate), and the high scenario (175 million metric tons) to represent 1 to 1.5 standard deviations above the mean.

Table 1 Information used to value Antamina

Scenario		Low case	Expected	High case
Reserves	(million metric tons)	100	127	175
Mine life	(years)	12	14	18
Copper production	(million lbs/year)	313	339	365
Zinc production	(million lbs/year)	155	168	181
Operating costs[a]	(millions of 1996 US$)	131	138	145
Copper treatment charge[b]	(1996 US$/lb)	0.28	0.28	0.28
Zinc treatment charge[b]	(1996 US$/lb)	0.22	0.22	0.22
Feasibility study[c]				
1996	(millions of 1996 US$)	6	6	6
1997	(millions of 1996 US$)	18	18	18
Capital expenditure[c]				
1998	(millions of 1996 US$)	54	55	55
1999	(millions of 1996 US$)	246	255	264
2000	(millions of 1996 US$)	281	292	303
Per year after 2000[c]	(millions of 1996 US$)	8.7	9.0	9.3
Closure costs[c]	(millions of 1996 US$)	45	45	45

Estimates of mine life, yearly production, capital expenditures and operating costs for three development scenarios corresponding to three potential outcomes of the quantity of ore in the deposit. The expected scenario can be thought to represent the mean or the median of the distribution of the possible ore content of the deposit, the low scenario to represent 1 to 1.5 standard deviations below the mean, and the high scenario to represent 1 to 1.5 standard deviations above the mean.

[a]While the operating costs are expressed in US dollars, less than half would be in set in Peruvian Sols. These costs are expressed in real (1996) amounts, and exclude the treatment charges paid to a smelter.

[b]The treatment charges were paid by the mining company to smelters (typically in Japan) and would be set and paid in US dollars. These treatment charges are set by contract with the smelter, and tend to adjust with the relative demand for smelting capacity, which tends to increase with the price of copper. From the smelter, the mining company received revenue equal to the amount of metal produced times the US dollar market price of copper (or zinc) minus the treatment charge.

[c]The feasibility study, the initial and continuing capital expenditures, and the closure costs are expressed in real (1996) amounts. These expenses would primarily be set and paid in US dollars.

VALUATION OF THE ANTAMINA INVESTMENT OPPORTUNITY

In a "typical" *all-cash* auction, interested firms would be asked to bid the amount of cash they would pay up-front for the right to develop the mineral resource. Some of the first applications of real options, such as Paddock, Siegel and Smith (1988) deal with this type of problem. In this section, we apply classic real option valuation to value the investment opportunity, as if bidders were asked to submit all-cash bids.

The economic value of the Antamina mine is the value of the discounted stream of expected future sales of copper and zinc, less any development and extraction costs. It is well known that the standard valuation technique of risk-adjusted DCF analysis fails to capture all sources of value associated with the investment, in that it assumes that the decision to invest is both *irreversible* and *inflexible*, ie, the investment cashflows are both committed and fixed for the life of the project. The main contribution of real options models is to incorporate managerial flexibility inherent in the project in its valuation. Added flexibility value, overlooked in DCF analysis, comes from the managerial decisions that can take advantage of such mineral price movements: operating flexibility and investment timing flexibility.

Operating flexibility includes any option to vary operating parameters, including shutting down, reopening, expansion, contraction, abandonment, optimisation of cut-off grades and varying production rates as prices change.[2] Investment timing flexibility is the ability to delay the start of property development.[3] Of course, both types of managerial flexibility can be exercised jointly. For example, Paddock, Siegel and Smith (1988) value, on unexplored, undeveloped offshore petroleum leases, the option to delay exploration and/or extraction for up to five years, and then the option to shutdown, reopen or abandon once the field has been developed.[4]

In this section, we value the *right to develop* the Antamina mineral reserves, focusing on the investment options available to the winning bidder. We value the winning bidder's right to develop the property as a call option, treating the developed mine as the underlying asset and the development expenses as the exercise price of this option. Following the extant literature, we consider the developed mine as a levered claim on a traded underlying asset, where the leverage comes from operating leverage and the traded underlying is a set of forward contracts on copper and zinc. By mapping the underlying asset (the developed mine) into traded assets (metals prices), we can use the risk-neutral valuation methodology that has been previously developed in Cox and Ross (1976) and Harrison and Kreps (1979).

To value Antamina, we use a Monte Carlo simulation model, implemented within the popular spreadsheet program, Excel. This decision was made for three reasons. First, the European style of this option makes Monte Carlo a feasible approach. Second, this model was developed for classroom or practitioner use, and its transparency allows users to see the assumptions used and the workings of the valuation model. Third, with this approach, we can vary both the optimal amount of investment and the mine life based on the amount of ore found. In particular, we calculate a series of possible price paths for copper and zinc prices, which are then used as inputs to a DCF model which establishes the distribution of values of the underlying asset, the developed mine. The model values the real option to develop the mine after the two-year feasibility study is completed and the related ore uncertainty is resolved. The model has a number of simplifying assumptions:

1. We focus on the investment timing option, not the operating options. In this model, we assume that once development of the project has begun, the options to delay or speed up development, temporarily shutdown, reopen, change the production profile (eg, speed up or slow down production) or the ore grade cannot be exercised.

2. The probability distribution of ore quantity can be adequately characterised by three discrete outcomes. We assume that there are three possible quantity outcomes (corresponding to the outcomes in Table 1), and each has a known probability of occurring.[5]

3. The risk-free discount rates are known and deterministic. We discount the mine revenues (derived from forward prices or simulated at a risk-neutral drift rate) at a constant risk-free rate, as in Brennan and Schwartz (1985a,b). They argue that if the mine owner could enter into forward contracts to sell the mine output in the future, the price risk would be eliminated, and the relevant discount rate would be the risk-free rate. In Brennan and Schwartz (1985b), they hypothesise known and certain costs, thus, they discount costs at the risk-free rate as well.[6] We follow this treatment. Other uncertain variables, such as exchange rates and inflation rates, are also assumed to be known and deterministic.[7]

4. Project data gathered from industry sources are representative. To arrive at the other information used to value the mine, given in Table 1, we interviewed industry experts and consultants. We acknowledge that the parameters we use are not necessarily those used by bidders for Antamina, although we have been told that they are representative of those used. In addition to those assumptions given in the table, we assume working capital of 25% of net revenue (gross revenue less treatment charges), five year straight-line depreciation starting in 2001, a tax rate of 30%, and a long-term inflation of 3.5% per annum for all costs.[8]

5. The copper and zinc prices follow a joint diffusion process. Applying the work of Gibson and Schwartz (1990) and Schwartz (1997), we assume that the copper and zinc prices and convenience yields follow a joint diffusion process. In particular, the copper and zinc prices follow geometric Brownian motion, while their convenience yields are mean-reverting. The prices of each commodity are negatively correlated with that commodity's convenience yield, while the copper and zinc returns are positively correlated with each other. More formally, the dynamics of prices and convenience yields are given by[9]

$$\frac{dP_c}{P_c} = (r - \delta_c)dt + \sigma_c dz_c$$

$$\frac{dP_z}{P_z} = (r - \delta_z)dt + \sigma_z dz_z \tag{1}$$

$$d\delta_c = k_c(\alpha_c - \delta_c)dt + \sigma_{\delta c}dz_{\delta c}$$

$$d\delta_z = k_z(\alpha_z - \delta_z)dt + \sigma_{\delta z}dz_{\delta z}$$

where the P's are prices, r is the risk-free rate, the δ's are convenience yields, and the subscripts c and z correspond to the copper and zinc processes, respectively. The α's are the long-term mean convenience yields, and the k's are the mean-reversion coefficients, which dictate the speed at which the convenience yields will approach their long-term means. The dz's are correlated increments to standard Brownian processes, and we allow for correlations among these noise terms:

$$\begin{aligned} dz_c dz_z &= \rho_{cz}dt \\ dz_c dz_{\delta c} &= \rho_{c,\delta c}dt \\ dz_z dz_{\delta z} &= \rho_{z,\delta z}dt \end{aligned} \tag{2}$$

where the ρ's denote the corresponding correlation coefficient between the relevant Brownian motions. The first equation relates price of zinc and copper, while the second and third equations relate the commodity prices to their respective convenience yields.

To obtain estimates for the relevant parameters, we used copper and zinc spot and futures market data from the London Metals Exchange for the period September 1991 to July 1996. We used the spot and three-month futures prices of zinc and copper to determine a convenience yield, given by (Gibson and Schwartz (1990), equation (9))

$$\delta = r' - 4\ln\left(\frac{F(S,3)}{S}\right) \tag{3}$$

where r′ is the three-month risk-free rate, S is the spot price, and F(S,3) is the three-month copper or zinc futures price. To determine the mean reversion parameters k and α, we ran the regression[10]

$$\delta_t - \delta_{t-1} = a + b\delta_{t-1} + \varepsilon_t \tag{4}$$

for both zinc and copper convenience yields, and calculated

$$\begin{aligned} \alpha &= -\frac{\hat{a}}{\hat{b}} \\ k &= -\ln(1 + \hat{b}) \\ \hat{\sigma} &= \hat{\sigma}_\varepsilon \sqrt{\frac{\ln(1 + \hat{b})}{(1 + \hat{b})^2 - 1}} \end{aligned} \tag{5}$$

Table 2 Metal price process parameters

(a) Convenience yield mean-reversion parameters

κ_c	κ_z	α_c	α_z
0.8	0.6	6.0%	–0.2%

(b) Copper and zinc volatilities

σ_c	σ_z	$\sigma_{\delta c}$	$\sigma_{\delta z}$
22%	18%	25%	23%

(c) Copper and zinc correlations

	P_c	P_z	δ_c	δ_z
P_c	1.00			
P_z	0.44	1.00		
δ_c	0.57		1.00	
δ_z		0.73		1.00

Daily data from the London Metal Exchange for the period September 1991 to July 1996 were used to determine the metal price process parameters used in equations (1) and (2). **(a)** shows the zinc and copper three-month futures convenience yield mean-reversion parameters k and α, obtained from the data and equations (3), (4) and (5). **(b)** shows the copper and zinc return annualised volatilities, calculated on the basis of the prior 90-day price changes. A year is assumed to have 250 trading days. **(b)** also shows the annualised convenience yield volatilities, from the third equation in equation (5). **(c)** shows the return correlations ρ for copper and zinc and their convenience yields. All correlations are for daily data for the period September 1991–July 1996, and are significant to 1%.

where $\hat{\sigma}_\varepsilon$ is the standard error of the regression. Table 2 shows the mean reversion parameters k and α, the volatilities (calculated over a 90-day window), and the relevant correlations.

We forecast the copper and zinc prices and convenience yields monthly for the first two years, using the discrete version of the diffusion processes of equation (1), and the relevant parameters from Table 2. The initial spot prices and convenience yields (prevalent at the time of the bidding in mid-1996) were US$1.00/pound and 20% for copper, and US$0.55/pound and –5% for zinc, respectively. Under the assumption that a forward market extends indefinitely into the future, and that the convenience yield is mean-reverting, the forward price of copper (or zinc) at time T after year 2 can be approximated by (Schwartz (1997), equations (18) and (20)):

$$F(P,\delta,T) = P \exp\left[-\delta \frac{1 - e^{-kT}}{k} + A(T)\right] \quad \textbf{(6)}$$

A(T) is given by:

$$A(T) = \left(r - \alpha + \frac{1}{2}\frac{\sigma_\delta^2}{k^2} - \frac{\sigma_\delta \sigma \rho_\delta}{k}\right)T + \frac{1}{4}\sigma_\delta^2 \frac{1 - e^{-2kT}}{k^3} + \left(\alpha\kappa + \sigma_\delta \sigma \rho_\delta - \frac{\sigma_\delta^2}{k}\right)\frac{1 - e^{-kT}}{k^2} \quad \textbf{(7)}$$

where σ is the commodity price volatility, δ is the (stochastic) convenience yield, α is the long-term mean convenience yield, k is the convenience yield mean reversion coefficient, ρ_δ is the correlation of the convenience yield with the commodity price, and σ_δ is the convenience yield volatility.[11]

The Monte Carlo simulation generated 10,000 price and convenience yield paths from year 0 to year 2 using monthly timesteps. For each run, the ending metal prices and convenience yields at year 2 were used as

Figure 1 Distribution of values of Antamina if winning bidder forced to develop or has option to develop

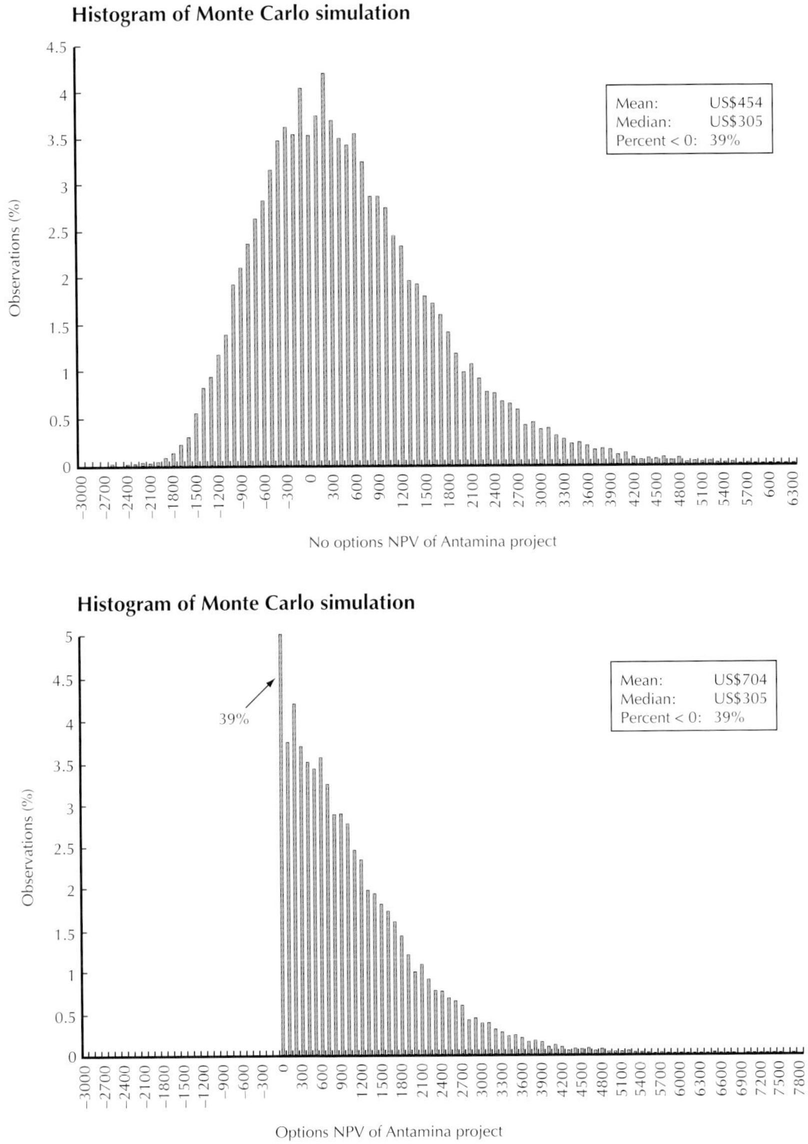

The top panel shows the results of a 10,000-run Monte Carlo simulation of the Antamina mine if the mine had to be developed at the end of year 2. The mean NPV is US$454 million. The project had a negative NPV in 39% of the trials. The bottom panel shows the effect of the option to abandon on the NPV of the project. In this instance, NPV-negative outcomes are not developed, resulting in a mean NPV of US$704 million.

inputs to equations (6) and (7) to generate the forward prices for copper and zinc in future periods. These forward prices were then used to derive the cashflows in three DCF models, which correspond to the high-, medium-, and low-ore outcomes. As a baseline, the top graph in Figure 1 shows the distribution of values of the Antamina mine if the mine *had to be developed* at year 2 (ie, there was no option to delay development, and a standard DCF model applied). The mean NPV of this "real forward contract", ie, commitment to develop at the end of two years, is US$454 million, with a wide distribution as shown by the figure.

This top panel shows that in about four of ten outcomes the value of developing the mine is negative at year two. The essence of the *real option* is that at this go/no-go point, managers could optimally choose not to proceed if the economic value of developing the mine was negative.[12] By walking away from the project these negative outcomes are truncated, as shown in the bottom graph in Figure 1.

As shown in Table 3, the decision not to develop is made in 23% of the high-ore scenarios, 39% of the medium-ore scenarios and 53% of the low-ore scenarios. The mine is not developed when metals prices are low, convenience yields are large (which means that forward curves are shallow or downward sloping) and the quantity of ore is low. Under these circumstances, the value of a developed mine is too low to justify the expenditure of the development charges.

By abandoning the project at year 2, management can eliminate all NPVs less than zero outcomes. If that were the case, the mean NPV of the mine would rise to US$704 million. The difference between the two, US$250 million, is the value of the development timing (or abandonment) option at year 2. This corresponds to 55% of the standard DCF valuation of US$454 million, which is on the high end of the findings of Davis (1996).

As with any valuation model, the outputs are only as valid as the parameters used as inputs to the model. We have been told by some executives in the mining industry that these valuations might be "too high", but without access to their private assumptions we cannot ascertain whether this difference is due to our parameter inputs or other factors. Certainly the choice of the diffusion process is an important modelling decision. Schwartz (1997) points out that the mean-reverting characteristics of metal prices have a first-order impact on the valuation and the decision of when to exercise real options. Our application of his methodology confirms his point. In our early modelling of the value, we calculated the mine value ignoring mean reversion after year 2 (ie, if the forward curve was given by the equation given in footnote 11). In this case, the NPV of the "real forward contract" was US$1.3 billion, while the development timing option makes the NPV rise to US$1.7 billion for a US$400 million value for the option to defer development. These values are 2.5 to 3 times as large as those that result from the analysis using Schwartz's (1997) specification for

Table 3 Characteristics of Antamina project, conditioned on whether the project is developed at the end of year 2

	Project developed	**Project not developed**
High ore:		
Copper price/convenience yield	US$1.04/5.5%	US$0.63/8.9%
Zinc price/convenience yield	US$0.64/–0.5%	US$0.47/0.5%
% of trials	77	23
Expected ore:		
Copper price/convenience yield	US$1.07/5.4%	US$0.67/7.9%
Zinc price/convenience yield	US$0.65/–0.6%	US$0.50/0.4%
% of trials	61	39
Low ore:		
Copper price/convenience yield	US$1.10/5.3%	US$0.70/7.4%
Zinc price/convenience yield	US$0.70/–0.7%	US$0.52/0.3%
% of trials	47	53
All outcomes:		
Copper price/convenience yield	US$1.07/5.4%	US$0.68/7.8%
Zinc price/convenience yield	US$0.65/–0.6%	US$0.50/0.3%
% of trials	61	39

Mean copper and zinc prices and convenience yields, conditioned on whether the project is developed at the end of year 2. The means are shown for the High, Expected, and Low ore outcomes, and for all outcomes combined. We also show the per cent of trials in which the mine was developed for all outcomes. All means are significantly different to better than $p = 0.01$.

the dynamics of the metals' prices. Were we to ignore all volatility in both the price and convenience yields of the metals, the value of the project would be about a quarter of that we found: US$112 million for the commitment, and about US$158 million with the option to develop.[13] While it is obvious that the level of volatility would affect the valuation of the real option, the "seemingly-technical" choice of model for the diffusion process for commodity prices has as large and important an impact on the project valuation.

The historical data used to generate the convenience yield process parameters can also have a material impact on the valuation. In the past, copper and zinc convenience yields have been highly unstable. For example, the 3-month convenience yield of zinc oscillated between –10% and 60% in the period 1991–96, with periods of relative stability followed by short bursts of high volatility (ie, the distribution of convenience yields is "long-tailed" and skewed). For this reason, the choice of the window for estimation of the convenience yield parameters will influence these parameters and the resulting mine valuation.

THE INCENTIVE IMPACTS OF REAL-OPTION BIDS

The valuation of the real option to develop the Antamina mine, while important to potential bidders, is a well-studied problem. Also, the role of information and bidding structure on bid prices has been extensively studied – at least under all-cash bidding rules.[14] However, the bidding rules established by the Peruvian government – and the incentives they created – are much more novel than those previously studied. The auction rules themselves recognised and embodied the development timing option, and had bidders specify both premium and exercise price of the option to develop the property. The details of the rules created strong, and arguably perverse, incentives for potential bidders, leading them to promise to invest large sums in developing the property and thereby increasing the likelihood that they would ultimately choose to walk away from the project.

The rules of the game

Bidders for Antamina were asked to submit two figures at the time they bid: an *initial payment* and an *investment commitment*:

- The *initial payment* is the amount of cash that is paid immediately. It must be at least US$17.5 million.
- For the next two years, the winning bidder could explore the property and, under the auction rules, was obligated to spend at least US$13.5 million on exploration. As exploration was projected to cost US$24 million, this constraint was not binding in a real sense, except if initial exploration were to rule out all subsequent development.
- At the end of two years, the winner must decide to either:
  (1) walk away from the project, have no more rights to the property, and owe the government no additional funds; or
  (2) develop the mine. If the winning bidder proceeds, and its investment by the end of year five falls short of the amount that it promised to invest (the pledged *investment commitment*), it owes the Peruvian government a penalty equal to 30% of this shortfall, to be paid at the end of year five. The minimum investment commitment is US$135 million. This penalty was presumably put in place to prevent firms from making "false promises" regarding their intended investment plans.
- The winning bidder would be the firm proposing the largest "bid". To establish the value of the "bid", the authorities would add the initial payment and 30% of the investment commitment. This weighting scheme seemed designed to give greater reward to cash payments over promises of future investment.

The time line in Figure 2 summarises the actions and decisions of the winning bidder.

Figure 2 Time line of managerial decisions in Antamina auction

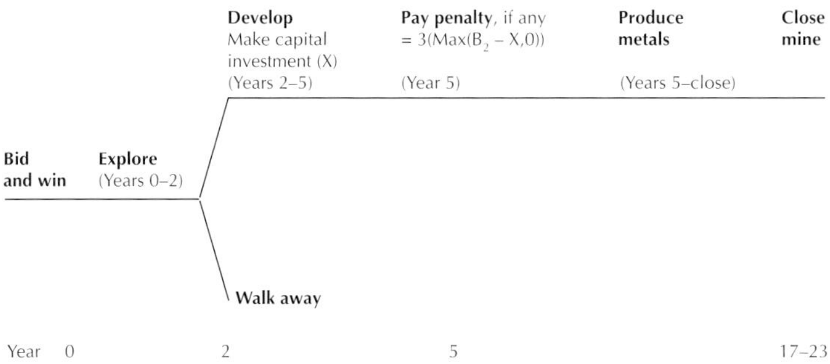

This figure shows the set of decisions facing a firm bidding on Antamina, focusing on the decision whether to develop the mine at the end of exploration (year 2). B_1 represents the initial cash payment made at year zero, B_2 represents the pledged investment commitment, and X represents the actual amount spent on development in years 2 to 5 (exercise price) if the firm proceeds. The penalty is paid in year 5, after development is completed. Closure takes place when the reserves are depleted, which is predicted to occur between years 17 and 23, depending on the ore found.

Analysing the government's bidding rules

To appreciate these bidding rules, it is useful to contrast this auction mechanism with a more traditional one, in which firms made *all-cash* bids, where firms propose only how much cash they would pay today for the right, but not obligation to develop in the future.

Under the all-cash bid, the expected value for each bidder, conditional on winning, is:

$$E[PV(Max(S - X,0))] - B \tag{8}$$

where the actual physical investment cost (X) is the exercise price of the real option to develop the property, the cash bid (B) is the premium paid for the option, and the underlying asset is the value of the developed reserves (S). PV[·] represents the present value operator, and E[·] represents the expectation operator.

By contrast, under the rules established to sell Antamina, the bidder bids both the initial payment B_1 and the promised investment commitment B_2, and its expected value is:

$$E[PV(Max(S - X - PEN,0))] - B_1 \tag{9}$$

where PEN is the penalty, given by:

$$PEN = 0.3Max(B_2 - X,0) \tag{10}$$

since bidders would always bid B_2 at least equal to X,[15] this simplifies to:

$$E[PV(Max(S - 0.7X - 0.3B_2,0))] - B_1 \quad \textbf{(11)}$$

Comparing the actual bidding structure with the all-cash scheme, the crucial difference is that rules under Antamina have firms propose not only the option premium B_1, but also a portion ($0.3B_2$) of the exercise price $0.7X + 0.3B_2$. Thus, the government effectively has bidders propose the two option terms simultaneously.

Incentives for bidding

In analysing this auction, we can ask how the all-cash and Antamina bidding rules would compare along a variety of dimensions: the bids submitted by the firms, the likely identity of bidders, the winner's subsequent decision regarding development of the property, and the possibility of renegotiating the contract.

In setting bids under an all-cash, first-price bid rules, bidders would be unwilling to offer a cash bid that would make their expected return negative, and their reservation price would equal the first term in equation (8) above. They face the classic winner's curse problem, in that the winning bidder confronts the distinct possibility of having the most over-optimistic valuation, and hence *ex post* overpaying for the property.[16]

The Antamina bidding rules provide a number of subtle complications, both with respect to a bidder's choice between the cash and investment commitment portions of their bid, and with respect to their subsequent decisions. In setting the bid, the firm must figure out how to allocate its bid between the initial payment B_1 and the investment commitment B_2. In understanding the firm's incentives, we consider the relative costs and benefits of adding another dollar to each of the portions of the bid. By bidding another dollar in the initial cash payment, the bidder gets an addition dollar in "bid points" under the government's weighting scheme, but must spend a dollar (conditional on winning.) By bidding another dollar in the investment commitment, the bidder gets only US$0.30 in "bid points", but must pay the present value of US$0.30, conditional on deciding to go forward at year 2. If we compare the bid points received to the expected cost, conditional on winning the bid, we see that there is a strong incentive for firms to increase their pledged investment commitments rather than the initial cash payment, as shown in Table 4.

Quite simply, it is "cheaper" to bid future promises – on which one can freely renege – than to bid cash. In the extreme, if the initial payment were zero (and exploration expenses were also zero), a firm could bid an extraordinarily high investment commitment B_2 and still be assured of not losing money or of not suffering from winner's curse. This is because a call option cannot have negative value, and with zero premium there is no

Table 4 Comparison of increasing initial payment or pledged investment commitment under antamina-style bidding rules

	If the bidder bids another US$1 on the initial payment	**If the bidder bids another US$1 on the investment commitment**
Number of *bid points* received	US$1.00	US$0.30
Expected cost, conditional on winning the bid	US$1.00 Must pay this immediately, regardless of ultimate decision to develop	E[PV(US$0.30)] < US$0.30 The penalty is not paid until later, and only if the project is ultimately developed
Ratio of *expected cost* to *bid points*	1	Less than 1

The table shows the number of "bid points" a firm would receive by adding an additional dollar to either the initial payment or investment commitment, under the rules set by the Peruvian government. A "bid point" is the measure used to judge alternative bids by the Peruvian government.

exercise price that could make the option worth less than zero. In this extreme hypothetical case, the bidder would only develop if ore prices were very high and there was a sufficiently large ore deposit on the property; otherwise it would walk away without spending any money on the project. It couldn't overbid, because it could walk away after learning whether the development was warranted but before spending any money on the project.

As the government's rules insisted that bidders invest at least US$17.5 in the initial payment plus another US$13.5 million in exploration, there can be some winner's curse. Supposing that firms bid the minimum initial cash amount and spend US$24 million on exploration, one can calculate the highest exercise price (investment commitment) that a firm could bid and still break even, or a reservation investment commitment. Figure 3 plots the the outcome to bidders as a function of the pledged investment commitment bid.[17] Given the low option premium (initial payment and exploration expenditures), bidders could bid up to about US$18 billion on the investment commitment before the value of the project turned negative. The notion that a firm might bid such a "high" investment commitment may seem puzzling. However, once expressed in option terms, the idea may seem less difficult to grasp: suppose that you could buy a two-year call option on a stock that is currently priced at US$10, but which is quite volatile ($\sigma = 0.5$).[18] The call-writer asks you, "How high an exercise price would you accept for paying a penny premium?" In principle, you would be willing to set the strike price as high as US$80, or eight times the current stock price in return for paying a US$0.01 for the option.

The bidding rules also affect the outcome in another way, by giving a comparative advantage to smaller, less-capitalised companies. Ultimately,

Figure 3 Net present value to winning bidder under Antamina auction, and the probability that the project would be developed versus size of pledged investment commitment

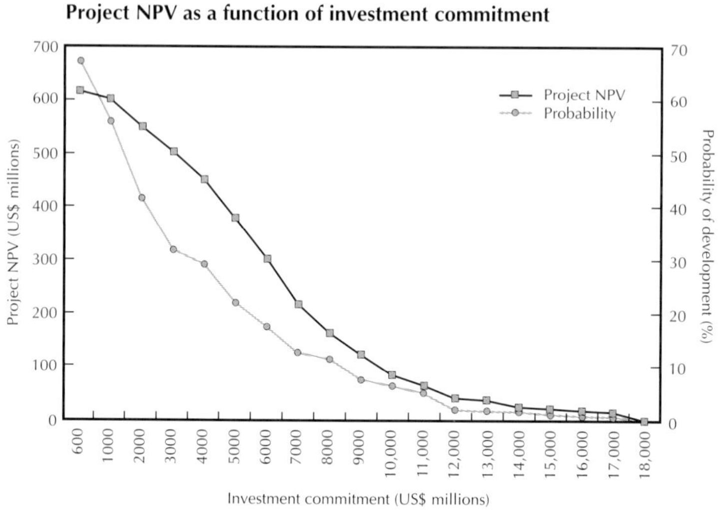

The left-hand scale shows the net present value of Antamina mine as a function of the promised investment commitment. The right-hand side shows the probability of mine development as a function of the investment commitment. The NPV graph intersects the x-axis (ie, the NPV of the project turns negative) at about US$18 billion, with a probability of development of 0%. Each point in the graph corresponds to the mean value for a 1,000-run Monte Carlo simulation.

only three firms submitted bids for Antamina. Two were among the largest and best-capitalised in the industry. The smallest of the three, a joint venture of Rio Algom and Inmet, had a combined market capitalisation of only US$1.5 billion, or 10% of the largest bidder, and 33% of the next bidder. Under an all-cash auction, the joint venture of the smaller mining companies might be unable to raise sufficient funds to match the bids of the larger firms.[19] However under the Peruvian bidding rules, smaller bidders can make future *promises* to develop the mine *without having to raise very much money today*. This type of bidding structure might encourage the smallest players to bid the most aggressively, because they have a comparative advantage in bidding future promises.

Incentives for subsequent investment

However, the level of the investment commitment – and hence penalty to be paid if the property is developed – will affect the subsequent development decision. Unlike the all-cash bid, in which the cash bid becomes an irrelevant sunk cost, under the Antamina-style rules, the investment commitment is not a sunk cost. The property will be developed only if it is economic to do so, which from equation (11), will occur when the net operating profits less 0.7 times the physical investment less 0.3 times the investment commitment is greater than zero, or if:

$$S - 0.7X - 0.3B_2 > 0 \quad \textbf{(13)}$$

As the pledged investment commitment (B_2) rises, it becomes less likely that the property will clear this hurdle, and the probability that the property will be developed falls. Figure 3 also plots the probability that the mine would be developed, conditional on the winning bid having the investment commitment shown along the horizontal axis.[20] As the investment commitment (exercise price) increases, the cost of building the mine and paying the penalty increases. As a result, the call option to develop is more likely to remain unexercised at maturity. Each point represents the percentage of trials for which the firm would optimally choose not to proceed, conditional on the stated level of investment commitment. Only when the investment commitment rises above US$18 billion does the probability that the project will be developed fall to 0%. Thus, competition among bidders would tend to lessen the probability that the mine would get developed, unlike the case of an all-cash bid.

In addition, the bidding procedure instituted by the Peruvian government might induce firms to make marginal investments that are uneconomic, rather than pay the penalty. Suppose the managers of the winning bidder found that it was optimal to proceed, despite the fact that the firm had bid a large investment commitment, and therefore owed a penalty to the Peruvian government. In particular, suppose that the company bid an investment commitment of US$900 million, while the necessary physical investment to actually develop Antamina was US$600 million. In this example, were the firm to invest only US$600 million, it would owe the Peruvian government a penalty of US$90 million, or 30% of US$300 million. Alternatively, suppose it could spend another US$300 million on the project, by buying additional trucks, helicopters or jets. Ideally, these "extra" investments would have some fungible value beyond that delivered to the Antamina project. As long as the losses from purchasing these assets were less than US$90 million, it would be optimal to purchase the extra "flexible" assets, claim them as capital investments for Antamina, avoid the penalty, and at some later date redeploy them to other uses to recoup their value. Given this incentive, the government might need to closely monitor the type of investments made by the winning bidder.

A related potential problem is that the winning bidder might hope to use inflated transfer prices to increase its recorded investment in the project. Were the winning bidder to hire a related firm to carry out some of the development, they could pay this subsidiary above-market rates on work performed on Antamina, and thereby reduce the amount of penalty owed to the government.[21]

Incentives for renegotiation

The existing rules seem to encourage a form of *ex post* renegotiation that would invalidate portions of the formal contract, in particular the penalty. At the end of two years, the winning bidder might conclude that the project was not economically feasible if it must pay the penalty, but that it would be feasible in the absence of the penalty. It could walk away from the project, or it could appeal to the government to forgo the penalty. In response to this offer, the government could either accept this proposal, or it could call the firm's bluff, take control of the property, and offer it again for sale. However, at that time, the most informed bidder would be the firm that had just completed the exploration, and others might be reluctant to bid against them. If the information collected by the winning bidder about the property was revealed, the government could not recover the penalty, because other bidders would know that the project would be unfeasible were they to invest the original investment commitment. Thus, the penalty may be unenforceable in a practical sense even when the true development option (without penalty) has positive intrinsic value.

This time-inconsistency type of problem plagued another recent auction: the Federal Communication Commission's spectrum auction of wireless phone licences.[22] The FCC auctioned off these spectrum rights in 1996, with 89 companies bidding US$10.2 billion for 493 licences. Bidders with less than US$40 million in annual revenues received attractive financing terms from the government, with no interest payable for six years. For a variety of reasons, the wireless business has failed to be as attractive as initially anticipated, and bidders requested the government to restructure their obligations, or else the winning bidders would default. While one solution would be to re-offer the licences for sale, these winning bidders could seek protection in bankruptcy courts, and it could be costly and time-consuming for the FCC to gain legal access to the assets. In essence, these bidders bid future promises, like the investment commitments in the Antamina example. When things didn't work out, they optimally chose not to exercise the real option to proceed with development, and at that point the original contracts ceased to be binding.

Does this chosen bidding structure meet the government's stated objectives?

Starting in the mid-1990s, the right-of-centre administration of President Alberto Fujimori sought to undo decades of state intervention, and to return many of its state-owned firms to private ownership. The objectives of such a large-scale privatisation programme were many, including (a) the development of the project that would generate some local employment, taxable revenue and hard currency for the country, and (b) higher cash proceeds for the government in the initial sale. Officials in the privatisation programme publically emphasised the former goal over the latter, stating in general terms, "At the moment the treasury is not in particular need of money, but what's imperative is that we allow private companies to carry out what state entities have been doing".[23] Speaking directly about the Antamina privatisation, one official said, "We obviously want to sell at an interesting price, but the principal objective is to maintain and develop the sector by attracting quality companies".[24]

Based on our analysis, we doubt whether the proposed rules would meet the government's stated objectives. As discussed above, the rules created incentives for weaker firms to bid large investment commitments, reducing the likelihood of development. Furthermore, the incentive to bid investment commitment over cash payments means that the government would receive very little money up-front in the form of initial payments. Given the incentives for potential renegotiation of the penalties to be paid, their future payment is suspect. Finally, the resulting disincentives seem to require substantial *ex post* monitoring of the types of investments made by the winning bidder. It appears that the government has set up a set of rules that not only deliver little money up-front, but also reduce the likelihood of development later on. These rules are difficult to reconcile with their stated goals.[25]

Alternative bidding rules

In order to evaluate alternative bidding rules, one must more carefully specify the government's objective function. In much of auction theory, one assumes that the seller seeks to maximise expected revenue. For example, the classic literature on auctions, going back to Vickrey's work (1961), analysed alternative auction mechanisms on the basis of their expected revenue.[26] In Antamina's case, the government's objective is not necessarily to maximise expected revenue, and the option character of the bid complicates the task of comparing bids even further.

Consider the simpler stylised example given earlier, where the owner of a share offered it up for "sale" via an auction where bidders could bid any pair of premiums and exercise prices on a call option. The share-owner is offering to write a call on the share rather than selling it outright, and gives potential bidders the right to select any contract type they chose. Were the seller's goal to maximise its expected revenue or value, it would select the bid for which the premium received exceeded the value of the call delivered, ie, $P - c(X)$, where P is the premium offered, and $c(X)$ is the value of the call with exercise price X. There are infinite combinations of these parameters to which the seller would be indifferent. These are diagrammed in Figure 4, which shows the set of bids which have zero expected value to both buyer and seller. While the bids graphed have the same value, they present the bidder with different risk exposures, as shown on the bottom graph, which plots the profit/loss to the seller for two proposed contracts. Given the sellers' risk preferences, we could select from among these equal-value choices.

Figure 4 Trade-off between the premium and exercise price for a call option

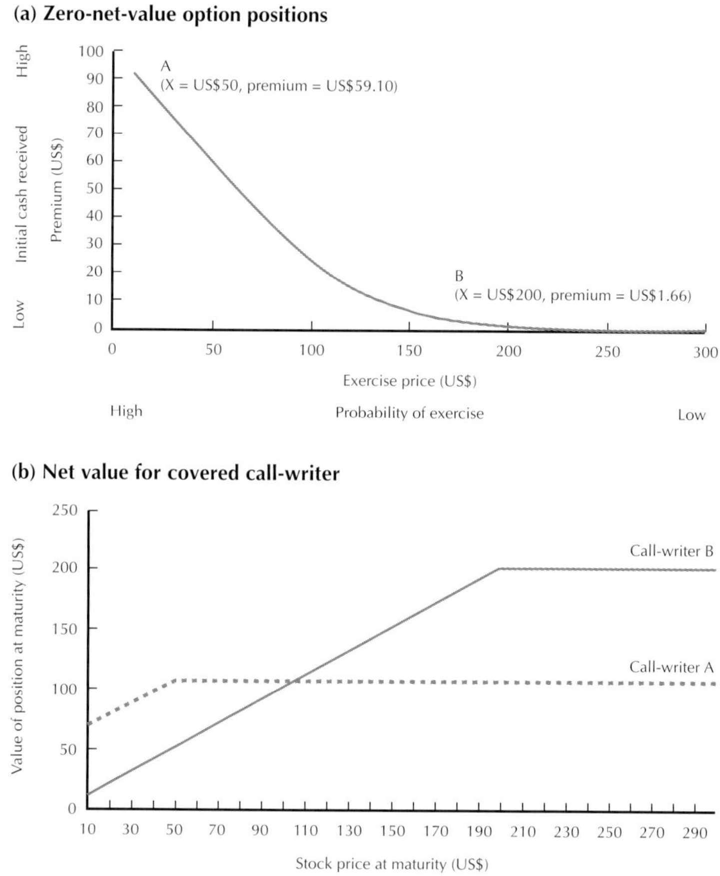

(a) Shows the combination of call option exercise prices and premiums for which the value of 2-year European-style call option struck at the given exercise price less the premiums equals zero, ie, those set of fairly priced calls. This plot is for a stock price of US$100, a maturity of 2 years, a risk-free rate of 10%, and a volatility of 25%. (b) Shows the risk exposure at maturity of two call-writers, each of whom wrote a fair-value option at exercise prices A and B from (A).

In the case of Antamina, the government has stated a preference for high probability of development as a primary goal and high cash proceeds as a secondary goal. The probability of development (exercise) is inversely related to the exercise price, and cash proceeds are directly related to the option premium. Thus, while points A and B on the figure have equal value, using these preferences, the option represented by A would dominate the one represented by B, in that it would be more likely to be exercised, as well as deliver a higher up-front cash price. Taken to the extreme, this would suggest that the government's interests would be best served by reducing the exercise price as low as possible by eliminating the penalty, which would mean that its should prefer an all-cash auction. This more traditional solution seems as if it would better meet the government's objectives than the actual bidding structure.

POSTSCRIPT

The Peruvian government used the proposed auction rules in the Antamina sale. It received three bids and announced the winner on July 12, 1996. The largest firm that submitted a bid, RTZ-CRA, Ltd, offered US$17.5 million as its initial payment, and promised a US$900 million investment commitment. Noranda, the second largest firm, also bid US$17.5 million up-front, and promised a more generous US$1,900 million investment commitment. The winner, the small Rio Algom/Inmet partnership which bid US$20 million up-front, and an investment commitment of US$2,500 million.[27]

At a press conference on July 12, Rio Algom Ltd and Inmet Mining Corp said that they intended to spend about US$1 billion to develop their new Peruvian copper purchase. "If it does not prove viable, we just lose our up-front investment", stated a Rio Algom spokesman.[28] Clearly, the joint venture partners seemed well aware of the options embedded in the bidding rules, since they signalled that they never planned to spend the investment commitment of US$2.5 billion, but rather a more reasonable US$1 billion (plus penalty, were they to proceed.). They also apparently wanted to demonstrate this understanding to security analysts, to make sure that they were not perceived as having "overpaid" for the mine. The winning bidder clearly understood the option-nature of the offering, and followed a bidding strategy consistent with the one that could have been predicted.[29]

The power of real option analysis – in conjunction with simple game theory – is that it promises to help us better understand not only how things are valued, but also how managers will behave when faced with flexibility and choices. In this case, a real options perspective can help us understand how firms would bid on a large capital investment project, as well as to predict what types of problems might crop up in the future.

1 The material in this section is excerpted with modifications from the Harvard Business School case study "Bidding for Antamina", No. 297-054, prepared by Peter Tufano and Alberto Moel. Copyright ©1997 by the President and Fellows of Harvard College; all rights reserved.

2 Models of the operating flexibility to shutdown, reopen and abandon a mineral resources project were first developed by Brennan and Schwartz for a gold mine (Brennan and Schwartz (1985a)) and for a copper mine (Brennan and Schwartz (1985b)). Palm, Pearson, and Read (1986) valued the option and to shutdown and reopen both low cost and high cost copper mines, while Cavender (1992) valued the option to shutdown and reopen for a small open-pit gold mine. The flexibility to adjust cut-off grade, and to stockpile work in process was modelled by Mardones (1993).

3 The value of the option to delay development of an oil field for a finite time has been studied extensively, by Brennan and Schwartz (1985a,b), McDonald and Siegel (1986), Paddock, Siegel and Smith (1988), Trigeorgis (1990), Bjerksund and Ekern (1990), Laughton (1993) and Laughton and Jacoby (1991).

4 Bjerksund and Ekern (1990) carry out a similar analysis, except the option to delay is perpetual. Trigeorgis (1991) values the option to expand, defer or cancel a non-fuel mineral project.

5 We assume that the Low and High case both have a 0.2 probability of occurring, while the Expected outcome has a probability of 0.6. In reality the quantity would be a continuous random variable (while the investment programme would be lumpy). We use a discrete version of the distribution to simplify the computational burden of the Monte Carlo model.

6 The assumption that costs should be discounted at the risk-free rate is appropriate if costs are riskless in the sense that they have no systemic risk.

7 The extension of commodity price modelling to include stochastic interest rates is discussed in Schwartz (1997).

8 The project value will be sensitive to these assumptions. For example, inflation affects the operating costs, investment expenditures and closure costs of the project: higher inflation will decrease the value of the project.

9 We make the additional simplifying assumption that the market price of convenience yield risk λ is zero. Schwartz (1997) finds that λ for copper was not significantly different from zero in the period 1988–95. Given our assumption that the mine can be expressed as a claim on copper, its λ would also be zero. While this assumption simplifies our analysis, we recognise its limitations. In particular, a portion of the project (the cost structure) is not a tradable asset, and therefore may have a market price of risk.

10 The methodology used is described in Gibson and Schwartz (1990) and Dixit and Pindyck (1994), p. 76.

11 Under the assumption that the term structure of convenience yields at the end of year two does *not* display mean-reversion, the forward curve is given by the usual expression

$$F_1(P,\delta,T) = P \exp[(r - \delta)T]$$

where the convenience yield (δ) is assumed constant. Schwartz (1997, esp. fig. 7), shows that the incorporation of stochastic convenience yields into the calculation of the term structure has a large effect on the shape of these curves, and the effect differs across different time periods. While the spot prices of the metals are correlated in our analysis, our calculation of the forward curves does not explicitly take metal price correlation into account in the construction of the term structure of the forward curves.

12 In reality the managers could postpone this decision by accepting the penalty, but we do not model this choice.

13 Even though there is no commodity price volatility, the option value is still positive since there is ore quantity uncertainty built into the model.

14 For an analysis of the role of private and public information on bids for oil and gas lease auctions using all-cash bids, see Porter (1995). The number and size of the bids received can also be influenced by elements of the bidding process itself, such as the auction mechanism, the speed of the auction process and the number or order of assets to be auctioned. For a review of this literature, see Lopez-de-Silanes (1996).

15 If X is the amount a bidder would spend to develop, conditional on proceeding, then the bidder can make an investment commitment promise of at least X without incurring any risk of penalties. By bidding B_2 less than X, the bidder would not reduce its penalty, but would reduce the likelihood of being selected as the winner of the auction. Thus $B_2 = X$ dominates $B_2 < X$, and thus the penalty can be expressed as $0.3^*(B_2 - X)$.

16 The first mention of winner's curse in the academic literature was in the context of bidding for oil and gas drilling rights by Capen *et al.* in 1971.

17 This calculation assumes that the bidder makes the minimal initial payment of US$17.5 million as well as the full US$24 million exploration expenditure. We assume that once development is begun, the bidder will complete the full development project. In reality, the bidder could abandon the project after partial development or underinvest (and pay a penalty if its development expenditures were less than the promised investment commitment.) These features would tend to increase the value of the project, because the holder would only exercise them when it was optimal to do so.

18 The risk-free rate is 5% (continuously compounded) for this example.

19 In this instance, a US$1+ billion cash bid by RioAlgom/Inmet might not be feasible, as it would represent two-thirds of their market capitalisation.

20 Technically, this is the risk-neutral probability that the mine would be developed, rather than the objective probability of its exercise.

21 This is merely a restatement of the general proposition that whenever contractual payoffs are determined by accounting figures, there is a temptation for firms to manipulate these reports to their advantage. There are many examples of this phenomenon, ranging from firms accelerating or delaying large write-offs to maximise managers' bonuses (Healy (1985)) or under-reporting earnings to avoid having to make payments to income bond holders (Tufano (1997)).

22 This situation has been covered widely in the press. For example, see Bryan Gruley and Quentin Hardy, "Wireless Bidders Ask to Restructure Debt", *Wall Street Journal*, June 26, 1997, page A3. For an academic discussion of the auction rules, see McMillan (1994).

23 Saul Hudson, "Interview – Peru upbeat on energy and mining ", *Reuters Financial Service*, June 14, 1996.

24 Saul Hudson, "Forty Firms Qualify for Peru Copper Prospect Sale", *Reuters Financial Service*, June 21, 1996.

25 The rules may be consistent, however, with a possible *unstated* goal of favourable public relations. A large investment commitment could be viewed positively as a signal of the government's privatisation offerings.

26 See Riley (1989) for an overview of this literature.

27 Sally Bowen, "Canadians in US$2.5 billion Peruvian Copper Plan", *Financial Times*, July 16, 1996, p. 33.

28 "Canadian Group Hopes to Double Antamina Reserves", *Reuters Financial Service*, July 12, 1996.

29 In the MBA classroom, students played the role of bidders for Antamina. Those who appreciated the nature of the real options bid apparently differently from those who did not with the former offering lower initial payments and investments commitments sometimes over US$15 billion.

BIBLIOGRAPHY

Bjerksund, P. and S. Ekern, 1990, "Managing Investment Opportunities Under Price Uncertainty: From 'Last Chance' to 'Wait and See' Strategies", *Financial Management*, Autumn, pp. 65–83.

Brennan, M. and E. Schwartz, 1985a, "A New Approach to Evaluating Natural Resource Investments", *Midland Journal of Corporate Finance* 3, pp. 78–88.

Brennan, M. and E. Schwartz, 1985b, "Evaluating Natural Resource Investments", *Journal of Business* 58(2), pp. 135–57.

Capen, E.C., R.V. Clapp and W.M. Campbell, 1971, "Competitive Bidding in High-risk Situations", *Journal of Petroleum Technology* 23, pp. 641–53.

Cavender, B., 1992, "Determination of the Optimum Lifetime of a Mining Project Using Discounted Cashflow and Option Pricing Techniques", *Mining Engineering* 44, pp. 1262–8.

Cox, J. and S. Ross, 1976, "The Valuation of Options for Alternative Stochastic Processes", *Journal of Financial Economics* 3(1/2), pp. 145–6.

Davis, G.A., 1996, "Option Premiums in Mineral Asset Pricing: Are they important?", *Land Economics* 72(2), pp. 167–86.

Dixit, A.K. and R.S. Pindyck, 1994, *Investment under Uncertainty*, Princeton, Princeton University Press.

Gibson, R. and E. Schwartz, 1990, "Stochastic Convenience Yields and the Pricing of Oil Contingent Claims", *Journal of Finance* 45(3), pp. 959–76.

Harrison, J. and D. Kreps, 1979, "Martingales and Arbitrage in Multiperiod Securities Markets", *Journal of Economic Theory* 20, pp. 381–408.

Healy, Paul, 1985, "The Effect of Bonus Schemes on Accounting Decisions", *Journal of Accounting Research*, pp. 71–83.

Laughton, D., 1993, "Reversion, Timing Options, and Long-term Decision Making", *Financial Management*, August, pp. 225–40.

Laughton, D. and H.D. Jacoby, 1991, "A Two-method Solution to the Investment Timing Option", in F.J. Fabozzi (ed.), *Advances in Futures and Options Research* 5, JAI Press, Greenwich, CT, pp. 71–87.

Lopez-de-Silanes, F., 1997, "Determinants of Privatization Prices", *Quarterly Journal of Economics* 112, pp. 966–1028.

Mardones, J., 1993, "Option Valuation of Real Assets: Application to a Copper Mine with Operating Flexibility", *Resources Policy*, pp. 51–65.

McDonald, R. and D. Siegel, 1986, "The Value of Waiting to Invest", *Quarterly Journal of Economics* 101(4), pp. 707–27.

McMillan, J., 1994, "Selling Spectrum Rights", *Journal of Economic Perspectives* 8(3), pp.145–62.

Paddock, J., D. Siegel and J. Smith, 1988, "Option Valuation of Claims on Real Assets: The Case of Offshore Petroleum Leases", *Quarterly Journal of Economics* August, pp. 479–508.

Palm, S.K., N.D. Pearson and J.A. Read, Jr, 1986, "Option Pricing: A New Approach to Mine Valuation", *CIM Bulletin* May, pp. 61–6.

Porter, R.H., 1995, "The Role of Information in US Offshore Oil and Gas Lease Auctions", *Econometrica* 63(1), pp. 1–28.

Riley, J.G., 1989, "Expected Revenue from Open and Sealed Bid Auctions", *Journal of Economic Perspectives* 3(3), pp. 41–50.

Schwartz, E., 1997, "The Stochastic Behavior of Commodity Prices: Implications for Valuation and Hedging", *Journal of Finance* 51, pp. 923–73.

Trigeorgis, L., 1991, "A Real-options Application in Natural Resource Investments", in F.J. Fabozzi (ed), *Advances in Futures and Options Research* 5, JAI Press, Greenwich, CT, pp. 153–64.

Tufano, P., 1997, "Business Failure, Judicial Intervention, and Financial Innovation: Restructuring US Railroads in the Nineteenth Century", *Business History Review* Spring, pp. 1–40.

Vickrey, W., 1961, "Counterspeculation, Auctions and Competitive Sealed Tenders", *Journal of Finance* 16, pp. 8–37.

9

*Miracle on Sixth Avenue: Information Externalities and Search**

Andrew Caplin and John Leahy

New York University; Boston University

After a long dormant period, Lower Sixth Avenue in New York has undergone a rapid revitalisation. We show that a simple search theoretic model with information spill-overs can explain both the period of under-use and the rapid turnaround. The model reduces to a simple equation, which allows us to do comparative statics on the duration of vacancies. We show how the solution reacts to changes in market structure and changes in search technology. The model is suggestive of difficulties that may occur in many markets in which there are changes over time in the optimal use of resources.

At one time Lower Sixth Avenue was one of the most fashionable retail centres in Manhattan, but the area collapsed as a retail centre when the affluent moved further uptown and to the suburbs. Today, Sixth Avenue is in the midst of a rapid turnaround, as reported in the *Sunday New York Times* by Claudia Deutsch (June 27, 1993):

> After decades of catering to virtually no one save the odd buyer of a quick lunch or piece of hardware, part of that once fashionable strip ... is again turning into a major retail area.

How can we account for the revitalisation of this depressed area in the middle of a recession?

The *New York Times* article provides an important clue. In 1992, Bed Bath & Beyond opened Manhattan's first superstore in this area. It was the success of this store that prompted other stores to move to the area. One local businessman is quoted as stating that, "Sixth Avenue was dead as a

*This chapter was first published in *The Economic Journal* 108, January, pp. 60–74 (1998) and is reprinted with the permission of Blackwell Publishers Ltd. The authors wish to thank Charles Freeman and two anonymous referees for their helpful comments and acknowledge research support from the National Science Foundation and the Sloan Foundation.

doornail before, but Bed Bath & Beyond has really opened it up", while another asserts that, "It was Bed Bath & Beyond that really legitimised the neighbourhood". The article goes on to point out that the arrival of the store changed local rents: "Bed Bath & Beyond is said to be paying in the low US$20s a square foot. Barnes and Noble will be paying in the US$30s. Current negotiations in the area are inching towards US$40."

There are two features of the Sixth Avenue experience that we wish to explain. The first is the long period of underuse, and the second is the rapid rise in the area's prospects following the arrival and success of the first superstore. In this chapter, we develop a search model to explore one possible explanation for these events: information externalities. Before Bed Bath & Beyond set up shop, all potential tenants faced the common problem of assessing whether or not New Yorkers would be willing to shop on Lower Sixth Avenue, which is a rather inconvenient location. The size of the crowds at Bed Bath & Beyond demonstrated to other retailers that many shoppers were indeed willing to make the trip. This knowledge led to a snowball effect as other tenants rushed to lease in the area.

We show that our search theoretic model with information spill-overs can indeed explain both the long vacant period and the rise in rents following the success of Bed Bath & Beyond. One important feature of the model is that the delay until first occupancy is sub-optimally long. This is due to the information externality. When deciding whether or not to accept a tenant, owners do not take into account the value to others of the information revealed in their match and are therefore too selective in equilibrium. The model shows that the failure of optimality may be extreme. We provide examples in which social optimality demands immediate occupancy, but the market produces an arbitrarily long delay.

Another possible explanation for the sudden change in fortunes on Sixth Avenue is a crowd externality, in which later arrivals rushed in to take advantage of the crowds generated by Bed Bath & Beyond.[1] In this chapter we also incorporate crowd externalities in the search model, and show that they differ in important ways from information externalities. In particular, in the crowd story the desire of one agent to locate increases the desire of others to locate and benefit from the crowds. Actions are strategic complements. In regard to the information externalities the desire of one agent to locate increases the incentive for others to wait and learn from this initial decision so that actions are strategic substitutes.

We believe that the information externality story fits the facts of Sixth Avenue better than the crowd story. One reason for this is that Bed Bath & Beyond succeeded before any other stores located in the area. It follows that Bed Bath & Beyond was not in need of external economies in order to be profitable. A second factor is the striking increase in rents that occurred in the two years after Bed Bath & Beyond arrived. Rents rose almost 100% over this period, which is hard to understand unless the information upon which rents were based had changed.[2] A third argument in favour of the information story is that the crowds are not all that large in comparison to other retail districts in New York nor are the spill-overs obvious. The shopping area extends only four blocks and has stores as diverse as Bed Bath & Beyond, Barnes and Noble, Tri-City Camera, Staples, TJMaxx, and Today's Man. It is not clear that towels, books, cameras, office supplies and clothes are natural complements.

A final point favouring the informational interpretation is the timing of subsequent lease signings. The stores that followed Bed Bath & Beyond apparently signed their leases after Bed Bath & Beyond had demonstrated success. The simplest agglomeration story would say that Bed Bath & Beyond would only have needed to sign a lease in order to prompt others who were already searching to follow suit.

So what is the information that was revealed? In the *New York Times* article, a local retail broker clearly states that there was scepticism concerning shoppers' willingness to return to Sixth Avenue:

> Let's face it, on a sunny afternoon Fifth Avenue is bustling and Sixth is deserted. Who knows to what extent bargain hunters will seek bargains that cost a US$20 round trip?

Sixth Avenue is in a rather inconvenient location. There is little parking. There is no subway. There are few residential high rises. New York city is already a congested, competitive market, so the lure of variety and value that makes stores like Bed Bath & Beyond so successful in the suburbs is not necessarily an attraction in New York. There was legitimate reason to doubt that large discount stores would succeed in the area. When the store succeeded, other stores, such as Barnes and Noble, that had previously taken a wait and see attitude, decided that Sixth Avenue was a good place to do business.[3]

RELATION TO LITERATURE

Our model contributes to the understanding of information externalities, which can be viewed more generally as examples of the private supply of a public good. Banerjee (1992) and Bikhchandani *et al* (1992) showed how information externalities give rise to herding; the information that agents learn from observing the actions of others may lead them to ignore their own private information. Rob (1991), Caplin and Leahy (1993a; 1994a), and Chamley and Gale (1994) showed how information externalities may cause agents to delay action inefficiently, since there is an incentive to wait for others to take costly action. All of the above models are set in the context of irreversible investment decisions, as opposed to the search theoretic setting of this chapter.

We believe that search is the most natural setting for an analysis of information externalities.[4] Whenever agents engage in the purposive search for information, their search behaviour will be greatly affected by what they can learn from the actions of other agents. Information externalities will not only result in herding and delays, but will also change the nature of the information that the economic system will end up gathering and incorporating in equilibrium.

Another valuable feature of the search setting is that it provides by far the simplest and most flexible framework for analysing information externalities. The model boils down to one equation with one endogenous variable, and this equation is easily amended to allow for variations in the market structure.

We believe that the insights of this model have applications to many different economic settings. Caplin and Leahy (1994b) use the model to study the natural rate of unemployment. They consider the search decision of a group of workers who are simultaneously laid off from their jobs, and show that the presence of information spill-overs raises the level of search unemployment. Huang and Shirai (1994) have used it to analyse foreign direct investment into China. They show that search externalities may explain the timing and magnitude of entry into that market. Davis and Haltiwanger (1992) and Davis *et al* (1994) suggest a role for search and information externalities in their analysis of labour and capital reallocation among manufacturing plants in the United States.

THE MODEL

The basic idea is as follows. The vacant buildings on Sixth Avenue can be seen as options available for a wide variety of alternative uses. Committing a vacant unit to any one of its possible uses is costly to the owner, since such a commitment will normally involve modifying the unit to fit the needs of the particular tenant. If owners can learn about the best use for their property from the experience of neighbouring properties, then they will have an incentive to delay renting. They will all play a waiting game, hoping that one of the other owners will be first to commit their property to a particular use and thereby generate the socially valuable information concerning whether or not the property is well matched to that use. Once a successful use is found the uncommitted owners will be able to use the information to improve the quality of their decisions.

To model the phenomenon of interest, we begin with a set of N identical vacant buildings, each with a distinct owner. We assume that there are two potential uses for each property, which we shall refer to as retail and office space. Owners must decide for which use their property is best suited. We assume that the payoff to using a building as retail is unknown and equal to $W > 0$. The value of using a building as retail is unknown and potentially greater. We use a search methodology to capture the time it takes for an owner to identify an appropriate retail tenant.

Every period owners make a series of decisions. First, each landlord decides whether or not to commit their building to office space and collect W or to search for a retail tenant. Each landlord that decides to search then meets a single potential retail tenant. These landlords then simultaneously and independently decide whether or not to offer these tenants leases. The tenants that are offered leases then occupy their buildings and the next period begins. To capture the cost of adapting a building to different uses, we make the extreme assumption that such transfers are impossible. Once a building is committed to office space or a retail lease is signed the building is committed to that use and tenant forever. This will mean that it will often be preferable to keep a unit vacant rather than commit it to an incorrect use.

We assume that both landlords and tenants are risk neutral and make their decisions to maximise the present value of profits, and that landlords discount future periods by $\delta \in (0, 1)$. The decisions that landlords make will, of course, depend on the payoffs to retail, the nature of the information process, and the bargaining between landlord and tenant. We now turn to these issues in turn.

The present value of leasing a building to a particular tenant depends on two factors. The first is the quality of the tenant that we denote by α. The second is the quality of the retail market that we denote by λ. For simplicity, we set total value generated by a match equal to the product of these two factors $\alpha\lambda$. Note that α is particular to an individual landlord-tenant match, where λ is characteristic of the market as a whole and therefore common to all matches. It is the desire to learn about λ that provides an incentive for landlords to learn from the experience of others.[5]

As concerns the bargaining problem between landlord and tenant, we assume that landlords can make "take it or leave it" offers. This means that landlords will extract all of the surplus from the match and that tenants will pay the expected value of $\alpha\lambda$ at the time of the lease. Placing all of the market power in the landlord's hands removes the potential externality arising from a division of the surplus and allows us to focus on the effect of information spill-overs.

The heart of the model lies in the information structure. We assume that tenants are identical *ex ante* and that all landlords and tenants have identical priors concerning the distributions of α and λ. We assume each α is drawn independently from a uniform distribution on $[0, 1]$ and is revealed to both the tenant and the owner when they meet and before a lease is signed. The critical assumption is that information concerning λ can only be released once a tenant occupies a building and that tenant's success can be judged. Initially, all agents know that λ is distributed uniformly on $[0, 2]$. This means that the first lease to be signed will stipulate a payment of α, which is the expectation of $\alpha\lambda$ when λ is unknown. We assume that all participants in the market have access both to the terms of any occupancy agreement between a landlord and tenant, as well as the actual level of payoff that an occupied unit gives rise to. These payoffs are realised and observed immediately after occupancy and before the next period begins. This implies that λ becomes public knowledge in the period after the first occupancy. The observation of the amount the first tenant pays will reveal the tenant's α type, while the observation of the payoff will reveal the product $\alpha\lambda$ and therefore pin down the market type. Subsequent tenants then pay $\alpha\lambda$. Finally, we initially assume that there is no recall, so that landlords have only one chance to lease to each potential tenant. The case with recall is considered below.

In order for the model to be interesting W must be small enough that owners find it profitable to search for retail tenants when λ is unknown. We assume that an owner would prefer search even if the owner were the only one searching. Solving the single-agent search problem leads to the bound

$$W < \frac{1 + x^2}{2}$$

where $x \in [0, 1]$ is the solution to $x = \delta(1 + x^2)/2$. Note that the presence of other searchers will only make search more profitable, so this is a sufficient condition that works for all parameter values.

The informational assumptions in the model imply that the market passes through two phases. In the first phase before any units have occupants, there is no information on the market type, and the owners must simply decide on an acceptance strategy: which α types to permit to take occupancy. The second phase begins in the period immediately after the first occupant has taken up residency, and involves a known value of λ. In this phase, there is a twofold decision facing the owners of vacant units. First, given the new information on λ, is it best to continue searching or to collect W? Second, if search is best, which α types should be permitted to take occupancy at cost $\alpha\lambda$ (the *ex post* value to an α-type)?

In looking to the model for an answer to these questions, we limit attention to equilibria in Markov strategies in which the decision rules of each agent depend only on information that is directly payoff relevant. In this case a strategy for an individual owner is a set of acceptable α types in the uninformed phase, a set of λ values that warrant continued search through potential occupants, and a set of acceptable α types corresponding to those values of λ that imply continued search. Given the information externality, the optimal strategies are interdependent, and we look for a symmetric Nash equilibrium.[6]

The simple nature of the search framework leads to simple strategies. It is clear that any stationary equilibrium will involve cut-off rules: a reservation type U in the uninformed phase, a cut-off level $\bar{\lambda}$, and a reservation type $I(\lambda)$ in the informed phase contingent on the realised value of λ. The

Nash equilibrium condition is that all decisions must be optimal given that others are using the same decision rules. We summarise this equilibrium below.[7]

Definition: *An equilibrium consists of a cutoff* $\bar{\lambda} \in [0, 2]$, *a cutoff* $U \in [0, 1]$, *and a function* $I(\lambda) = [\bar{\lambda}, 1] \rightarrow [0, 1]$ *with the property that the following strategy is optimal for each landlord on the assumption that all others are pursuing it:*

(a) *In the uninformed phase with* λ *unknown, the landlord accepts the occupancy of any tenant of type* $\alpha \geqslant U$ *upon payment of* α, *and rejects potential tenants with* $\alpha < U$.
(b) *In the informed phase in which* λ *is known, the landlord uses the property for office space if* $\lambda < \bar{\lambda}$, *and searches for a retail tenant if* $\lambda \geqslant \bar{\lambda}$. *When searching, the landlord accepts any tenant of type* $\alpha \geqslant I(\lambda)$ *upon payment of* $\alpha\lambda$ *and rejects potential tenants with* $\alpha < I(\lambda)$.

The main lines of a demonstration of existence and uniqueness of this equilibrium are presented in the next section and a qualitative analysis in the following one. For a formal analysis of these topics see Caplin and Leahy (1993b).

MODEL SOLUTION AND PROPERTIES

The existence and uniqueness of equilibrium

We solve the model backwards beginning with the informed phase. Once λ is known there is no longer any strategic interaction among the landlords. The problem of searching for the appropriate retail tenant reduces to a standard, single-agent search problem. Let $V(\alpha, \lambda)$ denote the value of an optimal strategy for an informed owner who is visited by a potential tenant of type α and knows that the market type is λ. The Bellman equation for this problem is,

$$V(\alpha, \lambda) = \max\left[\alpha\lambda, \delta\int_0^1 V(\alpha', \lambda)d\alpha'\right]$$

The first term inside the maximisation is the present value of renting to the tenant at hand. The second term is the discounted value of searching for another tenant. It follows immediately that there exists a reservation level of α, $I(\lambda)$, and that

$$\lambda I(\lambda) = \delta\int V(\alpha, \lambda)d\alpha = \lambda\delta\left[\frac{1 + I(\lambda)^2}{2}\right] \tag{1}$$

The left-hand side of (1) is the value of accepting the reservation type, whereas the centre and right-hand side are the expected value of rejecting this type in hope that a higher type will show up in the future. Equation (1) shows that $I(\lambda)$ is in independent of λ, $I(\lambda) \equiv I$. This independence arises from the multiplicative interaction between α and λ, which implies that the return to renting and searching are proportional to λ. Note also that I is an increasing function of the discount rate and has range (0, 1).

Search will only take place in the informed state if the value of search, which equals $\int V(\alpha, \lambda)d\alpha$, exceeds W. This pins down the cut-off $\bar{\lambda}$.

$$\bar{\lambda} = \frac{2W}{1 + I^2} = 2w \tag{2}$$

where we introduce $w = W/(1 + I^2)$, to ease notation. Note that the upper bound on W implies $\bar{\lambda} < 1$ and $w \leqslant 1/2$.

Using (1) and (2) we can integrate over the possible values of λ to derive the unconditional expectation of the value of being informed:

$$EV_I = \int_0^{\bar{\lambda}} \frac{W}{2} d\lambda + \int_{\bar{\lambda}}^2 \int_0^1 V(\alpha, \lambda) d\alpha d\lambda = \frac{1}{2}(1 + I^2)(1 + w^2) \tag{3}$$

This simple solution for the optimal strategy in the informed phase makes it easy to solve for the optimal strategy in the uninformed phase. Let $V_U(\alpha)$ denote the value of an optimal strategy for an uninformed landlord who is visited by a potential tenant of type α. Standard reasoning implies,

$$V_U(\alpha) = \max\{\alpha, \delta[pEV_U + (1 - p)EV_I]\}$$

where p is the probability that the landlord who waits will remain uninformed and EV_U and EV_I denote the expected value of being uninformed and informed respectively. Again, the optimal strategy is for each owner to keep searching for tenants until they meet a tenant with type $\alpha \geqslant U$, where U satisfies the indifference condition

$$U = \delta[pEV_U + (1 - p)EV_I] \tag{4}$$

Given U, we pin down the other unknowns in the model. First, the probability that the landlord remains uninformed is the probability that all other landlords will reject their tenants. Since we are looking for symmetric equilibria, $p = U^{N-1}$. Second, it is easy to show that $EV_U = (1 + U^2)/2$. Substituting these into (4) and using (1) to eliminate δ and (3) to eliminate EV_I gives us an expression for U in terms of I:

$$U = U^{N-1} I\left(\frac{1 + U^2}{1 + I^2}\right) + (1 - U^{N-1})I(1 + w^2) \tag{5}$$

This is the equation that will be the focus of the analysis.

Proposition 1: *For any data* $N \geq 2$, $\delta \in (0, 1)$ *and* $W \in (0, (1 + I^2)/2$, *define* I *as the unique solution to (1) with* $I \in (0, 1)$ *and* $w = W/(1 + I^2)$. *There exists a unique equilibrium of the form given in Definition 1, with* $I(\lambda) = I$, $\bar{\lambda} = 2w$, *and* $U \in (I, \min [I, (1 + w^2)/I])$ *uniquely determined by (5).*

Properties of the equilibrium

Proposition 1 indicates that the model conforms to the case of Sixth Avenue. It is easy to see that $U > I$. This follows because the only difference between search in the two states is the possibility of becoming informed. If p were equal to zero then U would equal I, but $p > 0$ provides an added incentive to wait. The fact that landlords are more selective during the uninformed phase tends to increase the time until the first unit is occupied relative to the time to fill subsequent vacancies. It is also easy to confirm that, if the first match is successful enough that other owners choose to search for retail tenants, then the expected rental rate increases. The expected lease payment in the uninformed phase is $(1 + U)/2$. The expected payment in the informed phase given that $\lambda > \bar{\lambda}$ is $(1 + w)(1 + I)/2$. That the latter is greater follows from (4).[8] Rents rise due to the improvement in λ, in spite of the fact that less desirable α types may receive leases. Therefore, the model predicts the extended initial period of vacancies and both the higher rents and the increased speed with which lots became occupied once Bed Bath & Beyond succeeded.

The basic comparative static results of the model are intuitively clear. It is easily shown that the equilibrium level of U is strictly increasing in W. An increase in W raises EV_I, the amount that an owner receives from waiting for the information to be revealed. Similarly, an increase in N with U fixed raises the probability that another owner will fill their vacancy this period. This increases the value of waiting for the information on the appropriate use of the property. Any force that increases the value of rejecting a current offer must by definition, raise the minimum acceptable current offer, implying that U itself must rise with N.

The fact that U is increasing in N raises the question of how the expected waiting time until first occupancy, $1/(1 - U^N)$, varies with N. If U were independent of N then we know that adding enough vacancies would ultimately raise the probability of the information escaping in the first period to the limiting value of 1. It turns out, however, that it is indeed possible for delay to survive in the limit, as U(N) rises toward 1 quickly enough to preserve a strictly positive probability that all vacancies remain unfilled. The result is summarised in Proposition 2.

Proposition 2: *The limiting value of* U(N) *as* N *increases depends on the value of* $(1 + w^2)I$:

(a) *If* $(1 + w^2) I \leq 1$, *then* $U(N) \to (1 + w^2)I$, *and* $U(N)^N \to 0$.

(b) *If* $(1 + w^2)\,I > 1$, *then* $U(N) \to 1$ *and* $U(N)^N \to \tilde{p}$, *where*

$$\tilde{p} = \frac{[(1 + w^2)I - 1](1 + I^2)}{[(1 + w^2)(1 + I^2) - 2]I} \in (0, 1)$$

There is a good economic reason for the importance of the product $(1 + w^2)I$ in determining the nature of the free rider problem in large markets. It is readily confirmed that $(1 + w^2)I = \delta EV_I$ is the expected value of a vacant property in the next period should someone else's match reveal the true state today. If $(1 + w^2)I > 1$, there can be no equilibrium in which the information is practically guaranteed to come out, since this requires some landlord to accept an offer of no higher than one, while guaranteeing an expected payoff above one for anyone that turns down such an offer. This means that as the number of firms increases, the probability that each firm rejects a tenant must rise toward 1 fast enough to maintain aggregate uncertainty concerning whether someone else will accept an offer.

Proposition 2 shows how severe the free rider problem can become. Note that with w at its maximal value of $1/2, \tilde{p}$ approaches 1 as I increases to 1. What this means is that we can find large markets in which the expected wait until first occupancy, which heads toward $1/(1 - \tilde{p})$ as N increases, is arbitrarily long.

Recall that I is related to the discount rate δ by (1). Hence an I close to one implies a δ close to one. This suggests another experiment: we can think of an increase in δ as an increase in the search speed and ask what happens to the extent of delay as search becomes instantaneous. To address this question, fix the discount rate δ in terms of a period of calendar time, for example one year. Suppose that a period in the model corresponds to a fraction $1/n$ of a year. The discount rate for a period is therefore $\delta^{1/n}$. As n heads to infinity, the time between searches falls until search becomes effectively instantaneous. We saw above that in a large market if $EV_I > 1$ the expected number of periods until the first occupancy is $1/(1 - \tilde{p})$. Given n, this corresponds to $1/[n(1 - \tilde{p})]$ units of calendar time. It is easy to show that as n heads to infinity this delay approaches $EV_I/\ln\delta$, which is positive and finite. Again, the delay must be positive, since, if it were not, each agent would have an incentive to wait and act after learning the true state of the market.

Our final experiment concerns the social optimality of the equilibrium of the model. It is straightforward to see that a social planner subject to the same information constraints as the private market will face a greater incentive to accept offers in the uninformed period. This is because the social planner will take into consideration the potential value of the information to other market participants. It is also intuitive that as N increases to infinity the probability that some offer will be accepted in the first period will rise to one since the social value of the information rises rapidly with N. In contrast with the rapid release of information in the social optimum, Proposition 2 shows that the private market equilibrium may result in delay even as N rises without bound.

EXTENSIONS

Recall in the search market

Barnes and Noble decided against taking a lease in 1991 but changed its mind once it saw the success of Bed Bath & Beyond. This suggests that recall may be an important option available to landlords. To allow for recall we amend the model by, if in any period, each landlord can receive a new offer they may also recall any previous offer.

Let R denote the reservation type in the uninformed phase with recall. A landlord that meets a tenant of type R will be indifferent between leasing to his tenant and waiting until the next period. As before in a symmetric equilibrium the agent who does not accept will become informed with probability $1 - R^{N-1}$, in which case the agent faces the choice of immediately accepting W or searching. Note that because $R > I$, if this agent searches in the informed state, the agent already has a satisfactory offer in hand.[9] The agent will therefore only meet one more tenant and choose between this tenant and the tenant of type R. The expected value of informed search in therefore $\lambda(1 + R^2)/2$, and informed search occurs only if $\lambda > 2W/(1 + R^2)$.

The indifference condition that determines R is therefore:

$$R = \delta R^{N-1}\left(\frac{1 + R^2}{2}\right) + \delta(1 - R^{N-1})\left(\frac{1 + R^2}{2}\right)\left[1 + \left(\frac{W}{1 + R^2}\right)^2\right] \quad (5)$$

The left hand side is the return to accepting the reservation type. The first term on the right hand side represents the probability that the agent remains uninformed in the next period multiplied by the expected value of remaining uninformed, whereas the second represents the expected value if the information does become public multiplied by the probability of that event.

It is straightforward to show that (5) has a unique solution $R \in (U, 1)$, and that there is a simple symmetric equilibrium in which in the uninformed phase, any landlord accepts a type α tenant if and only if $\alpha \geqslant R$. Once the information has been revealed, any landlord with maximal offer $M \geqslant I$ allows in one more tenant and accepts the maximum offer, provided $\lambda \geqslant 2W/(1 + M^2)$ and otherwise sells immediately for W. For a searcher with maximal offer below I, the optimal strategy once the information has been revealed is identical to the optimal strategy in the case without recall.

The model with recall fits the timing of events on Sixth Avenue even better than the model without recall. Recall lengthens the time until first occupancy and gives rise to a burst of activity following the first success. Intuitively, recall is important in this setting because offers that would be unacceptable in the uninformed phase may become acceptable once the state of the market becomes known. The ability to come back to these offers makes the informed state more valuable than it would be in absence of recall, thereby increasing the minimum acceptable offer in the uninformed period. Given that $R > U$ the time to first occupancy is greater with recall. Given the backlog of acceptable offers, an initial success leads to an immediate burst of completed deals.

Ownership structure

A monopoly owner of all vacancies would be able to radically improve the functioning of the market by looking at the entire set of offers for the different units and accepting the best one. This would enable the monopolist to release the information early and reduce the amount of waste associated with committing units for incorrect use. To understand the monopolist's optimal strategy, note that once the information has been released, the monopolist follows exactly the same strategy as the competitive market. The difference occurs in the uninformed phase. Provided that the monopolist accepts at least one offer in a period, (3) gives the expected value of any units that remain unfilled: $\delta EV_I = (1 + w^2)I$. Any offers above this should also be accepted. The only issue is therefore whether the maximum offer should still be accepted, even if it is below this level. The answer is yes if and only if the maximal offer exceeds some lower bound B. The entire issue of computing the monopoly optimum reduces to the identification of B, and it is straightforward to show that $B < I$ reflecting the value that the monopolist places on learning market information, (see Caplin and Leahy, 1993b).

The monopoly optimum also has very different comparative static properties from the competitive outcome. Consider the effects of an increase in the value of office space W, with δ and N fixed. In economic terms, this makes the information on λ more valuable, since the alternative is more attractive. As we have seen, this makes competitive private agents less willing to accept an offer in the uninformed phase, since they are made to give up an increasingly valuable option, but for the monopolist however, the increased value of all vacant units makes it more important to release information early so that the proper use can be found for the vacancies. This suggests that an increase in W should lower B, and this is indeed the case.

It is not surprising that where information externalities are important, there are certain advantages to monopoly ownership. The shopping malls that have sprung up in so many places indicate that these incentives have not gone unnoticed in the market. The case with competitive owners appears more likely to characterise areas that are long established and have seen multiple use, such as Sixth Avenue. In complex urban settings, the nature of the correlations between payoffs at different locations changes over time, and so the relevant peer group also changes. This means that without monopoly ownership of the entire city, it is very likely that any new developments will cut across units that have different owners. This does not mean that there will not be efforts at consolidation and takeover once the common interest is noticed, but there is no evidence that such a process can take place without itself involving the expenditure of considerable time and effort, especially if hold-up problems arise. Even without common ownership, the various landlords may write contracts with side payments aimed at improving asset allocation in the spirit of Coase. However this may be, the model points to the importance of ownership structure in determining the speed with which large numbers of vacant spaces become filled up.

Information and payoff structure

There are several conceptually straightforward amendments to the model that would reduce the extent of the information externality without fundamentally changing the message. If the payoffs at the various locations were less than perfectly correlated, then there would be less to be gained by waiting for another store's results. If there was a smaller cost of switching the unit among its various uses, then the loss caused by an incorrect decision would fall which would also reduce the free rider problem. In addition, allowing for landlord heterogeneity may greatly reduce the extent of the free rider problem in large markets.

There are other changes that may operate to worsen the problem. One of these is the additional externality introduced when there are crowd spillover effects. As we show in the next subsection, this also tends to give rise to a sub-optimal delay in filling the first vacant unit. Another such change would involve endogenising the extent of the search effort. In this case, the presence of other searches would slow down the privately optimal search speed. Information is a public good and would be underprovided by private agents. The reduction in search speed would add another force tending to delay information revelation, especially in large markets.

Another set of issues arise in considering changes in the information structure. There are many ways to change the information flow that all end up reducing the extent of the learning that takes place from a single observation of a store's success. One way to assure this is to make the reasonable assumption that observations of rents and profitability are highly noisy. One could also imagine changing the structure of information and the protocol in individual bargains in a manner that would impede the flow of information among agents. While these amendments should be expected to reduce the discontinuity in market outcomes, there is little reason to believe that this would change the fundamental form of the externality.

Crowd effects

We now amend the model to allow for crowd spill-over effects. To keep things simple, we get rid of the information spill-over. To capture the crowd effects in the simplest possible manner, we assume that the payoff to an α type in any period is $\alpha(1 - \delta)$ while there are no other units occupied, and rises to $\lambda\alpha(1 - \delta)$ per period as soon as there is a single other unit occupied, with $\lambda > 1$. This means that one store alone does not draw the crowd, but that the presence of two stores draws in the whole crowd. In all other respects we retain an identical model structure, although there is no role for the outside option W.

In this framework just as in the model of information spill-overs, there is a two-phase market history, with a reservation type C with no other units occupied, and I once one or more unit is occupied. The model is identical to the information model in the second phase, so that the second cut-off is the familiar value of I identified in (1) above.

To calculate C we use the condition for indifference between acceptance and rejection. Note that both of these values are endogenous to the model and depend on the equilibrium strategies themselves. To see how much to charge an α type with no other occupants in place, the landlord must compute the expected present value of occupancy for this α type: it will be above α but below $\alpha\lambda$. The expected surplus to an α type who takes occupancy in a period with no prior occupants depends both on the probability that some other unit gets occupied in this period, and the probability that units get occupied in subsequent periods. Hence the value depends on both C and I: the value can be written as $\alpha K(C, I)$, where a little algebra shows.

$$K = \lambda - (\lambda - 1)C^{N-1}\left(\frac{1 - \delta}{1 - \delta I^{N-1}}\right) \tag{7}$$

A little further manipulation yields the indifference condition as between accepting and rejecting the offer when no units are occupied as,

$$CK(1 + I^2) = C^{N-1} I(1 + C^2) K + (1 - C^{N-1}) I(I + I^2)\lambda \tag{8}$$

Together (7) and (8) determine C and K.

It is straightforward to establish that an equilibrium exists, and that any solutions satisfy $C > I$, so that owners are more selective when the market is small, and $K(1 + C) < \lambda(1 + I)$, so that rents rise with market size. This means that the crowd story fits many of the same qualitative aspects. In particular, early entry encourages later entry. There is, however, a big difference, in that the crowd effect is in many ways less strong. In economic terms the reason for this is that the cost of leading is far higher in the model of information externalities, in which entry may be a permanent mistake.

One reflection of this difference between the information model and the crowd model is shown by considering what happens as the number of vacancies grows. In contrast to the information model, in the crowd model one can show that in large markets as N leads to infinity, any sequence of equilibria C(N) falls toward I. The reason for this is that with crowd spillovers, the more you trust someone else to accept a tenant, the more you wish to accept a tenant yourself. In a large market, if all tenants of type $\alpha \geqslant I$ are accepted in the first period, then it is practically guaranteed that there will be a crowd right away. There is no need to wait. The more general issue is that while a large number of players may help to resolve the crowd externality problem, large numbers play a less essential role in resolving informational problems. In the information model, the fact that another landlord is accepting a tenant encourages delay, so that the acceptance decision of distinct landlords are strategic substitutes, while in the crowd model they are strategic complements.

CONCLUSIONS

We have developed a theory of search with information externalities, and have shown how this may help account for the regeneration of Sixth Avenue in the middle of a recession. In qualitative terms, the key assumption is that the first store that makes the costly decision to locate must do so without knowing the level of some market risk, and that their act of locating will help to reveal this to those who have waited. The result is an incentive for all to free ride off the decisions of someone more anxious to locate than themselves.

We feel that similar forces are at work in the many markets in which there are changes over time in the optimal pattern of usage of given resources, and in which there are costs of moving the resources between uses. This makes it important to explore the applicability of analogous frameworks in the labour market and the capital market as well as the property market. At an even broader level, the model points to the importance of learning from the actions of others in the process of search and adjustment. We have argued elsewhere that such patterns of learning are pervasive in many markets (see Caplin and Leahy, 1995), and we are actively exploring these issues in a variety of different settings.

1 The crowd externality story is implicit in many models of strategic complementarily and is discussed in detail by Rauch (1993) in the context of shopping malls.
2 It may be possible to develop a monopoly version of the crowd externality in which the first tenant gets a low rent in order to attract future tenants. Monopoly ownership, however, appears extremely unlikely in Manhattan.
3 In her article, Deutsch points out that Barnes and Noble had backed out of a possible deal in 1991.
4 In contrast to the current work, the externalities that are most often discussed in the search literature deal with the effects that the behaviour of other searchers has on the matches

available to a searcher. Congestion externalities (Tobin, 1972) shift the distribution of matches one way, while thin market externalities (Diamond, 1981) shift it the other direction. Mortenson (1982) discusses the externalities that arise through the division of the surplus generated by a match. The papers most closely related to ours are Bolton and Harris (1993), who consider the effect of information externalities on experimentation in a multi-armed bandit setting, and King (1995), who analyses herding equilibria in a search setting in which one person searches while others wait and free-ride upon the searcher's decision.

5 In what follows we will be comparing the present values of various options. These present values, such as $\alpha\lambda$, will, in general, depend on the discount rate δ. As changes in the discount rate affect all present values proportionately, we will not need to take account of this effect when calculating the effect of changes in δ on agents' decisions.

6 There may also be asymmetric equilibria to our model as in King (1995).

7 Note that this definition of equilibrium applies only to values of W low enough to ensure that search is worthwhile in the uninformed phase.

8 Equation (4) shows that $U < \delta EV_I = I(1 + w^2)$. This and the restrictions on I and w imply $1 + U < (1 + I)(1 + w)$.

9 Note that the presence of recall does not affect I. Also $R > U$ as discussed below.

BIBLIOGRAPHY

Banerjee, A.V., 1992, "A Simple Model of Herd Behavior", *Quarterly Journal of Economics* 107, pp. 797–818.

Bikhchandani, S., D. Hirshleifer and I. Welch, 1992, "A Theory of Fads, Fashions, Customs, and Cultural Change as Informational Cascades", *Journal of Political Economy* 100, pp. 992–1025.

Bolton, P. and C. Harris, 1993, "Strategic Experimentation", Mimeograph, Oxford University.

Caplin, A. and J. Leahy, 1993a, "Sectoral Shocks, Learning, and Aggregate Fluctuations", *Review of Economic Studies* 60, pp. 777–94.

Caplin, A. and J. Leahy, 1993b, "Miracle on Sixth Avenue", Harvard University Discussion Paper 1665; reprinted as Chapter 9 of the present volume.

Caplin, A. and J. Leahy, 1994a, "Business as Usual, Market Crashes and Wisdom after the Fact", *American Economic Review* 84, pp. 548–65.

Caplin, A. and J. Leahy, 1994b, "Mass Layoffs and the Natural Rate of Unemployment", Working Paper 4766 (NBER).

Caplin, A. and J. Leahy, 1995, "The Economics of Adjustment", in R. Cross (ed.), *The Natural Rate of Unemployment Twenty Five Years On* (Cambridge University Press).

Davis, S. and J. Haltiwanger, 1992, "Gross Job Creation, Gross Job Destruction, and Employment Reallocation", *Quarterly Journal of Economics* 107, pp. 819–64.

Davis, S., J. Haltiwanger and S. Schuh, 1996, *Job Creation and Destruction* (Cambridge: MIT Press).

Deutsch, C., 1993, "From the Carriage Trade to the Category Killers", *Sunday New York Times,* June 27, p. 13.

Diamond, P., 1981, "Mobility Costs, Frictional Unemployment, and Efficiency", *Journal of Political Economy* 89, pp. 798–812.

Huang, D. and S. Shirai, 1994, "Information Externalities Affecting Foreign Direct Investment: The Case of China", Mimeograph, Columbia University.

King, S., 1995, "Search with Free Riders", *Journal of Economic Behavior and Organization* 26, pp. 253–71.

Mortensen, D. 1982, "The Matching Process as a Non-cooperative Bargaining Game", in J. McCall (ed.), *The Economics of Information and Uncertainty* (University of Chicago Press).

Rauch, J., 1993, "Does History Matter only when it Matters Little? The Case of City-Industry Location", *Quarterly Journal of Economics* 108, pp. 843–67.

Rob, R., 1991, "Learning and Capacity Expansion under Demand Uncertainty", *Review of Economic Studies* 58, pp. 655–75.

Tobin, J., 1972, "Inflation and Unemployment", *American Economic Review* 52, pp. 1–18.

Part IV

Corporate Securities as Strategic Options

10

*Warrant Valuation and Exercise Strategy**

David C. Emanuel

The University of Texas at Dallas

This chapter demonstrates that warrant valuation and exercise strategy differ fundamentally from call option valuation. Simultaneous exercise of warrants is shown to be suboptimal and a monopolist owning all warrants can achieve a higher value. Unless warrants are perfectly divisible, no satisfactory equilibrium exists for the valuation and exercise of widely held warrants. The problems encountered appear to be quite general and stem from necessary assumptions about future corporate dividend policy and capital structure. Such assumptions are necessary for any model of corporate security valuation.

On the face of it, the theoretical valuation of warrants should be trivial given an adequate theory of option valuation, as pioneered by Black and Scholes (1973). Apart from differences in lifetime, the effect of which may initially be ignored, the sole difference is that a warrant is a security issued by the firm and a call option is not, being instead a marketable side-bet. Because of this, there is some dilution of equity and dividend when the warrant is exercised. However, this difference would seem trivial and it would appear that a minor modification of the boundary conditions to reflect equity dilution at exercise would be sufficient to model the difference.

This turns out not to be the case. The theoretical valuation of call options and warrants involves making assumptions about the future dividend

* This chapter was first published in *Journal of Financial Economics* 12, 1983, pp. 211–35 and is reprinted with the permission of Elsevier Science. It is based on Chapter 3 of the author's PhD dissertation, "Extensions and Developments of Option Pricing", unpublished, Stanford University (1980). The author would like to thank his dissertation committee, J.C. Cox, D.M. Kreps and R.A. Litzenberger, all of Stanford University, for their helpful guidance and patience. He would also like to thank M.J. Brennan of The University of British Columbia for helpful suggestions and C.W. Smith of the University of Rochester for suggesting the approach used in the second section.

policy and capital structure of the firm. Subject to these and other assumptions, the calls can be uniquely priced and the price is independent of the number of calls written, as each call is a separate bet. Given rational participants, all call options will be exercised simultaneously. However, when multiple warrants are outstanding, the exercise of some of the warrants leads to the creation of new shares and to concomitant changes in the dividend policy and capital structure of the firm. The value accruing to one warrant holder is therefore not independent of what the other warrant holders do. In general, it ceases to be optimal to exercise all warrants simultaneously.

In this chapter, the implications of this observation are examined in detail. For the most part, the discussion focuses on the problem of valuing perpetual warrants. Even in this simple case, the task of valuing multiple warrants is of course intimately connected with the problem of optimal warrant exercise, which has no obvious solution.

The discussion begins with an analysis of the situation where all warrants are to be exercised simultaneously. This is called the "naive" exercise strategy. Valuation functions are obtained in both the certainty and uncertainty cases and are found to belong to the same one parameter family of functions. Because of this fortunate result, one may often usefully refer to the problem of valuation in a certain world without loss of generality.

In general, a higher aggregate value of the warrants outstanding is obtained if they are not exercised simultaneously, but are instead exercised optimally. This is the strategy a rational monopolist owing all warrants would follow.

Of considerable interest is the question of whether a competitive equilibrium valuation function and exercise strategy with widely held warrants exists. A concept of a non-simultaneous exercise equilibrium is introduced and developed. However, this "equilibrium" has the disturbing property that if fair prices for all securities exist, then apparent arbitrage opportunities exist, although any attempt to exploit the arbitrage opportunity leads to instant revision of expectations and prices and makes the arbitrage no longer attractive.

If the possibility of a monopolist owing all warrants is excluded, it is not clear what will happen. Analysis of zero-sum games among many players without communication presents insurmountable difficulties, particularly in a continuous-time context. If limited communication is allowed, one suggestion, although hardly satisfactory, is the concept of a "pre-emptive strategy" solution whereby players may pre-empt one another by announcing their intended exercise strategies.

Since the dividend and capital structure policy of the firm is crucial, considerable discussion is given to this matter. In particular, the assumptions are examined to determine whether the results are specific to the particular set of assumptions made. This is not the case. Similar problems also exist with convertible bonds and other securities issued by the firm. There is some discussion of the "neutral dividend policy". This refers to a hypothetical adjustment of the firm's dividend policy as warrants are exercised in such a way that simultaneous exercise becomes an optimal policy. There exists a unique satisfactory policy of this type.

While the analysis and discussion presented herein suggest that using inferior strategies reduces warrant value by only a modest amount in most normal circumstances, it should be realised that analogous problems exist in almost every economic context.

NAIVE WARRANT VALUATION IN A WORLD WHERE ASSET RETURNS ARE DETERMINISTIC

The following sufficient set of assumptions is used:

(A.1) The short-term riskless interest rate r is constant.

(A.2) All assets and securities are continually tradeable in any amount without transaction costs or taxes.

(A.3) All assets are priced to yield a deterministic rate of return identically equal to r. (These assumptions are stronger than they need be. Given the sure rate of return, a requirement that there be no arbitrage opportunities would imply that all rates of return equal r.)

(A.4) The firm consists of assets whose market value equals V(t) at time t. Out of assets a continuous proportional dividend is paid to shareholders at a constant rate α. That is to say, shareholders receive dividends totalling $V\alpha\Delta t$ in the infinitesimal period of time between t and $t + \Delta t$.[1]

(A.5) Claims on the firm consist of n_1 identical shares and n_2 identical warrants with an exercise price of X per warrant. Warrants may be exercised at any time in perpetuity and receive no dividends until exercised. Money received from warrant exercise is used to purchase more assets and leads to an increase in the scale of the firm.

(A.6) All securities are rationally priced to yield a rate of return r subject to the stated assumptions about warrant exercise being correct.

(A.7) All warrants will be exercised simultaneously at the optimal time so that their value is maximised.

Assumption (A.2) implies that, in order to avoid arbitrage possibilities, $V = n_1S + n_2W$. This is essentially the Modigliani and Miller (1958) irrelevance proposition for capital structure. Liberal use is made of this result throughout the analysis here.

Assumption (A.4) is somewhat stronger than the analogous assumptions used in the derivation of the Black–Scholes (1973) argument for option

Table 1 Definition of symbols used

Symbol	Definition
t:	Calendar time
V(t):	Value of firm at time t
n_1:	Number of shares of firm outstanding
n_2:	Number of outstanding warrants to purchase one share of the firm
X:	Exercise of price of a warrant
r:	Interest rate
σ^2:	Variance rate of return on the firm
μ:	Expected rate of return on the firm
α:	Proportional dividend rate being paid continuously by the firm
$W_N(V,n_1,n_2)$:	Average warrant value given that the naive exercise strategy is followed[a]
$S_N(V,n_1,n_2)$:	Rational share price given that it is known with certainty that the naive warrant exercise strategy will be followed
$V_N^*(n_1,n_2)$:	Value of the firm at which to exercise the next warrant given that the naive exercise strategy is followed
$A_N(n_1,n_2)$:	Constant coefficient in average warrant value formula
W_0,S_0,V_0^*,A_0:	Analogous definitions to the above subject to the warrant exercise strategy that maximises aggregate warrant values being followed
W_C,S_C,V_C^*,A_C:	Analogous definitions to the above subject to the competitive "equilibrium" exercise strategy being followed
ξ:	Exponent of V in warrant value formula (also known as "XI")

[a]Note that, in general, warrant value has the form $W = AV^{\xi}$. Also note that the use of terms such as $W(V,n_1,n_2)$ in equations can be very cumbersome. Therefore, the parenthesised arguments are often omitted where their values should be obvious from the context. Whenever a comma appears inside parentheses, the contents of the parentheses should be read as an argument string. Otherwise, the contents should be read as an arithmetic expression, if possible.

pricing. In addition to having perfect markets in the relevant securities, the underlying assets of the firm must be traded in a perfect market. Alternatively, one could instead require that the aggregate value of securities outstanding equal the theoretical value V. The fact that slightly stronger sufficient assumptions are used in deriving a hedging argument for warrant value should not be viewed as being particularly disturbing. The comments made here apply equally to the problem of valuing preferred shares or any other corporate securities.

One might ask why warrants would be issued in the first place if their existence does not affect the market value of the firm. This objection cannot readily be answered under the perfect market assumptions listed above. Existence of warrants would be neither helpful nor harmful. When the assumptions of market perfection are relaxed, warrants need no longer be redundant securities. This would then appear to imply that the value of the firm is dependent on its capital structure and to contradict the irrelevance proposition of Modigliani and Miller (1958). However, as the discussion by Miller (1977) indicates in the case of corporate debt, the Modigliani–Miller proposition may be false for all firms in aggregate and, at the same time, true for each individual marginal firm. In this context, one should also reference Jensen and Meckling (1976) and Smith and Warner (1979). Without further discussion of the issue, Assumption (A.2) is therefore made and its necessary implications for irrelevance of capital structure are gratefully utilised.

The assumptions made here and in the next section are, in most important respects, equivalent to those made by Smith (1977). The key difference is that Assumption (A.7) implies that the warrants are indivisible, whereas Smith assumed infinite divisibility. Smith also assumed that all traders were price takers. The subsequent development of this chapter shows the interesting consequences of making warrants indivisible, regardless of whether or not warrant holders are price takers.

Given the perpetual stationary setting of this problem it is sufficient to determine $V_N^*(n_1, n_2)$, the value of the firm at which all n_2 warrants should simultaneously be exercised when n_1 shares are outstanding. Subject to the assumptions of constant rate of return and constant dividend rate, the value of the firm will grow at a steady rate $r - \alpha$ (or decline steadily if α exceeds r).

If V_N^* is less than $V_N^*(n_1, n_2)$, it will then grow to $V_N^*(n_1, n_2)$ in time t, where

$$Ve^{(r-\alpha)t} = V_N^*(n_1, n_2)$$

or

$$e^{rt} = (V_N^*(n_1, n_2)/V)^{(r/(r-\alpha))} \tag{1}$$

At that point in time, by assumption, all warrants will be exercised. The exercise price obtained by exercising n_2 warrants raises the firm's value to $V_N^*(n_1, n_2) + n_2X$. Post-exercise each share will then be worth $(V_N^*(n_1, n_2) + n_2X)/(n_1 + n_2)$, since there will be $n_1 + n_2$ shares of identical value. Immediately prior to exercise, each warrant will be worth X less than this amount because the exercise price is yet to be paid. That is to say, each warrant at time t will be worth $(V_N^*(n_1, n_2) + n_2X)/(n_1 + n_2) - X$. Discounting this value at rate r back to time 0 one obtains

$$\begin{aligned} W_N(V, n_1, n_2) &= e^{-rt}W_N(V(t), n_1, n_2) \\ &= e^{-rt}((V_N^*(n_1, n_2) + n_2X)/(n_1 + n_2) - X) \\ &= e^{-rt}(V_N^*(n_1, n_2) - n_1X)/(n_1 + n_2) \end{aligned}$$

Substituting from equation (1) one obtains

$$W_N(V, n_1, n_2) = (V/V_N^*)^{(r/(r-\alpha))}(V_n^* - n_1X)/(n_1 + n_2) \tag{2}$$

Given Assumption (A.7), it remains to determine at what value of V_N^* this expression is maximised. Simple calculus reveals that

$$V_N^* = rn_1X/\alpha = \xi n_1X/(\xi - 1) \qquad \textbf{(3)}$$

where

$$\xi = r/(r - \alpha) \qquad \textbf{(4)}$$

This is the optimal naive strategy as long as $r > \alpha$. (If $r < \alpha$, the optimal strategy is to exercise immediately if and only if V exceeds n_1X.)

Substituting back into equation (2),

$$W_N(V, n_1, n_2) = (V(\xi - 1)/(\xi n_1X))^{\xi} n_1X/((n_1 + n_2)(\xi - 1)) \qquad \textbf{(5)}$$

$$= (V\alpha/(rn_1X))^{(r/(r-\alpha))}(r - \alpha)n_1X/((n_1 + n_2)\alpha) \qquad \textbf{(6)}$$

At this point some pertinent observations are in order:

First, it should be noted that ξ is a single parameter sufficient to describe the combined effects of r and α. This will also be evident in the next section.

Second, the number of shares and warrants enter into equations (3) and (5) in a very trivial way. Rewriting equation (5) as

$$W_N(V, n_1, n_2) = (V/V_N^*)^{\xi} n_1X/((n_1 + n_2)(\xi - 1))$$

it can be seen that the equity dilution factor $n_1/(n_1 + n_2)$ enters simply as a multiplier in the formula. This is hardly surprising since the warrants have value solely through being exercised and since the payoff on the exercise depends linearly on value per share. The fortunate consequence is that, in essence, one needs only to calculate naive warrant values and exercise strategies in one case (or, for example, in the analogous case involving call options) to have effectively obtained an answer for all combinations of n_1 and n_2. This is true not only for the perpetual warrants discussed there, but also for finite-lived warrants.

Third, if the naive warrant exercise strategy is followed, then the rational share price (ie, based on the assumption of that exercise strategy) will be

$$S_N(V, n_1, n_2) = (V - n_2W_N(V, n_1, n_2))/n_1$$

and

$$S_N(V, n_1, n_2) - W_N(V, n_1, n_2) = (V - (n_1 + n_2)W_N(V, n_1, n_2))/n_1$$

$$= (V - (V(\xi - 1)/(\xi n_1X))^{\xi} n_1X/(\xi - 1))/n_1$$

Examination of this expression indicates that if $\xi > 1$ (ie, if $r > \alpha$) and if $V \leqq V_N^*$ (which, by assumption, it is as exercise occurs at V_N^*) then the value of this expression is $\leqq X$, with strict equality only occurring when $V = V_N^*$. This is reassuring as the condition

$$S - W \leqq X$$

is normally regarded as a globally applicable "no-arbitrage" constraint for a warrant.

Fourth, in the only interesting case (ie, where $\xi > 1$), reference to equation (3) shows that

$$\Delta V_N^* / \Delta n_1 > X$$

This observation is crucial to this chapter, as shall be seen in the fourth section.

NAIVE WARRANT VALUATION WHEN ASSET RETURNS ARE RISKY

As in the previous section, warrant valuation theory under uncertainty can be treated analytically subject to a suitable set of assumptions. These assumptions are (A.1), (A.2), (A.4), (A.5), and (A.7) as defined in the previous section.

(A.8) Securities are rationally priced so that the return on no combination of securities or assets strictly dominates the return on any other combination. Dominance here is defined, in the manner of Merton (1973) to mean that return on the dominant combination surely at least equals and with some probability exceeds return on the dominated combination.

(A.9) The return on the assets of the firm, conditional on their initial value, has a lognormal distribution with constant rate of variance. More explicitly, the instantaneous increment may be defined by

$$dV/V = (\mu - \alpha)dt + \sigma dz \qquad (7)$$

where μ is the expected rate of return on the firm, α is the constant dividend rate and σ is the square root of σ^2 the constant variance rate of return. dz is the instantaneous increment in z(t), the sample path of a Wiener process with zero mean drift and unit variance rate. V naturally jumps considerably when some or all warrants are exercised. However, this is treated as a boundary condition and is of no immediate concern.

Since, by assumption, the firm's value and the warrants are tradeable, it is possible to construct a riskless hedge. This is done by holding $\partial W/\partial V$

times the firm long and one warrant short. If one argues that this hedge is riskless and must therefore earn the riskless rate of interest, then one obtains the differential equation

$$(rV - \alpha V)(\partial W_N/\partial V) - rW_N + (\sigma^2/2)V^2(\partial^2 W_N/\partial V^2) = 0 \qquad \textbf{(8)}$$

where the partial derivatives may be taken to be ordinary derivatives, since there is no time dependence and n_1 and n_2 only change at the boundary $V_N^*(n_1, n_2)$. The analysis leading to equation (8) exactly parallels Black and Scholes (1973). All solutions to this equation have the following form:

$$W_N = AV^{\xi_1} + BV^{\xi_2} \qquad \textbf{(9)}$$

where ξ_1 and ξ_2 are roots of the quadratic equation

$$(r - \alpha)\xi - r + (\sigma^2/2)\xi(\xi - 1) = 0 \qquad \textbf{(10)}$$

Examination reveals that the lower root ξ_2 of (9) must be negative and that ξ_1 must be greater than one if the risk free rate, r, the dividend rate, α and σ, the standard deviation of the rate of return, are strictly positive.

Since $W(0,n_1,n_2) = 0$, it follows that the coefficient B in (9) is zero. Therefore,

$$W_N(V, n_1, n_2) = A_N(n_1, n_2)V^{\xi}$$

where

$$\xi = (-(2r - 2\alpha - \sigma^2) + \sqrt{((2r - 2\alpha - \sigma^2)^2 + 8r\sigma^2)})/2\sigma^2 \qquad \textbf{(11)}$$

(At this point it should be noted that all stationary exercise strategies will lead to valuation equations of the form shown in (11), as long as the strategy has the reasonable property that the warrant is exercised immediately $V \geqq V^*(n_1,n_2)$. Given this property, then whenever some warrants remain unexercised after exercise of some warrants, V will still be in the interior of the region bounded above by V^* and below by 0.)

From Assumption (A.7) requiring naive simultaneous exercise, the following boundary condition is obtained at $V_N^*(n_1, n_2)$:

$$W_N(V_N^*(n_1, n_2),n_1, n_2) = (V_N^*(n_1, n_2) - n_1X)/(n_1 + n_2) \qquad \textbf{(12)}$$

This relationship comes from the identical argument immediately following equation (1). Combining equations (11) and (12),

$$\begin{aligned} A_N(n_1, n_2) &= W_N(V_N^*(n_1, n_2),n_1, n_2)/V_N^{*\xi} \\ &= (V_N^*(n_1, n_2) - n_1X)/(n_1 + n_2)V_N^{*\xi} \qquad \textbf{(13)} \end{aligned}$$

Therefore,

$$W_N(V, n_1, n_2) = (V/V_N^*)^{\xi}[(V_N^*(n_1, n_2) - n_1X)/(n_1 + n_2)] \qquad \textbf{(14)}$$

This is identical to equation (5) obtained in the previous section and consequently all the subsequent analysis of that section is valid, except that ξ should be read as being defined by (11) instead of (4). The formal similarity of the solutions obtained in a world of certainty and a world of uncertainty is thus established. If understanding or intuition is aided one may feel at liberty to consider oneself living in the certain world with the equivalent value of ξ. (This was suggested by C.W. Smith.)

THE SUBOPTIMALITY OF SIMULTANEOUS EXERCISE

Using the analysis of either the second or third section, if the warrants are to be exercised simultaneously, then exercise should occur when

$$V = V_N^*(n_1, n_2) = \xi n_1 X/(\xi - 1)$$

where, with the exception of an uninteresting special case, ξ is strictly greater than one. Given $\xi > 1$, it is relatively simple to demonstrate the suboptimality of simultaneous exercise when Assumption (A.7) is dropped and warrants may be exercised separately. (For the moment, the criterion of optimality is undefined but may be thought of as maximising aggregate warrant value.)

Lemma 1. If m *warrants are exercised when* $V = V_N^*(n_1, n_2)$, *then a rational holder of the remaining* $n_2 - m$ *warrants, knowing that fact, will not immediately follow suit and exercise all his warrants simultaneously.*

Proof. After the exercise of m warrants at the point assumed, the value of the firm will be increased to $V_N^*(n_1, n_2) + mX$,

$$\begin{aligned} V_N^*(n_1 + m, n_2 - m) &= \xi(n_1 + m)X/(\xi - 1) = \xi n_1 X/(\xi - 1) + mX + mX/(\xi - 1) \\ &> V_N^*(n_1, n_2) + mX \quad \text{(since } \xi > 1\text{)} \end{aligned}$$

Therefore, after the exercise of m warrants, the value of the firm will be below the point at which the holder of the remaining $n_2 - m$ warrants would exercise them simultaneously. Q.E.D.

It is not easy to give Lemma 1 or the subsequent lemmas any intuition. The results stem mechanically from the functional form of V_N^*. Essentially, exercising some warrants changes the character of the firm. In particular, while α the dividend rate on the firm is unchanged, the dividend rate per share declines as warrants are exercised. Since dividends provide the main reason for exercising warrants (unless they are about to expire), the incentive to exercise a warrant might well decline if the rate of dividends per share declines due to other warrants being exercised. A later section explores the dividend policy assumption more thoroughly.

Lemma 2. Given the conditions of Lemma 1:

(1) *The remaining* $n_2 - m$ *warrants are worth more than* $W_N(V^*_N, n_1, n_2)$ *each.*
(2) *Ex post, the* m *warrants initially exercised were worth less than* $W_N(V^*_N, n_1, n_2)$ *each.*
(3) *Total value of all* n_2 *warrants is greater than* $n_2 W_N(V^*_N, n_1, n_2)$.

Proof of (1). The holder of the remaining $n_2 - m$ warrants could immediately exercise with the effect that all n_2 warrants are exercised simultaneously. This would guarantee a value of just $W_N(V^*_N, n_1, n_2)$. However, this has been shown by Lemma 1 to be a suboptimal strategy. Therefore, his warrants must be worth more than $W_N(V^*_N, n_1, n_2)$. Q.E.D.

The intuition here is simply that the holder of the remaining $n_2 - m$ warrants has freedom of choice. All things being equal, the rational exercise of choice can only benefit him.

Proof of (2). If the holder of the remaining $n_2 - m$ warrants at once exercises them, the shares would each be worth $S_N(V^*_N, n_1, n_2)$. But he chose not to do this, thereby increasing the value of his warrants. This gain in value is at the expense of the owners of the $n_1 + m$ shares. Each share is therefore worth less than $S_N(V^*_N, n_1, n_2)$. Therefore, *ex post*, the exerciser of the m warrants obtains shares worth less than $S_N(V^*_N, n_1, n_2)$ each. Their net value is less than $S_N(V^*_N, n_1, n_2) - X$. But

$$W_N(V^*_N, n_1, n_2) = S_N(V^*_N, n_1, n_2) - X$$

Therefore, the value of his warrants is less than $W_N(V^*_N, n_1, n_2)$ each. Q.E.D.

By assumption, the participants are in a zero-sum game. Any choice which benefits one group must hurt the other group, made up of shareholders.

Proof of (3). The aggregate value of all claims on the firm must equal V (from Assumption (A.8) for example). Also

$$n_1 S_N(V^*_N, n_1, n_2) + n_2 W_N(V^*_N, n_1, n_2) = V$$

But in the proof of (2) above, it was shown that each share was worth less than $S_N(V^*_N, n_1, n_2)$. Therefore, in order for the aggregate of all claims to be worth V, the n_2 warrants must collectively have a value greater than $n_2W_N(V_N^*, n_1, n_2)$. Q.E.D.

Lemma 2.3 simply says that since all shareholders were hurt by the change in warrant exercise strategy, the class of all warrant holders must have gained from the change. Some warrant holders may have been hurt, but the wealth gain to other warrant holders outweighs that loss.

The basic features of the problem have now been identified. The naive assumption of simultaneous exercise at $V_N^*(n_1, n_2)$ does not seem reasonable. There is a distinct advantage to letting someone else exercise his warrants at $V_N^*(n_1, n_2)$ and simply waiting. This raises the question of who rationally would exercise his warrants first.

THE OPTIMAL STRATEGY FOR A MONOPOLIST CONTROLLING ALL THE WARRANTS

Quite often, the outstanding warrants and convertible securities issued by a corporation will be closely held or controlled by a single individual. This situation can arise in a number of ways. First, the senior management may have issued warrants to themselves. The ostensible reason for this might be to deter unwelcome takeovers. Second, where a venture capitalist has provided substantial financing, he may also have obtained a large number of warrants. In this case, the intention would be to allow the venture capitalist to seize control of the corporation if necessary. A third possibility arises in connection with underwritten equity issues. Smith (1977) reports that warrants are quite often issued to the underwriters as an additional form of compensation. In each of these cases, the warrant holders might well be expected to act in a coordinated way, much as a monopolist would.

A rational person owning claims will seek to maximise their total value. Since the aggregate value of all claims on the firm equals the value of the firm and since exercising warrants cannot *ex ante* make the firm more or less valuable, a rational person will then attempt to exercise his warrants in such a way as to minimise the aggregate value of those claims on the firm that he does not own.

A monopolist owning all warrants and perhaps some shares as well (but less than 100%) will not own all the shares. Therefore, his exercise strategy will seek to minimise the value of the shares he does not own. In other words, he will seek to minimise share value at all times. This, of course, is the same as seeking to maximise aggregate warrant value at all times. However, it was necessary to establish that the monopolist's strategy would not depend on his stock position. For, in the course of exercising warrants or hedging, the monopolist will acquire a position in the common stock of the firm.

Furthermore, the monopolist will have no incentive to divest himself of any of his warrants. If he does so a suboptimal exercise strategy will probably be used resulting in share value not being minimised. It is clear then that a stable situation will result if a monopolist owns all the warrants. He has no incentive to deviate from a strategy that minimises share value. The other rational shareholders will recognise this and the share price will therefore reflect this strategy. The monopolist, if he chooses, will be able to hedge his position by shorting the correctly priced shares. The warrant price will not, of course, be observable, there being no market in the warrants. However, an average warrant value will be readily computable.

It is known from the analysis of the previous section that the monopolist will not exercise his warrants simultaneously. An interesting result obtainable by virtually identical analysis is that as each marginal warrant is exercised by the monopolist, it is worth less than the average warrant yet to be exercised. Consequently, as each warrant except the last is exercised, the share price at exercise will be less than the average warrant value plus the exercise price. Therefore, use of the difference between share price and average warrant value will be a poor guide to determining the monopolist's strategy.

The optimal exercise strategy for the monopolist may be defined by the following recursive method:

First note that

$$A_0(n_1,1) = A_N(n_1,1)$$
$$= (V_N^*(n_1,1) - n_1X)/(n_1+1)V_N^{*\xi}(n_1,1) \quad \text{[from (13)]} \tag{15}$$
$$V_0^*(n_1,1) = V_N^*(n_1,1) \tag{16}$$

since with only one warrant outstanding, the optimal and naive strategies must be identical.

The recursive optimisation step can be defined with the help of the property that the rational share price will not change as a warrant is correctly exercised. That is,

$$\begin{aligned} S_0(V_0^*(n_1,n_2),n_1,n_2) &= S_0(V_0^*(n_1,n_2)+X,n_1+1,n_2-1) \\ &= (V_0^*(n_1,n_2) - n_2W_0(V_0^*(n_1,n_2),n_1,n_2))/n_1 \\ &= (V_0^*(n_1,n_2) - n_2A_0(n_1,n_2)V_0^{*\xi}(n_1,n_2))/n_1 \end{aligned}$$

(recalling that A was introduced in the list of definitions so that average warrant value $W = AV^{\xi}$),

$$\begin{aligned} &S_0(V_0^*(n_1,n_2)+X,n_1+1,n_2-1) \\ &= (V_0^*(n_1,n_2)+X-(n_2-1)W_0(V_0^*(n_1,n_2)+X,n_1+1,n_2-1))/(n_1+1) \\ &= (V_0^*(n_1,n_2)+X-(n_2-1)A_0(n_1+1,n_2-1)(V_0^*(n_1,n_2)+X)^{\xi})/(n_1+1) \end{aligned}$$

Therefore,

$$(V_0^*(n_1, n_2) - n_2 A_0(n_1, n_2) V_0^{*\xi}(n_1, n_2))/n_1$$
$$= (V_0^*(n_1, n_2) + X - (n_2 - 1) A_0(n_1 + 1, n_2 - 1)(V_0^*(n_1, n_2) + X)^{\xi})/(n_1 + 1) \qquad \textbf{(17)}$$

Since the intention is to obtain the values of A_0 and V_0^* recursively, it should be noted that if $A_0(n_1 + 1, n_2 - 1)$ is known, equation (17) has two unknowns, $V_0^*(n_1, n_2)$ and $A_0(n_1, n_2)$. Clearly, the monopolist's criterion is to maximise aggregate warrant value and is equivalent to maximising average warrant value or $A_0(n_1, n_2)$. In other words, $V_0^*(n_1, n_2)$ should be chosen to maximise $A_0(n_1, n_2)$. More formally,

$$A_0(n_1, n_2) = \max_{0 \leqq y \leqq V^*(n_1+1, n_2-1) - X} \times \left[\frac{(n_1 + 1)y - n_1(y + X - (n_2 - 1)A_0(n_1 + 1, n_2 - 1)(y + X)^{\xi})}{(n_1 + 1)(n_2 y^{\xi})} \right] \qquad \textbf{(18)}$$

and $V_0^*(n_1, n_2)$ is equal to the value of y that maximises (18).

Knowing $A_0(n_1 + n_2 - 1, 1)$ and $V_0^*(n_1 + n_2 - 1, 1)$ as defined in equations (15) and (16), recursive application of (17) and (18) allows $A_0(n_1,n_2)$ and $V_0^*(n_1,n_2)$ to be obtained eventually.

Needless to say, further discussion of the analytical properties of this solution seems pointless. However, a numerical example would seem quite useful:

Example 1. Let $\xi = 2$ [a value that can be obtained by substituting quite reasonable values in equation (4) or equation (11)]. Also, let $X = 1$. For simplicity, let there be only one share and two warrants in existence. Then, from (3),

$$V_N^*(1,2) = \xi n_1 X/(\xi - 1) = 2$$

ie, naive strategy indicates exercising both warrants when $V = 2$. From (10) and (13),

$$W_N(V,1,2) = (V_n^*(n_1, n_2) - n_1 X)V^{\xi}/(n_1 + n_2)V_N^{*\xi} = V^2/12$$

On the other hand, from equations (15) and (16),

$$V_0^*(2,1) = V_N^*(2,1) = 4$$

and

$$W_0(V,2,1) = A_0(2,1)V^{\xi} = V^2/24$$

Substituting into equation (18) we obtain

$$A_0(1,2) = \max_{0 \leqq y \leqq 3} (y^2 + 26y - 23)/96y^2$$

which is maximised when y = 23/13.

This means that $A_0(1,2) = 2/23$, $V_0^*(1,2) = 23/13$ and $W_0(V,1,2) = 2V^2/23$. Thus, in this example, it appears that the average warrant value is increased by slightly more than 4% as a result of following the monopolist's strategy. The first warrant is exercised somewhat earlier and the second somewhat later than under the naive strategy.

More complex examples have been solved by the use of computer programming techniques. The recursion defined in equation (18) can thereby be rapidly solved exactly. Some results are presented in Figure 1. The figures give graphs of V_0^*/V_N^* and A_0/A_N as n_1 and n_2 are varied complementarily.

Figure 1 demonstrates the effect of varying ξ (or XI) from 1.1 to 5.0 for a firm with n_1 shares and $20 - n_1$ warrants outstanding. A_0/A_N (the graphs marked A^*) is seen not surprisingly always to have a value greater than 1. However, when the number of warrants equals the number of shares (a situation corresponding to warrants with the potential to dilute the equity by a factor of 2), this ratio is at most 1.04. This indicates that, at modest levels of potential dilution and at reasonable values of ξ, the gain from use of the

Figure 1 Ratio of V_0^*/V_N^* and A_0/A_N as n_1 and n_2 are varied

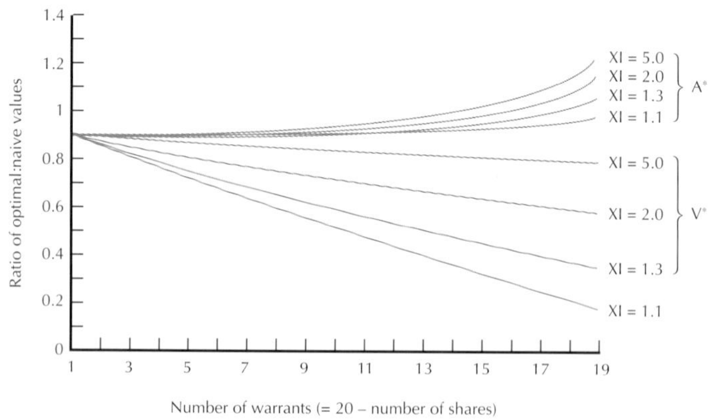

optimal strategy is not great. However, the relative gain rises sharply as the potential dilution increases (ie, $n_2 \gg n1$) to extreme levels.

Increasing ξ also increases the relative value of the optimal strategy. As ξ approaches 1.0, the gain becomes very small. Using the certainty world definition of $\xi(= r/(r - \alpha))$, this can be seen to be reasonable. A value of ξ close to 1.0 implies α is much smaller than r. This in turn means that immediate dividend yield is relatively unimportant. Therefore, the value of a warrant will be quite insensitive to the exercise strategy followed. On the other hand, the graphs of V_0^*/V_N^* show that the optimal exercise strategy deviates from the naive most strongly when ξ is close to 1.0. This is not surprising, given the form of V_N^* in equation (3). As n_2/n_1 and ξ both tend to infinity, A_0/A_N approaches the limit e (2.718). Increasing the total number of shares and warrants above twenty has almost no effect on V_0^*/V_N^* and A_0/A_N calculated at comparable potential dilution ratios. The reason for this is that the optimal strategy gains over the naive by the use of non-simultaneous exercise. With more than a few warrants, the advantage of non-simultaneous exercise can be substantially collected.

Figure 2 demonstrates the analogous results when the warrants are finite-lived and not perpetual. Since there is no known analytical form for the value of finite-lived American call options or warrants that pay proportional dividends, the functions have been obtained purely by numerical methods. Essentially, the situation is as described in the third section except that now the time to expiration of the warrants must be specified. In the example presented here $X = 40$, $r = 0.1$, $\sigma^2 = 0.05$, $\alpha = 0.075$ and the time to expiration varies from zero to sixteen years. In the perpetual case, these values would give ξ a value of 2.0. In Figure 2, A_0/A_N and V_0^*/V_N^* are plotted as before. (In actual fact, since the functional form is now unknown, instead of A_0/A_N the ratio of optimal to naive warrant values is plotted as computed when $V = n_1X$, ie, when the firm is worth X per share outstanding.) The plots closest to 1.0 correspond to a time to expiration of one unit. Those plots further away represent time to expiration increasing by one unit per plot. An interesting feature of Figure 2 is that V_0^*/V_N^* converges almost immediately to the ratio that is optimal in the perpetual case. Convergence is so rapid that the curves are hard to distinguish. Convergence of A_0/A_N as time to expiration increases is somewhat slower but still very rapid.

Figure 2 Moving away from the horizontal line at 1.0, the lifetime incereases by one unit per plot. The parameters were chosen to correspond to a XI value of 2.0

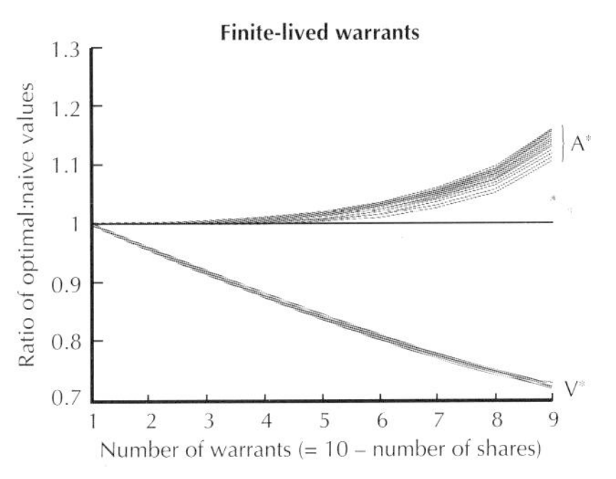

Convergence is faster at higher levels of potential dilution than at moderate levels. Even with unit time to expiration, A_0/A_N diverges considerably from unity.

The conclusion to be drawn from the analysis of this section is that modest increases in warrant values can be obtained by use of the monopolist's strategy. In extreme cases, the gain in value can be quite large. The same comments also apply to warrants that are relatively short-lived. The analysis of finite-lived warrants can be made computationally simpler by assuming that V_0^*/V_N^* is always equal to the ratio for the perpetual warrant situation.

ON THE EXISTENCE OF EQUILIBRIUM EXERCISE STRATEGIES FOR WIDELY HELD WARRANTS

The discussion of the previous section strongly suggests that normally a monopolist should purchase all warrants and exercise them optimally. If, for some reason, this possibility is excluded the holders of widely distributed warrants must do the best they can. In any event, it is of interest to determine what type of equilibrium might exist in those circumstances.

If such an equilibrium exists, it may involve simultaneous warrant exercise or not. However, if warrant exercise does not occur simultaneously, one wonders how non-communicating market participants determine the order in which such warrants are exercised. In such a case the first warrant exercised will be worth either more than, less than or the same as the average unexercised warrant. It if is worth less, then there seems to be no reason for any warrant owner to be that first self-sacrificing individual. If it is worth more, then warrant holders will trample themselves in the rush to be first. Therefore, without communication, such an equilibrium will necessarily require that the value obtained by the first holder to exercise equal the average value of the unexercised warrants.

If such an equilibrium can be established, holders will be indifferent about the sequence in which they exercise their warrants. In order to determine non-simultaneous exercise sequence, holders will have to know how many warrants have actually been exercised. When the time comes to exercise the next warrant, anyone could step forward and proffer his.

However, equality of warrant value is only a necessary condition for equilibrium. All simultaneous exercise strategies satisfy this condition but are not necessarily equilibria. For example, it has already been established that the naive strategy offers an incentive for an individual to defect and defer his exercise decision. An equilibrium should also be stable, providing no incentive for an individual to deviate from the strategy by either late or early exercise.

In order for a stable competitive equilibrium to exist, it is required that the following three conditions be met:

$$V_C^*(n_1,1) = V_N^*(n_1,1) = \xi n_1 X/(\xi - 1) \tag{19}$$

(since the exercise of the last warrant is simple).

$$S_C(V_C^*(n_1, n_2), n_1, n_2) = W_C(V_C^*(n_1, n_2), n_1, n_2) + X \tag{20}$$

(since all warrants are equally valuable if the equilibrium exists and exercising when $V = V_C^*$ converts a warrant to a share).

$$\forall V < V_C^*(n_1, n_2): \quad S_C(V, n_1, n_2) \leqq W_C(V, n_1, n_2) + X \tag{21}$$

(otherwise, the early exercise incentive or "arbitrage" opportunity exists).

Assuming that the first two conditions are met, the problem of exercising the penultimate warrant can be analysed. First, the value W_C of the final warrant to be exercised can be obtained,

$$W_C(V, n_1 + 1, 1) = \frac{(V(\xi - 1)/\xi X(n_1 + 1))^{\xi}(n_1 + 1)X}{(n_1 + 2)(\xi - 1)} \tag{22}$$

This is just the same as W_N from equation (14), with V_N^* from equation (19) substituted in. If the penultimate warrant is exercised when $V = y$, the warrant value immediately before exercise is

$$S_C(y + X, n_1 + 1, 1) - X$$

where

$$S_C(y + X, n_1 + 1, 1) = (y + X - W_C(y + X, n_1 + 1, 1))/(n_1 + 1)$$

Therefore, when the firm is worth V (< y), the penultimate warrant is worth

$$\left(\frac{V}{y}\right)^{\xi}\left(\frac{y + X}{n_1 + 1} - \frac{((y + X)(\xi - 1)/\xi X(n_1 + 1))^{\xi} X}{(n_1 + 2)(\xi - 1)} - X\right)$$

Whereas the final warrant is worth

$$\left(\frac{V}{y}\right)^{\xi}\frac{((y + X)(\xi - 1)/\xi X(n_1 + 1))^{\xi}(n_1 + 1)X}{(n_1 + 2)(\xi - 1)}$$

This is just equation (22) adjusted by a factor $((y + X)/y)^{\xi}$ to reflect the fact that firm value will jump by X when it reaches y. In equilibrium, these two values are equal. In other words, rearranging and simplifying,

$$(y - n_1 X) = \frac{((y + X)(\xi - 1)/\xi X(n_1 + 1))^{\xi}(n_1 + 1)X}{\xi - 1} \tag{23}$$

Examination of this equation reveals that it is of the form

$$y - n_1X = K(y + X)^{\xi}$$

Clearly, there can be no more than two positive real roots of equation (23). Substitution of the value

$$y = \xi(n_1 + 1)X/(\xi - 1) - X$$

into equation (23) reveals that this value (corresponding to exercising the penultimate warrant simultaneously with the final warrant) is a root. Differentiation of both sides of equation (23) at this value of y indicates that it is a unique, multiple root. Therefore,

$$V_C^*(n_1,2) = \xi(n_1 + 1)X/(\xi - 1) - X$$

but

$$V_N^*(n_1,2) = \xi n_1X/(\xi - 1) < V_C^*(n_1,2)$$

and

$$W_N(V,n_1,2) > W_C(V,n_1,2) \quad \forall V \leqq V_N^*(n_1,2) \tag{24}$$

$$S_N(V,n_1,2) < S_C(V,n_1,2) \quad \forall V \leqq V_N^*(n_1,2) \tag{25}$$

since the naive strategy applied to the exercise of the last two warrants yields a higher warrant value than any other simultaneous exercise strategy. However, at the naive strategy exercise point,

$$W_N(V_N^*(n_1,2),n_1,2) + X = S_N(V_N^*(n_1,2),n_1,2) \tag{26}$$

Therefore, combining equations (24), (25) and (26),

$$W_C(V_N^*(n_1,2),n_1,2) + X < S_C(V_N^*(n_1,2),n_1,2) \tag{27}$$

Since $V_N^*(n_1,2) < V_C^*(n_1,2)$, it is thus evident that the third condition necessary for a stable equilibrium [equation (21)] is violated. In other words, a stable competitive exercise equilibrium cannot exist when two warrants are outstanding. Since the situation with more than two warrants necessarily devolves to the case with two warrants, the non-existence of equilibrium has been generally established.

However, when warrants are perfectly divisible rather than exercisable only in discrete lumps, it turns out that a competitive equilibrium does exist. The demonstration of this result is lengthy. It is shown in Emanuel (1980) and more rigorously, more broadly and at greater length in subsequent work by Constantinides (1981). Only the main characteristics of this equilibrium are now discussed here.

The strategy and value functions implied by the competitive equilibrium with perfectly divisible warrants are

$$V_C^*(n_1,2) = V_N^*(n_1,2) \tag{28}$$

$$W_C(n_1, n_2) = W_N(n_1, n_2) \tag{29}$$

While this appears to imply that the competitive strategy is identical to the naive strategy, it must be interpreted differently. Under the naive strategy, whenever V_N^* is attained all outstanding warrants are simultaneously exercised. Under the competitive strategy, only an infinitesimal number of warrants should be exercised at V_N^*. Subsequently, the number of shares n_1 changes and so does V_N^*. However, even though the competitive strategy involves non-simultaneous exercise, the warrants all have equal value which is identical to that obtained under the naive strategy.

Demonstrating this result is lengthy and is omitted here. The best intuition for the result can be obtained by reference back to the certainty framework. It can then be shown that the competitive strategy causes V to grow at rate r while warrants are being continuously exercised. This rate is a combination of the rate $r - \alpha$ from retained earnings and growth in V due to the proceeds of warrant exercise. Share value, calculated by discounting the infinite stream of dividends, can then be shown to be identical to that obtained in the second section under the naive strategy. Finally, the value of the Kth warrant to be exercised under the competitive strategy can be computed. That value is the same as given for the naive strategy in equation (6) and clearly does not depend upon K. These results together imply that the competitive strategy described in equation (28) for infinitely divisible warrants is a genuine equilibrium strategy, although it need not be a unique equilibrium.

Example 2. Returning to the discussion of indivisible warrants, the example in the previous section of the optimal strategy for a monopolist controlling all the warrants can be continued. Following the analysis between equations (19) and (23) we obtain the results

$$V_C^*(1,2) = 3 \quad \text{and} \quad W_C(V,1,2) = 2V^2/27$$

This warrant value is substantially lower than $V^2/12$ or $2V^2/23$ obtained for the naive and optimal strategies, respectively. Moreover, the no-arbitrage condition (21) is clearly violated.

The fundamental problem here lies in the requirement that contemporaneous prices exist for all securities. This section has shown that such prices cannot exist for indivisible warrants unless a monopolist owns all of them, in which case the price is moot.

Figure 3 Ratio of V_C^*/V_N^* and A_C/A_N as n_1 and n_2 are varied and XI is fixed at 2.0. Total of n_1 and n_2 = 20

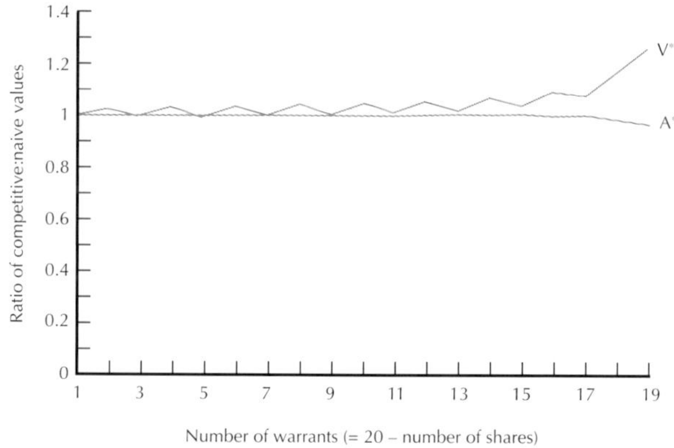

Figure 3 is analogous to Figure 1 and shows A_C/A_N and V_C^*/V_N^* (labelled as A^* and V^*, respectively) for a firm with n_1 shares and $20 - n_1$ warrants. Figure 4 is based on n_1 shares and $5 - n_2$ warrants. In both cases ξ is assumed to be 2.0. Clearly, the competitive strategy involves later exercise than the naive strategy, because no one wants to be first. Figure 3 indicates almost negligible impact on warrant value of non-simultaneous exercise compared to simultaneous exercise under the naive strategy. 20 potential shares are almost divisible enough. Figure 4 shows more significant effects, because the securities are more "chunky". (Recall that the values plotted for 4 warrants in Figure 4 correspond to the same dilution factor as 16 warrants in Figure 3.) As in Figure 1, increasing ξ causes the A^* graph to deviate more and the V^* graph to deviate less, although this is not depicted here.

Figure 4 Ratio of V_C^*/V_N^* and A_C/A_N as n_1 and n_2 are varied and XI is fixed at 2.0. Total of n_1 and n_2 = 5

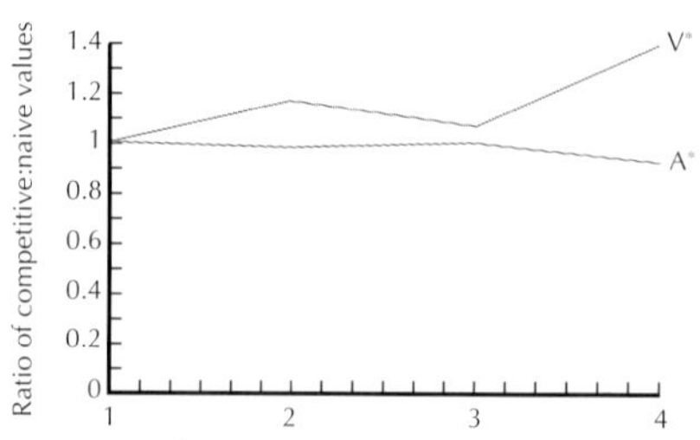

DISCUSSION OF DIVIDEND POLICY ASSUMPTIONS

The discussion so far has indicated the existence of non-trivial problems in the valuation of warrants when the assumption of simultaneous exercise is abandoned. However, any honest examination of the problem must carefully scrutinise the pertinent assumptions to determine whether the whole problem stems from unreasonable assumptions.

The pertinent assumptions here concern future dividend policy and capital structure. Such assumptions are almost universal in security and option valuation models and it is hard to see how one could proceed without them. However, one might seriously question whether a problem that existed only under very special assumptions was a problem at all.

The dividend policy assumption that was made required α, the rate at which assets were paid out as dividends, to be constant. This does not seem unreasonable. However, an alternative assumption that also seems reasonable would require that α be adjusted, as shares were exercised, in such a way that dividend per share immediately before and after exercise was unchanged. In other words, a proportional dividend policy would still be followed, but the dividend per share would be protected against dilution.

Quantitatively, this condition may be expressed as follows:

$$\alpha_{post} = \alpha_{pre}(n_1 + 1)V^*/n_1(V^* + X) \tag{30}$$

where α_{pre} and α_{post} are the values of α before and after exercising a warrant to create the $n_1 + 1$st share. Referring back to equation (3), which gives the optimal naive exercise strategy under certainty, one observes

$$V_N^*(n_1, n_2) = rn_1X/\alpha_{pre}$$

and

$$V_N^*(n_1 + 1, n_2 - 1) = r(n_1 + 1)X/\alpha_{post} = rn_1X/\alpha_{pre} + X$$

Thus, in the case of certainty, maintaining the dividend rate per share unchanged as warrants are exercised eliminates the problem as

$$\Delta V_N^*/\Delta n_1 = X$$

However, in the case of uncertainty,

$$V_N^*(n_1, n_2) = \xi n_1 X/(\xi - 1)$$

$$= \frac{-(2r - 2\alpha_{pre} - \sigma^2) + \sqrt{((2r - 2\alpha_{pre} + \sigma^2)^2 + 8\alpha_{pre}\sigma^2)}}{-(2r - 2\alpha_{pre} + \sigma^2) + \sqrt{((2r - 2\alpha_{pre} + \sigma^2)^2 + 8\alpha_{pre}\sigma^2)}} n_1 X \tag{31}$$

and

$$V_N^*(n_1 + 1, n_2 - 1)$$

$$= \frac{-(2r - 2\alpha_{post} - \sigma^2) + \sqrt{((2r - 2\alpha_{post} + \sigma^2)^2 + 8\alpha_{post}\sigma^2)}}{-(2r - 2\alpha_{post} + \sigma^2) + \sqrt{((2r - 2\alpha_{post} + \sigma^2)^2 + 8\alpha_{post}\sigma^2)}} (n_1 + 1)X \tag{32}$$

Depending on the values of r, α and σ^2, $V_N^*(n_1+1, n_2-1)$ exceeds $V_N^*(n_1,n_2)$ by more or less than X. Generally, one would expect the incentive for deferred exercise to be reduced but rarely exactly eliminated.

It is thus clear that the particular dividend policy is not crucial to the existence of the problem. Equally, the particular capital structure policy is not crucial. It was reasonably assumed that all the proceeds from warrant exercise would be reinvested in the firm. An alternative assumption would be that all proceeds would be used to repurchase shares. In that case, warrant exercise would not affect the value of the firm, merely the number of shares and warrants. As before,

$$V_N^*(n_1,n_2) = \xi n_1 X/(\xi - 1)$$

Given a share price of S, X/S shares would be repurchased. There would then be $n_1 + 1 - X/S$ shares and $n_2 - 1$ warrants after the exercise of one warrant. The number of shares outstanding will exceed n_1 if and only if $S > X$. Since V_N^* is an increasing function of n_1, $V_N^*(n_1 + 1 - X/S, n_2 - 1)$ will exceed $V_N^*(n_1,n_2)$ if and only if $S > X$. Since firm value is unchanged by warrant exercise, firm value will therefore be below $V_N^*(n_1 + 1 - X/S, n_2 - 1)$ after the exercise of one warrant, assuming $S > X$. However, it is inconceivable that the exercise of warrants will rationally occur when $S \leqq X$. Therefore, the problem persists under the alternative capital structure assumption.

Of some interest is "neutral dividend policy". That is to say, how should dividend rate be adjusted, as warrants are exercised, to make simultaneous exercise optimal. One is tempted to say that one requires that

$$V_N^*(n_1 + 1, n_2 - 1) \leqq V_N^*(n_1,n_2) + X$$

(assuming, once again, that the proceeds of warrant exercise are to be reinvested in the firm). Satisfying this condition would imply that, after the exercise of one warrant, it would then be optimal to exercise the remaining warrants immediately. However, strict inequality would not be very satisfactory because, in that case, the shareholders would have an incentive to purchase a few warrants and to exercise them early so as to force the other warrants holders into earlier exercise than is optimal. Consequently, the only strictly satisfactory neutral dividend policy would be one that adjusted α so that

$$V_N^*(n_1 + 1, n_2 - 1) = V_N^*(n_1,n_2) + X \qquad \textbf{(33)}$$

Adjusting α also changes ξ which will then also be a function of n_1.

Since $V_N^*(n_1,n_2) = \xi n_1 X/(\xi - 1)$, condition (33) is equivalent to

$$(\Delta/\Delta n_1)(\xi n_1 X/(\xi - 1)) = X$$

or

$$\xi/(\xi - 1) - (n_1/(\xi - 1)^2)(\Delta\xi/\Delta n_1) = 1 \quad \Delta\xi/\Delta n_1 = (\xi - 1)/n_1$$

The solution to this difference equation is

$$\xi - 1 = Kn_1$$

where K is an arbitrary constant determined by the initial value of ξ. Substituting back into $V_N^*(n_1,n_2)$, one observes that

$$V_N^*(n_1,n_2) = (Kn_1 + 1)n_1X/Kn_1 = X/K + n_1X \tag{34}$$

which clearly shows that V_N^* has the required property defined in equation (33). Under certainty, $\xi = r/(r - \alpha)$ and it can readily be verified that a policy of recursively changing α according to equation (30) satisfies conditions (33) and (34). This is hardly surprising, since we have previously determined that a policy of protecting dividend per share against warrant exercise leads to simultaneous exercise under certainty. The uncertainty case is more complicated and the neutral dividend policy is therefore just left as that unique policy which maintains the ratio $\xi - 1{:}n_1$ at a constant value.

IMPLICATIONS AND CONCLUSIONS

The results of the analysis presented in this chapter are empirically testable, in principle. One prediction is that warrants should not be exercised simultaneously and this can readily be observed. In principle, one could observe V^* functions empirically and compare them with the theoretically derived V^*. Another prediction would be that one would tend to observe monopolistically or closely held warrants whenever the gains from the use of the optimal exercise strategy are substantial. However, the gains are only substantial when potential dilution effects are extreme.

An examination of firms with warrants outstanding as listed in *Moody's Industrials Manual* has indicated that rarely does potential dilution of equity from all sources exceed a factor of two. There are a few exceptional cases where potential dilution is extreme. If data on warrant ownership were available, the prediction could be tested. However, such a test producing evidence consistent with this hypothesis would not necessarily be accepted as empirical proof of the theory. For it seems very likely that the population of firms with extreme potential equity dilution is, in any case, unusual, and one could well argue that such firms exhibit extreme potential dilution because they are closely held rather than vice versa. Empirical testing is therefore left as an unresolved issue.

The analysis here has focused exclusively on warrants. Analogous problems of timing of conversion exist for convertible bonds, although with smaller magnitude. The assumptions made here are not designed to create the problem and similar exercise timing problems exist for quite general classes of securities and assets.

In the economy, many real claims have characteristics similar to warrants. For example, the owner of a mine or an oilfield possesses a warrant on the market for his product whose exercise price is the cost of extraction. Likewise, owning letters patent or licences conveys a warrant on a technology or product. Problems caused by suboptimal sequencing of the exercise of such real claims will have a real economic impact. There is no reason to suppose that the gains from a coordinated strategy will always be small. However, it must be conceded that cost differences or other differences will distinguish real claims and make the optimal sequencing more obvious. Since uncertainty has very little to do with the basic problem, it does not appear that the existence of futures prices will help to resolve the situation either.

The analysis presented here indicates that the first exercised warrant is less valuable than a later exercised warrant. This may explain the apocryphal reluctance of firms to be the first to enter a new market. The analysis also suggests that monopoly control has its advantages. Strong patent law may be justified for that reason, as may government sponsorship of research and development. Firms with patents should be somewhat reluctant to create more warrants by issuing licences to use the patented technology. It would appear that, for the kind of problem considered here, a monopoly is most suitable and oligopoly the least suited to cope.

It could be argued that none of these extensions and conjectures tells us anything new. The existence of natural monopolies is certainly not a new idea, although they are normally rationalised on grounds of economy of scale rather then better coordination of investment decisions as is suggested here. However, the point of this concluding discussion has been to show that the theory developed here has some relevance in a wider sphere than warrant exercise.

1 All symbols are defined in Table 1.

BIBLIOGRAPHY

Black, F. and M. Scholes, 1973, "The Pricing of Options and Corporate Liabilities", *Journal of Political Economy* 81, pp. 637–59.

Constantinides, G.M., 1981, "Warrant-exercise and Bond Conversion in Competitive Markets", working paper, Graduate School of Business, University of Chicago, Chicago, IL.

Emanuel, D.C., 1980, "Extensions and Developments of Option Pricing", unpublished PhD thesis, Graduate School of Business, Stanford University, Stanford, CA.

Jensen, M.C. and W.H. Meckling, 1976, "Theory of the Firm: Managerial Behavior, Agency Costs, and Capital Structure", *Journal of Financial Economics* 3, pp. 305–60.

Merton, R.C., 1973, "Theory of Rational Option Pricing", *Bell Journal of Economics and Management Science* 4, pp. 141–83.

Miller, M.H., 1977, "Debt and Taxes", *Journal of Finance* 32, pp. 261–75.

Modigliani, F. and M.H. Miller, 1958, "The Cost of Capital, Corporation Finance and the Theory of Investment", *American Economic Review* 48, pp. 261–97.

Smith, C.W., Jr., 1977, "Alternative Methods for Raising Capital: Rights versus Underwritten Offerings", *Journal of Financial Economics* 5, pp. 273–307.

Smith, C.W., Jr. and J.B. Warner, 1979, "On Financial Contracting: An Analysis of Bond Covenants", *Journal of Financial Economics* 7, pp. 117–61.

11

*Design and Valuation of Debt Contracts**

Ronald W. Anderson and Suresh Sundaresan

Université Catholique de Louvain; Columbia University

This chapter studies the design and valuation of debt contracts in a general dynamic setting under uncertainty. We incorporate some insights of the recent corporate finance literature into a valuation framework.

The basic framework is an extensive form game determined by the terms of a debt contract and applicable bankruptcy laws. Debtholders and equityholders behave non-cooperatively. The firm's reorganisation boundary is determined endogenously.

Strategic debt service results in significantly higher default premiums at even smaller liquidation costs. Deviations from absolute priority and forced liquidations occur along the equilibrium path. The design tends to stress higher coupons and sinking funds when firms have a higher cash payout ratio.

By incorporating some insights of the recent corporate finance literature into a valuation framework, we obtain a model that seems promising for the empirical study of pricing of risky debt claims and which gives insights into the question of why certain contractual provisions are selected in some situations but not in others. Beyond this we believe our framework suggests a general approach incorporating strategic considerations in a valuation setting which can be usefully explored in future work.

*This was first published in *Review of Financial Studies* 9, 1996, pp. 37–68 and is reprinted with the permission of Oxford University Press. The authors would like to thank seminar participants at the Stockholm School of Economics, ECARE, IGIER, Baruch College, Carnegie-Mellon University, and the ESF Finance Network Symposium in Gerzensee, Switzerland for their comments, and Hayne Leland, John Parsons and William Perraudin for their discussions. The authors also thank Yonghua Pan and Pierre Tychon for computational assistance and comments. This chapter has been improved by the comments of the referee and the editor, Franklin Allen. This research has been supported through an ESF Network in Financial Markets travel grant. Responsibility for views expressed and for errors is the authors' alone.

DEBT VALUATION THEORY

Merton (1974) was the first to apply the valuation insights of Black and Scholes (1973) to the pricing of corporate debt contracts. He takes two key contractual provisions exogenously: first, the lower reorganisation boundary (which is the threshold value of the firm at which the control of the firm transfers from the stockholders to the bondholders) is specified. Second, the compensation to be received by creditors upon reaching the lower reorganisation boundary is taken as given. Jones, Mason and Rosenfeld (1984) extended the Merton model to coupon and callable debt, however, maintaining the assumption that the lower reorganisation boundary is the minimum of the firm's value at maturity and the promised face amount. This implies that there are no enforcements of indentures prior to maturity. When applied to a sample of 305 bonds of various ratings, their model resulted in prices systematically in excess of those observed in the market, that is, their model underestimated observed yield spreads. They interpreted their results as "establishing research priorities in what will be a large and complex task" of applying contingent claims pricing models to risky debt. Kim, Ramaswamy and Sundaresan (1993) pursued this agenda by extending the contingent claims pricing model to incorporate (1) enforcement of bond covenants on intermediate coupon payments and (2) a stochastic term structure. Their results indicated that yield spreads were sensitive to imposing cashflow checks prior to maturity and that they are an increasing function of the bankruptcy costs (parameterised in their model as the recovery rate). Surprisingly, yield spreads were found to be relatively insensitive to the volatility of interest rates. Maloney (1992) has extended this approach to allow for more general stochastic processes for the firm cashflows and interest rates.

All these studies take as exogenously given the firm's lower reorganisation boundary as well as the payoffs when that boundary is met. These models are also not consistent with the empirical regularities reported in studies on financial distress by Asquith, Gertner and Scharfstein (1991), Franks and Torous (1989, 1993) and Weiss (1990). These regularities are (1) bankruptcies are costly both because of direct costs and because of disruptions of the firm's activities; (2) bankruptcy procedures give considerable scope for opportunistic behaviour by the various parties involved; (3) deviations from absolute priority of claims are common; and (4) despite the incentives to do so, in practice it often proves impossible to renegotiate claims so that formal bankruptcy and liquidation often result. Thus costly liquidations represent a possible source of inefficiency associated with debt contracts.

CORPORATE FINANCE AND DEBT DESIGN

Historically, corporate finance has been concerned with other possible inefficiencies. Jensen and Meckling (1976) emphasised the agency costs of equity finance in the face of managerial moral hazard. In another vein, Myers (1977) identified the agency costs of debt by pointing out the underinvestment problem, that is, the possibility that firms will forego positive net present value investment opportunities because benefits are likely to accrue to debtholders rather than shareholders. Recent theory has reexamined these issues from the perspective of security design. This approach considers how both control rights and cashflow claims may be structured so as to minimise contracting inefficiencies. Aghion and Bolton (1992) and Zender (1991) emphasise that contracts that grant control to one class of agents exclusively may not be efficient because they fail to give the controlling agent the incentives to make first-best decisions, or because contracts sold to outsiders will not be sufficiently valuable to permit raising the required outside finance. They show that contracts with *contingent transfer of control rights* may minimise inefficiencies; this provides a rationale for debt contracts.

The properties of debt contracts have been studied in a number of theoretical models. For example, Hart and Moore (1991) examine the problem of designing optimal debt contracts in a dynamic setting under certainty. Bolton and Scharfstein (1993) study the trade-offs involved in making debt more or less easy to renegotiate. A number of other models examine the consequences of alternative bankruptcy models.[1]

SYNTHESIS OF VALUATION AND CORPORATE FINANCE

These corporate finance models tend to be very stylised and not truly dynamic. Furthermore, they have no valuation consequences or direct testable implications for corporate bond yields spreads. On the other hand, the valuation theory is truly dynamic but fails to endogenise contractual provisions.

Our article attempts to fill the gap by integrating these two strands of research. The basics of our approach are summarised as follows. For us, the terms of a debt contract and bankruptcy law establish the extensive form game in which debtholders and equityholders interact. The allocation of cashflows and the firm's reorganisation boundary are determined *endogenously* as the non-cooperative equilibrium in this game. Given this solution, we are then able to address the question of the design and the valuation of multiperiod debt contracts under uncertainty.

Two additional studies address some of the questions that we do and obtain results that are in some ways similar to our own. Leland (1993) reconsiders the contingent claims framework in continuous time. By assuming that all debt service must be met by issuing new equity he endogenises the firm's bankruptcy point. Specifically, it is the point where the value of equity is zero. He restricts his study to the case of perpetual debt essentially in order to obtain closed-form solutions permitting the calculation of a variety of interesting comparative statics. His approach is similar to ours in that it endogenises the lower reorganisation boundary of the firm. Otherwise the approaches are quite different. In particular, there is no modelling of the process for resolving financial distress and there is no scope for strategic debt service. Furthermore, since the framework is constrained to using perceptual coupon bonds, it cannot be used to address issues of financial contract design.

Like Leland, Mella-Barral and Perraudin (1993) also work in continuous time and confine their attention to perpetual coupon bonds. They explicitly model the shutdown decision of a firm operating with constant flow costs but with a stochastically varying output price. In this way they are able to study the effect of leverage on the operations of the firm. Their analysis allows for strategic debt service, and in this sense is closer to our model than is any other. However, they do not explicitly model the bankruptcy game. Their analysis is simplified considerably by the fact that they work with perpetual debt. This reduces its relevance to pricing actual bonds with finite maturities and specific (sometimes complicated) contractual features. Furthermore, this means that they cannot address the question of the design of debt contracts.

In the next section we set out model and derive the equilibrium values of corporate liabilities. Using a simple model of the bankruptcy process, we find the equilibrium often will result in renegotiations with deviations of priority in the favour of equity. In the second section, we compare our results to Merton (1974) and subsequent applications of the contingent claims approach to pricing risky debt. We find our model produces risk premiums more in line with levels observed in the market for reasonable parameter values. The third section is devoted to the design of debt contracts. We show how high cash payout rates, low leverage and low liquidation costs lead to higher coupon debt, in general. Sinking fund provisions are generally used to trade off the tax advantages of interest payments with risks of costly liquidations. The chapter concludes with suggestions of avenues for future work.

THE MODEL

Preliminaries

The setting we consider is simple: there is one owner-manager who has "access" to a technology and is endowed with a technology-specific human capital. His human capital is imbedded in a project which, if undertaken, will give rise to a stream of rents indefinitely into the future. The project requires financing in the amount of capital D which the owner-manager does not have. We assume that the financing is to be arranged through the issue of a debt contract, which we take to be characterised by the priority of claims and contingent transfer of control.[2]

We restrict attention to a single homogeneous group of creditors to abstract from holdout problems. The terms of the contract call for a payment of CS_t in period t up to the maturity date T. It is assumed that debt service is met out of cashflows and that asset sales or the issue of new securities would require the explicit agreement of creditors. Finally, we assume that all contracting parties have full information about the states of the nature.

The technology

The ongoing project is represented as a stochastic process for the value V_t, which is the present value of current and all future cashflows. The value is assumed to follow a simple binomial process as illustrated[3]

$$V_t \begin{array}{l} \nearrow uV_t \\ \searrow dV_t \end{array}$$

where $d = 1/u$. V_t may be interpreted as the cum dividend value of the firm were it to be financed entirely by equity. This is also the value of assets of the firm under alternative financing arrangements. This specification allows us to nest as a special case of Merton's (1974) analysis of zero coupon debt. Cashflows are assumed to be proportional to the value of the project; thus $f_t = \beta V_t$, where β is the payout ratio.

The mapping between the cashflows and value processes is based on martingale probabilities p which are time and state invariant and which are given by $r(1 - \beta) - d/u - d$.[4] We assume $u > r(1 - \beta) > d$. It will be noticed that the cash payout ratio β enters the martingale probability of an "up move" with a negative sign. This allows us to model within the same framework "growth firms" (low β projects for which a large proportion of current value is accounted for by expected future growth) and "cash cows" (high β projects for which a large proportion of current value is accounted for by near-term cashflows). Our main restriction is that in taking the cashflow process as exogenous we are not able to study explicitly the possible feedback of the financial contracting onto future production decisions.

Once under way, control of the project can be transferred only at a cost. This may in part take a direct form, such as legal costs, that reflect costly verification of collateral values. In part this may reflect a loss of project-specific human capital: after the transfer of control, it takes time and effort for the creditor to find another management team that will be able to produce the cashflows from the technology at its full efficiency. In what follows we assume these costs are all summarised in constant liquidation cost of K so that the collateral value equals $V - K$. The model could easily accommodate other functional forms for collateral values possibly at the cost of greater complexity. In other respects the markets are assumed to be frictionless. We examine later a variant of the model that allows for corporate income taxation.

Figure 1 The subgame at t

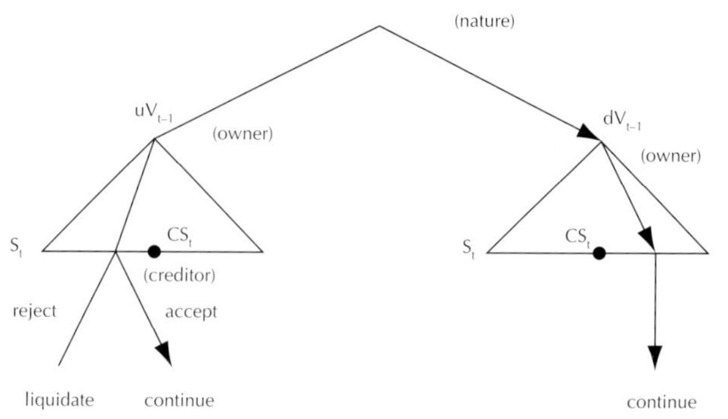

The arrows in Figure 1 indicate a decision. The event illustrated is a "down move" to dV_{t-1}. From that point the owner chooses a debt service that lies within the range indicated by the base of the triangle. Here he is shown as offering a debt service greater than the contracted service CS_t which leads to a continuation to date t + 1.

All cashflows from the project are paid out in the form of dividends, debt service or to cover bankruptcy costs. Thus the model satisfies the value preservation property, $V = E + B + L$, where E is the value of equity, B is the value of debt and L is the present expected value of future bankruptcy costs. Thus the value of the levered firm V_l, given by the value of its financial claims, is the asset value of the firm less the expected bankruptcy costs, $V_l = V - L$. We now proceed to discuss the interaction of the owner-manager and the creditor taking the financial contract as given. The optimal design of financial contracts is discussed below.

The game

The technology, the provisions of the debt contract and the law establish the environment within which the owner and the creditor interact. We represent this with an extensive form game which, though very simple, has interesting dynamic implications. Our general approach can readily accommodate modifications to this game form.

At any given time while the project is ongoing, there will be a realisation of the cashflow, f_t. Given this, the owner chooses a level of debt service, $S_t \in [0, f_t]$. If the debt service is equal to or greater than the contracted amount, CS_t, the game continues to the next period. If it is less than the contracted amount, the creditor has the choice of not initiating legal action (ie accepting the service) or initiating legal action (ie rejecting the debt service). If the debt service is accepted the project continues to the next date. If the debt service is rejected, the project is liquidated, leaving the creditor with the cum dividend value of the project less the liquidation cost.

The subgame originating from a state V_{t-1} is depicted in Figure 1.

At each node the player taking the action is indicated in parentheses. (A random variable is viewed as an action taken by nature.) The arrows indicate a decision. Thus in the figure the event is a "down move" to dV_{t-1}. From that point the owner chooses a debt service that lies within the range indicated by the base of the triangle. Here he is shown as offering a debt service greater than the contracted service CS_t, which leads to a continuation to date $t+1$. Had there been an "up move" to uV_{t-1} the owner is shown as offering a debt service below the contracted amount. Since a covenant of the debt contract is violated, this creates a decision node for the creditor. The figure indicates that he accepts the service rather than electing to liquidate the project. In Figure 1 the indicated actions of the owner and creditor are not necessarily their best responses. Below we will discuss the *equilibrium* in the game.

In our formulation we have modified the modelling of the firm bankruptcy process fundamentally as component to the approach of Merton where bankruptcy is determined simply by the state of the nature and the contract, leaving no scope for either the owner or the creditor to take any initiative. Our model gives the owner the scope for choosing to underperform his debt contract, even when the health of the project would enable him to fully meet his obligations. We will see that the costliness of liquidation means that the owner can exploit this opportunity to his own advantage. At the same time our model gives the creditor the scope for choosing how to approach a potential bankruptcy. He need not force bankruptcy when it is not in his interests to do so. This can be viewed as a simple representation of a negotiation between the owner and the creditor that arises whenever a contractual provision is violated. Admittedly this is very stylised and gives a first-mover advantage to the owner. Other ways of modelling bankruptcies could be considered that might allow more elaborate interactions, possibly to the creditor's advantage.

In our formulation an important loan indenture is meeting the currently scheduled payment of interest and principal. This is a basic and widespread feature of debt contractors. It should be noted, however, that violating this does not automatically throw the firm into bankruptcy. It merely creates a decision node for the creditor in which he must decide whether or not to initiate legal action against the owner-manager.

Requiring debt service to lie in the interval $[0, f_t]$ is important and requires some comment. First, it implicitly assumes that the owner cannot issue additional debt once the project is under way. Loan indentures very frequently do forbid the issue of additional debt with priority equal or superior to that of the original issue. Here we also exclude the issue of junior debt since the incorporation of subordinated debt valuation is a complication best approached after our basic model is fully understood. Similarly, we do not allow debt service to be funded through asset sales or equity issues. Loan indentures commonly do restrict asset liquidation since this might be exploited by owner-managers as a means of extracting value at the expense of undermining collateral values. Furthermore, the assets in place may be discrete.

The assumption that no new securities are issued to service existing debt simplifies our analysis and arguably comes close to reality in many circumstances. There may be fixed costs associated with security issues that may be higher for a firm in financial distress than for a healthy firm. Furthermore, adding new classes of claimants may make subsequent renegotiations more difficult so that the firm liquidation may simply be delayed but not prevented. Therefore, we believe working with a model where new security issues are excluded is a good starting point for the analysis of strategic debt service. Nevertheless, we would view enriching the game form to allow for securities issues in some circumstances to be an interesting avenue for future work.

Equilibrium

We assume that, once the debt contract has been established, the owner and the creditor each choose actions in their own self-interest. We assume complete information in the sense that the game as just described, as well as the payoffs to the two sides, are both common knowledge. As a result our attention is restricted to the subgame perfect equilibriums of the game. That is to say that an equilibrium is constructed under the assumption that at each possible decision point, including those that are never actually observed in equilibrium, the agent makes a maximising decision. As is typical in such settings, we construct the equilibrium recursively.

At maturity T, V_T is observed, and the owner selects a debt service S_T. If this is not less than the contracted amount, $S_T \geq CS_T$, the game ends. Otherwise, $S_T < CS_T$, the creditor must decide whether to accept or reject. If debt service fulfills the contract or if the creditor accepts, the payoffs to the debtholder and the owner, respectively, are

$$(S_T, V_T - S_T)$$

whereas if a debt service is rejected the payoffs are

$$(\max(V_T - K, 0), 0)$$

The assumption that the liquidation value of the firm cannot be negative reflects the fact that liquidation costs are deducted from remaining firm value.

Equilibrium in this subgame is formed by the decision rules of the creditor and owner that constitute the best responses in light of the payoffs. Given underperformance, $S_T < CS_T$, the best response of the creditor is to accept if $S_T \geq \max(V_T - K, 0)$ and to reject otherwise. The best response of the owner is to set $S_T = CS_T$ if $V_T - K > CS_T$. Otherwise he sets $S_T = \max(V_T - K, 0)$. Thus if the value of the firm is relatively high so that the liquidation value exceeds the contracted debt service, the owner is best off simply honouring the contract. For relatively low values of the firm, the owner is best off by making the minimum debt service payment which just leaves the creditor indifferent between accepting or liquidating the firm.

The payoffs for the creditor and owner conditional on the realisation of V_T are given by the equilibrium values which are

$$B(V_T) = \min(CS_T, \max(V_T - K, 0)) \tag{1}$$

for debt, while the value of the equity is

$$E(V_T) = V_T - B(V_T) \tag{2}$$

Notice that these payoffs reflect the possibility of strategic debt service: in terminal states where contracted payment is less than the project value but greater than its liquidation value, the owner-manager underperforms the debt contract, but by an amount insufficient to provoke the creditor to take legal action.

For periods prior to the term of the debt contract, the best responses of the agents will be based on similar reasoning. That is, the owner-manager in some states will try to reduce debt service to the point where the creditor is just indifferent between accepting the service and taking legal action. The important complication in these earlier periods is that, in calculating payoffs, the agents must take into account the values of the continuation subgames. Since these depend on the future realisations of the project values, they are uncertain. Under the assumption that the markets are dynamically complete, both the debtholder and the owner-manager will evaluate the continuation payoffs using the same martingale probabilities (see Harrison and Kreps (1978)). This preserves the complete information character of the game.

Suppose at time t the project is ongoing, and there has been a realisation V_t. The owner selects a debt service S_t. If this at least equals the contracted amount, CS_t, the game continues. If this underperforms the contract, the creditor can reject the payment, which results in the liquidation value

$$\max(V_t - K, 0)$$

Alternatively, the creditor can accept, which results in the payoff

$$S_t + \frac{pB(uV_t) + (1 - p)B(dV_t)}{r}$$

where r is one plus the risk-free interest rate per period. Thus, in the face of underperformance, the best response of the creditor amounts to selecting the greater of these two values. The owner determines his best action in light of this. For relatively high realised values of the firm, $(V_t - K) > CS_t$, the owner simply meets the contractual debt service. For relatively lower values of the firm, he elects to underperform the contract to the point that leaves the creditor indifferent between liquidating or not.

To find the equilibrium value of the debt contract, note that the owner will try to reduce the debt service but seeks to avoid provoking the creditor to liquidate the firm. If liquidation does not occur, then in state V_t^j the level of debt service is

$$S(V_t) = \min\left(CS_t, \max\left(0, \max(V_t - K,0) - \frac{pB(uV_t) + (1 - p)B(dV_t)}{r}\right)\right) \tag{3}$$

The value of debt thus is

$$B(V_t) = S(V_t) + \frac{pB(uV_t) + (1 - p)B(dV_t)}{r} \tag{4}$$

The corresponding value of equity is

$$E(V_t) = f_t - S(V_t) + \frac{pE(uV_t) + (1 - p)E(dV_t)}{r} \tag{5}$$

In some states, however, cashflows will be insufficient to pay an amount acceptable to the creditor. Since in this framework asset sales and new security issues are excluded, liquidation will result. Thus, forced liquidation occurs if

$$S(V_t) > f_t$$

In the case of forced liquidation, the value of debt is

$$B(V_t) = \max(0, \min(V_t - K, CS_t + P_t)) \tag{6}$$

where P_t is the principal of loan outstanding at time t. The value of equity is

$$E(V_t) = V_t - K - B(V_t) \tag{7}$$

This completes the construction of a subgame perfect equilibrium. This is a unique equilibrium if the liquidation costs are strictly positive. For in that case, even though S_t is chosen to leave the creditor indifferent between accepting and rejecting the service, rejecting never occurs in equilibrium. The owner's best action in anticipation of rejection is to slightly raise S_t, which would make the creditor strictly prefer acceptance.

This equilibrium merits a few comments. First, in contrast with the traditional approach in contingent claims valuation, the states in which contract default occurs have been determined endogenously from the primitives of the model. Furthermore, control of the project is transferred to creditors only in a subset of default states where the firm is illiquid. Second, the equilibrium may involve strategic debt service, which is consistent with the observed opportunism of contracting parties under many bankruptcy procedures. Our strategic debt service can be interpreted as the outcome of a negotiation process with a deviation from absolute priority in favour of equity. The size of the deviation depends on the health of the asset in place and the costliness of liquidation. Third, costly forced liquidations can occur in equilibrium since asset sales are infeasible. This kind of *ex post* inefficiency implies that the debt capacity of the project, that is, the maximum amount that can be raised to finance the project, may be less than its full asset value. This implies a possible *ex ante* inefficiency as well: positive net present value projects may not be undertaken if the debt capacity is less than the financing requirement of the owner-manager. Thus the model reflects an underinvestment problem that can be interpreted as an agency cost of debt. However, it should be recognised that since we do not allow feedback onto future investment decisions we do not incorporate the form of agency cost originally identified by Myers (1977).[5] Fourth, our model has been set out in discrete time. This has the advantage of making the modelling of the bankruptcy process transparent. Our particular model has the feature of giving a first-mover advantage to the owner-manager. A number of alternative models of the bankruptcy process that have been suggested in the literature could be incorporated as rather direct extensions of our framework. In contrast, modelling non-cooperative games in continuous time can present subtle difficulties. Indeed, Fudenberg and Tirole (1985) have convincingly argued that for continuous time games it is preferable to find the equilibrium in a discretised version of the game and then take limits as the time interval goes to zero. Fifth, the equilibrium holds for quite general stochastic processes. Furthermore, it is consistent with risk averse utility maximising owners and creditors.

VALUATION

In this section we use our model to evaluate straight debt contracts, that is, contracts offering nominally a fixed coupon and full principal reimbursement at maturity. This involves a promised payment of a coupon interest rate of c per period and a principal P to be reimbursed at the maturity date T. Thus we have $CS_t = cP$ for $t < T$ and $CS_T = (c + 1)P$. We first consider the

Table 1 Discount debt

(a) Time to maturity (T) = 2.00 years
V = 1, β = 0.00, c = 0.00, r = 1.05[1]

σ^2	d	Risk premium, R – r (%)[2] K = 0	K = 0.1	K = 0.2
0.030	0.200	0.000	0.000	0.000
0.030	0.500	0.012	0.106	0.513
0.030	1.000	5.129	8.396	12.734
0.030	1.500	20.578	25.556	31.429
0.100	0.200	0.002	0.050	0.378
0.100	0.500	0.817	2.031	4.138
0.100	1.000	9.738	13.365	17.779
0.100	1.500	23.022	27.581	32.845
0.200	0.200	0.118	0.796	2.702
0.200	0.500	3.087	5.682	9.358
0.200	1.000	14.266	18.404	23.310
0.200	1.500	26.601	31.232	36.345

(b) Time to maturity (T) = 10.00 years

σ^2	d	K = 0	K = 0.1	K = 0.2
0.030	0.200	0.004	0.031	0.113
0.030	0.500	0.376	0.629	0.994
0.030	1.000	2.430	2.988	3.680
0.030	1.500	4.969	5.678	6.568
0.100	0.200	0.474	1.049	1.956
0.100	0.500	2.113	2.860	3.799
0.100	1.000	4.826	5.622	6.568
0.100	1.500	7.116	7.919	8.869
0.200	0.200	1.872	3.435	5.522
0.200	0.500	4.379	5.695	7.356
0.200	1.000	7.351	8.494	9.878
0.200	1.500	9.546	10.609	11.867

[1]V refers to the initial value of the firm, d is the quasi-debt to value ratio, c is the coupon, β is the cash payout ratio (cashflow divided by the value of the firm), r is the riskless rate, K is the fixed liquidation cost and T is the time to maturity in years.
[2]R – r is the risk premium. The results are based on discrete time binomial valuation procedure in which the time to maturity is divided into 1,000 time steps.

case of a zero coupon (pure discount) debt to serve as a benchmark for comparison with Merton (1974). Later we consider the more important case of coupon bonds.

Discount debt

A useful benchmark for our model is Merton's (1974) analysis of zero coupon, risky bonds. Our model in the special case of a zero coupon ($c = 0$) and zero cost of liquidation ($K = 0$) is a discretised version of Merton's. Strategic considerations in the bankruptcy decision only enter into play for strictly positive bankruptcy costs ($K > 0$). Table 1 presents the risk premium of debt (ie the different between the internal rate of return on the risk bond, R, and the risk-free rate, r) for zero coupon bonds under a variety of assumptions about the remaining parameters. For ease of comparison we have parameterised our model as in Merton with respect to the volatility of value process, σ^2, the firm's quasi-debt firm value ratio, $d = Pe^{-rT}/V_0$, and have assumed $\beta = 0$.

The results for $K = 0$ correspond to those reported by Merton (1974) in his Table 1. The slight discrepancies between our results and his are due to our working with a discrete version of his continuous time model. Thus we see that, in the case of zero bankruptcy costs, the premium is increasing in the leverage, d, and the volatility of the underlying asset, σ^2. The effect with respect to time to maturity depends crucially on the degree of the firm's leverage. For example, for $d = 0.5$ the premiums for $T = 2$ are less than those when $T = 10$, whereas the reverse holds for $d = 1.5$. The interpretation is that, for firms with a low degree of leverage, default will occur only if the firm value declines substantially, a prospect

Figure 2 Effect of strategic debt service on default premiums

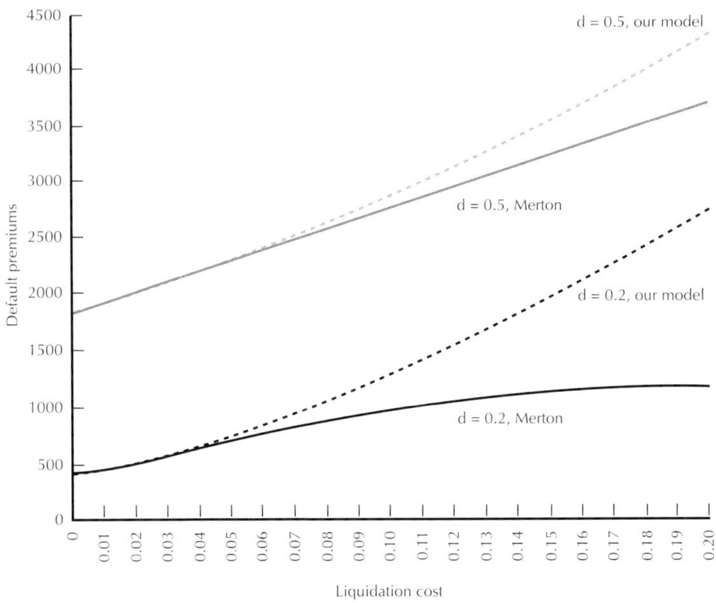

Starting firm value is assumed to be 1. The debt has a time to maturity of 10 years. A zero coupon bond, with a firm beta of 0.04 and a variance of 10% is considered. Riskless rate is assumed to be 5%.

that is more likely for long maturities than for short maturities. For highly leveraged firms, default will be avoided only if the firm value improves significantly, a prospect that is more likely for higher maturities.

The effect of positive bankruptcy costs is indicated in Table 1 by the yield spreads calculated under the assumption that K = 0.1, 0.2. For example, for a firm with $\sigma^2 = 0.03$ and d = 0.5, 2-year debt requires an insignificant spread (about 1 basis point) if liquidation is costless, whereas it carries spreads of 10 basis points and 51 basis points, respectively, when liquidation costs are 10% and 20% of value.[6]

Thus Table 1 demonstrates an important property of our model – properly accounting for costly bankruptcy can better explain observed premiums on traded risky debt than does the standard contingent claims framework.[7] Much of the increase in calculated spreads is accounted for by strategic debt service in our model. To see this we have also computed the implied spreads if Merton's (1974) model is modified by deducting liquidation costs K whenever the exogenous reorganisation boundary $\min(V_T, P)$ is met. For the case of $T = 2$, $\sigma^2 = 0.03$, $d = 0.5$ and $K = 0.2$ this ad hoc adjustment results in a yield spread of only 8 basis points. Thus most of the 51 point spread found in our model reflects strategic debt service in states where $V_T \geq P$. In order to get additional insights on the impact of strategic debt service in our model, we provide Figure 2 in which the liquidation costs are varied from 0 to 0.20 and the default premiums implied by our model are compared with those implied by Merton with ad hoc adjustment. Note that for a lower debt level of $d = 0.2$ the possibility of strategic debt service contributes much more to the default premiums even when the liquidation costs are relatively low. To see why, note that in the range of states $V_T \in (P, P + K)$ our model involves strategic debt service, whereas the modified Merton model repays debt in full. In the range $V_T < P$, debt payments in the two models are identical. For initially low leverage firms the former range is relatively more likely; therefore, a higher proportion of yield spreads are accounted for by strategic debt service.

Careful inspection of Table 1 reveals that the amount that bankruptcy costs increase the yield spreads depends systematically on the degree of leverage and the volatility of the underlying asset. This is shown clearly in Figures 3 and 4, which plot yield spreads on zero coupon debt in the Merton model ($K = 0$) and in our model with $K = 0.1$ with respect to d and σ^2.

In Figure 3 we see that yield spreads in the face of costly bankruptcy exceed those with costless bankruptcy for all degrees of leverage. The difference in yield spread grows with the degree of leverage, particularly in the case of short-term debt ($T = 2$). Similarly in Figure 4 we see that the premiums for bonds with costly bankruptcy are greater than premiums without bankruptcy cost by an amount that is an increasing function of the underlying asset volatility. Increasing the volatility increases the chances that asset values will erode sufficiently after 2 years so that shareholders will default in the payment of principal outstanding either opportunistically or through forced liquidation.

Having established some of the properties of our model for the case of pure discount instruments we now turn to the empirically more important case of coupon paying debt.

Coupon paying debt

A widely observed feature of risky debt contracts such as corporate bonds or syndicated bank loans is that loans indentures specify the regular payment of interest. Violation of this covenant can lead to bankruptcy and possible liquidation of the firm. This common practice reflects the fact that such flow payments are more readily verifiable than are firm values. Our model incorporates this feature when we specify a strictly positive coupon payment ($c > 0$). When liquidation costs are zero ($K = 0$), equilibrium debt service involves the payment of the scheduled coupon whenever the firm is able to pay ($\beta V_t \geq cP$); otherwise, the firm is liquidated. This is a direct extension of the Merton (1974) model.[8] When $K > 0$ equilibrium debt service in our model may fall short of scheduled coupon payment without provoking liquidation.

Figure 3 Effect of debt (d) on risk premium

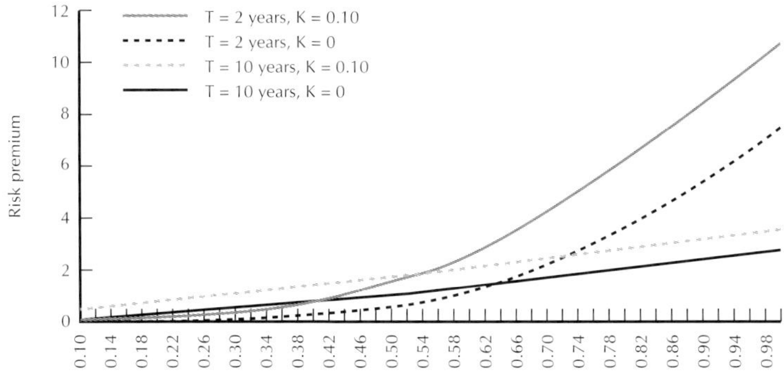

This figure shows the effect of debt level on risk premium. A zero coupon bond, with a firm beta of 0 and a variance of 3% is considered. Riskless rate is assumed to be 5%.

Figure 4 Effect of volatility on risk premium

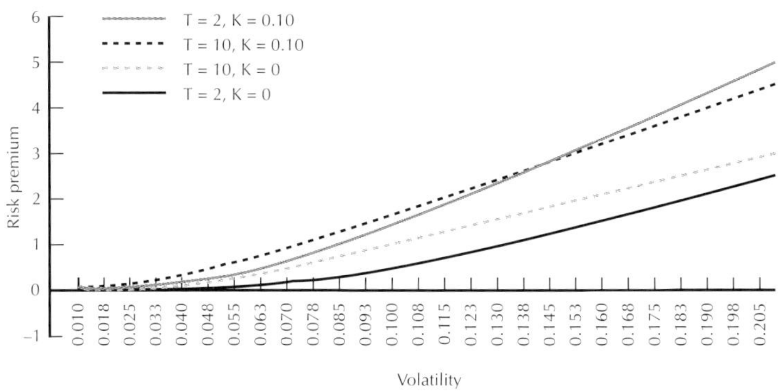

The effect of volatility on risk premium is plotted for debt contracts with 2 years and 10 years maturities. Two liquidation costs (0 and 10%) are used. We have assumed a zero coupon bond, with a firm beta of 0. Riskless rate is assumed to be 5%.

Results pertaining to coupon paying debt are presented in Table 2. Here we have taken the face value of the debt, P, to be the indicator of leverage.[9] The table reports results for a 10% per annum coupon bond to be paid continuously. The payout ratio, β, is also set 10% per annum. Other parameters are set as in the case of zero coupon debt.

The results are reported with respect to three levels of liquidation costs, K = 0, 0.1, 0.2. Required yield spreads are increasing in K, as expected. In fact, the premiums appear quite sensitive to changes in liquidation costs.

Table 2 Coupon debt

Time to maturity (T) = 2.00 years
V = 1, β = 0.10, c = 0.10, r = 1.05

σ^2	P	Risk premium, R – r (%) and bond price, B					
		K = 0		K = 0.1		K = 0.2	
		R – r	B	R – r	B	R – r	B
0.030	0.200	0.0000	0.2190	0.0000	0.2190	0.0033	0.2190
0.030	0.400	0.0005	0.4381	0.0467	0.4377	0.3931	0.4349
0.030	0.600	0.1154	0.6557	1.1265	0.6437	3.2311	0.6196
0.030	0.800	1.3613	0.8547	4.9593	0.8006	9.4939	0.7376
0.030	1.000	5.2333	0.9958	11.0074	0.8972	17.5217	0.7983
0.100	0.200	0.0006	0.2190	0.1135	0.2186	0.8058	0.2158
0.100	0.400	0.1188	0.4371	1.4226	0.4269	3.7347	0.4093
0.100	0.600	1.0055	0.6452	4.3400	0.6072	8.4881	0.5633
0.100	0.800	2.5711	0.8361	7.5786	0.7635	13.2828	0.6890
0.100	1.000	5.5684	0.9897	11.3698	0.8913	17.9093	0.7928
0.200	0.200	0.0430	0.2189	1.1839	0.2144	3.9532	0.2038
0.200	0.400	0.4974	0.4341	3.9863	0.4074	8.5708	0.3750
0.200	0.600	1.7529	0.6364	6.8005	0.5807	12.6833	0.5223
0.200	0.800	3.7057	0.8190	9.3993	0.7388	15.8690	0.6578
0.200	1.000	5.8894	0.9840	11.7317	0.8856	18.3205	0.7871

Time to maturity (T) = 10.00 years

σ^2	P	K = 0 R – r	K = 0 B	K = 0.1 R – r	K = 0.1 B	K = 0.2 R – r	K = 0.2 B
0.030	0.200	0.0321	0.2780	0.3602	0.2718	1.1849	0.2569
0.030	0.400	0.5155	0.5378	1.4209	0.5056	2.6057	0.4670
0.030	0.600	1.6407	0.7471	2.9029	0.6868	4.3726	0.6243
0.030	0.800	3.1564	0.9006	4.6492	0.8178	6.3662	0.7344
0.030	1.000	5.0813	0.9946	6.7689	0.8957	8.7301	0.7968
0.100	0.200	0.4765	0.2696	2.1688	0.2404	4.2609	0.2096
0.100	0.400	1.4857	0.5033	3.2769	0.4468	5.4490	0.3887
0.100	0.600	2.7824	0.6923	4.5286	0.6181	6.5894	0.5433
0.100	0.800	4.0296	0.8508	5.7473	0.7631	7.7546	0.6752
0.100	1.000	5.2222	0.9857	6.9302	0.8869	8.9183	0.7881
0.200	0.200	1.1050	0.2582	4.2667	0.2095	7.4505	0.1719
0.200	0.400	2.3809	0.4740	1.8206	0.4045	7.9177	0.3343
0.200	0.600	3.4482	0.6627	5.5296	0.5802	8.0542	0.4975
0.200	0.800	4.1523	0.8442	5.9937	0.7515	8.1681	0.6588
0.200	1.000	5.5106	0.9681	7.2566	0.8696	9.2954	0.7711

For example, for the parameters $\sigma^2 = 0.03$, $P = 0.06$ and $T = 2$ the spread ranges from 11 basis points to 323 basis points for $K = 0$ to $K = 0.2$, respectively. The sensitivity of coupon debt is the result of the fact that the frequent cashflow checks implied by coupon debt increase the scope for opportunistic debt service.

Figure 5 Risk structure of interest rates

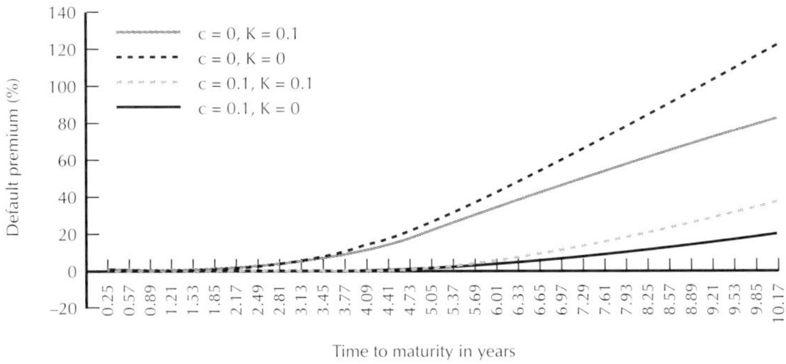

Starting firm value is assumed to be 1. The risk structure is plotted when the firm beta is 0.1, face is 0.3, and the variance is 3%. Riskless rate is assumed to be 5%.

The results for coupon debt are more directly comparable to risky debt traded in the market place. In the United States yield spreads on long-term corporate bonds of firms rated AAA or AA have averaged approximately 125 basis points in recent years (see Salomon Brothers (1992)). The median debt-to-capital ratio for firms in this category is approximately 20% (see Standard and Poor's (1982)). For a firm with this degree of leverage an asset variance of 10% corresponds to an equity volatility that is consistent with this firm being rated AAA or AA.[10]

From the results for the case of $P = 0.2$, $\sigma^2 = 0.1$, and $T = 10$, we see that these levels of yield spreads are produced on our model with $K = 0.045$. Thus, with coupon debt observed spreads are consistent with low levels of asset variance and bankruptcy costs of less than 5% of initial values.

The results in Table 2 reveal that the sensitivities with respect to the basic parameters are similar to those found in our model for discount debt. Yield spreads increase with increases of volatility (σ^2). They are almost always increasing in the degree of leverage used. The only exception appears in the case of $T = 10$, $\sigma^2 = 0.2$, and $K = 0.2$, where an increase in principal from 0.2 to 0.4 entails a decrease in the yield spread. The reason that increasing leverage can at times be beneficial to bondholders in our context is that it can make forced liquidations more frequent, which in turn can result in a higher net payoff to bondholders than if they were exposed to opportunistic debt service subsequently. The sensitivity with respect to time to maturity depends on the degree of leverage. The spread is increasing in T if leverage is low and decreasing if leverage is high.

Figure 5 presents the yield spreads as a function of time to maturity for bonds with face value of 0.3 when the underlying project has a payout rate of 10% per year and a volatility of 0.03. Four cases are plotted, corresponding

Figure 6 Effect of beta on risk premium

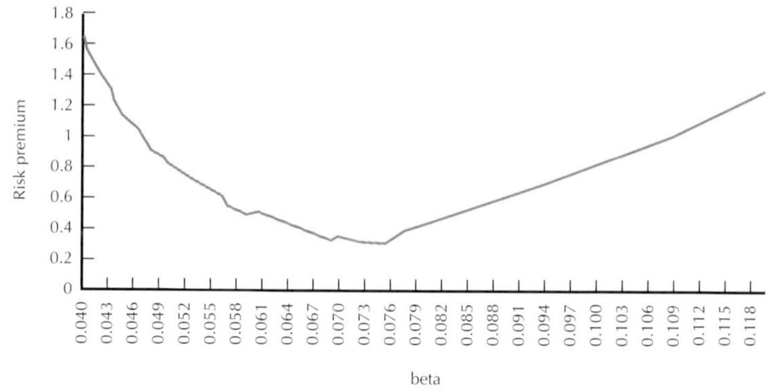

The effect of cash payout ratio (beta) on risk premium is illustrated for T = 10 years, face = 0.30 and coupon = 10%. The liquidation costs is assumed to be 10% and the variance is 3%. Riskless rate is assumed to be 5%.

Figure 7 Coupon effect on risk premium

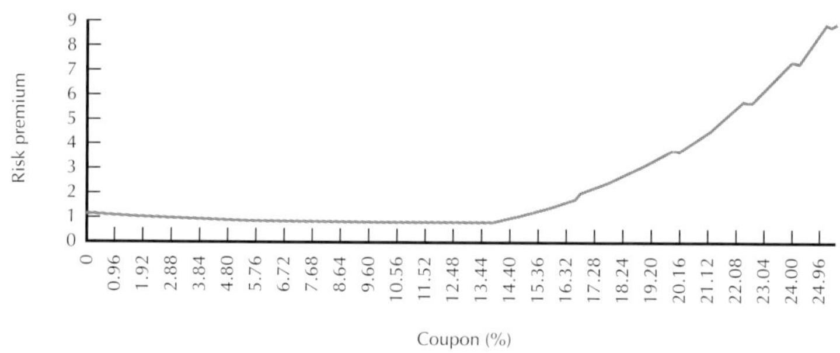

This figure examines the effect of increasing the coupon rate on risk premium. We assume that the debt has 10 years to maturity, beta = 0.10 and face = 0.30. The liquidation costs is assumed to be 10% and the variance is 3%. Riskless rate is assumed to be 5%.

to two values of liquidation costs (K = 0, 0.1) and two coupons (c = 0, 0.1). In each case the yield spread is increasing in the time to maturity, reflecting the low leverage involved. Furthermore, spreads are higher when the liquidation costs are positive. For the case presented, spreads are higher for the zero coupon bond than for the coupon bearing bond, holding liquidation costs constant. We explore this effect more explicitly in what follows.

Further subtleties of bond valuation in the face of opportunistic debt service are illustrated in Figures 6 and 7. Figure 6 depicts the relationship between the yield spread and the cash payout ratio, β. It is striking that yield spreads are initially decreasing and then increasing in this payout rate. Sufficiently low levels of β combined with a positive coupon interest lead to a quick forced liquidation. Even if liquidation costs are not so high as to prevent the bondholder from fully recovering the principal, this means the bond carries a risk premium, since the risk-free bond would be valued more than its face value when coupons exceed the risk-free rate. Thus an increase in β from low levels reduces the chances of this occurring and thus reduces required yield spreads. However, increases in β also imply that the growth prospects of the project are worsened.[11] Thus a high β project will not be cash constrained initially; however, it has a relatively high possibility of falling in value, pushing the project into the range where owner-managers will engage in opportunistic violation of the terms of the debt contract. In other words, extending high levels of long-term, fixed interest credit to today's cash cows is a risky business that should command a significant yield spread.

Figure 7 represents the relationship between required spreads and coupon level for debt with given principal and maturity. Again yield spreads at first decrease and later increase as the coupon is increased from 0 to 25%. This reflects two effects that operate in different directions. First, increasing the coupon increases the range of states in which a violation of an indenture is likely to occur. However, instead of unambiguously working in favour of the bond holder, the violation of the covenant presents the bondholder with a difficult choice. Either he initiates legal action and thereby incurs a deadweight loss, or he accepts the debt service the owner is willing or able to pay. Since accepting is often the better choice, the owner can take advantage of this to default opportunistically, that is, by increasing the range of outcomes where the debt will not be fully serviced. The second, opposite effect is that a positive coupon in some states results in bankruptcy for reasons of illiquidity. These forced liquidations in part can help the bondholder in cutting off a possible future of opportunistic debt service by the owner-manager. In effect the positive coupons combined with a no-asset-sale clause serves as a credible commitment by bondholders to require debt service.

To summarise, in comparison to the previous models in the valuation literature surveyed in the introductory section, our model produces higher risk premiums due to the opportunistic debt service and the resulting deviations from the absolute priority rule. Empirically observed yield spreads can be replicated with lower levels of bankruptcy costs and asset volatilities in our model.

Our discussion has underlined the ways that a costly bankruptcy process can directly and indirectly affect bond valuations. One general thread that runs through this is that a more costly bankruptcy process provides greater scope for opportunistic debt service since the bondholder will hesitate to initiate legal action. The corollary of this is that increasing liquidation costs which depress bond values can increase equity values. Of course, this effect depends crucially on the incidence of bankruptcy costs which in our model fall on the bondholder. The effect is also tempered to the extent there is a positive probability of forced liquidation. We explore these issues below in our treatment of the design of contract terms. Before doing so, however, we consider how valuation is affected by the introduction of corporate taxes.

Corporate taxes

We have already noted that the framework that we have introduced could be modified without changing the essential valuation methodology. In particular, literature on financial distress provides a rich array of one-shot bankruptcy models that could substitute for that in the second section. We view understanding the dynamic implications of such models a very worthwhile line of inquiry, but we will not attempt this here. Instead, we consider only a slight modification of our model. In particular, we model an extremely simple tax code where corporate earnings are taxed at a flat rate of τ but where all interest payments are tax deductible. In this context, equity valuation in the absence of forced liquidation involves replacing equation (5) with

$$E(V_t) = (f_t - S(V_t))(1 - \tau) + \frac{pE(uV_t) + (1 - p)E(dV_{t+1})}{r} \tag{8}$$

for $t < T$. At maturity, $E_T = (1 - \beta)V_T + (f_T - s_T cP)(1 - \tau) - s_T P$ where $s_T = S(V_T)/(1 + c)P$. In the case of forced liquidation, tax liabilities on current earnings are deducted before calculating liquidation values. This is the only way bond valuations are modified. Since this tax liability $\tau\beta V_t$ is small relative to V_t, introducing taxes will have little effect on equilibrium debt service strategies. Taxes will mean, however, that modifications of the terms of the debt contract can have different impacts on equity values depending on the tax rates applied. Thus taxes change bond valuation little but can be an important consideration in determining optimal contract terms.

DESIGN OF DEBT CONTRACTS

A problem of security design emerges when an inappropriate choice of contract features can result in inefficiencies, that is, a loss of aggregate value for the contracting parties. An optimal financial contract minimises these inefficiencies subject possibly to certain constraints. The literature on security design has considered several alternative sources of contracting inefficiency. Examples include incomplete risk sharing (Allen and Gale (1988)), inefficient information acquisition (Boot and Thakor (1993)), and inefficient allocation of control rights (Aghion and Bolton (1992)).

In our framework, the security design problem originates from inefficiencies in the process for managing financial distress. Specifically, they arise through costly liquidations which we have seen many be possible outcomes in equilibrium. This feature of our model is similar to other treatments emphasising control rights (see, in particular, Aghion and Bolton (1992) and Hart and Moore (1991)). In this related literature and in our model, debt is chosen despite its possible inefficiencies because the alternative forms of financial contracting, such as all outside equity, will imply inefficiencies or agency costs of their own which may exceed those of debt. However, unlike these models, our framework allows us to explore in detail the choice of the terms of the debt contract, taking into account their full dynamic implications in a stochastic setting. Consequently, we are closer to the practical design problem faced by companies issuing debt.

Formally, a change in the contract specification affects the game form faced by the contracting parties. The problem is to choose the contractual features so as to minimise the inefficiencies that emerge in the associated equilibrium. With many periods and possible states of the world, the design problem is likely to admit a great multiplicity of solutions unless some structure is placed on the set of possible contracts.[12] We consider contracts where contractual debt service is time dependent, but not state dependent, and all control rights remain with the owner-manager unless there is a default on contractual terms that induces the debtholders to proceed with liquidation. Further, we focus on four features: (1) coupon, (2) maturity, (3) face amount and (4) the amortisation schedule. These features allow us to model a large variety of contract forms encountered in practice. However, the treatment of convertible and callable bonds is left for future work.

Contract terms are modelled as follows. Let P_t be the outstanding principal amount of the bond in period t, T be the maturity date of the bond, c be the coupon, A_t be the principal amount that is amortised in period t and g be the grace period in years, that is, the number of years before amortisation is included in scheduled debt service. The index t refers to the number of years elapsed since the issue date. Then the contracted payments to the creditors are $CS_t = cP_t + A_t$, where $P_t = P_{t-1} - A_{t-1}$. The amortisation is computed as follows: $A_t = 0$, if $t \leq g$ and $A_t = P/(T - g)$, otherwise. By setting $g = T - 1$, we get straight coupon debt, which has zero coupon debt as a special case. Generally, this specification permits sinking fund schedules with different grace periods and purely amortising debt.[13]

In the design problem the owner-manager selects the vector (c, T, P, g) so as to maximise the value of the equity subject to constraint that the value of the debt be at least equal to the funding requirement for undertaking the project. This choice will be guided by the primitives of the firm which are (1) the payout factor β, (2) the liquidation costs K, (3) the amount of money that the firm needs to raise in the debt market D, (4) the volatility of the firm σ^2 and (5) the tax rate τ. Thus the problem is

$$\max_{c,T,P,g} E(V_0; \sigma^2, \beta, r, K, \tau)$$

such that

$$D \leq B(V_0; \sigma^2, \beta, r, K, \tau)$$

Here both E and B are given by the equilibrium values in the game implied by the contract. As seen above, the equilibrium generally will not be first-best efficient because it will typically result in forced liquidations in some future states. The solution to the security design problem may also be *ex ante* inefficient in the sense that positive net present value projects may not be undertaken. This will occur when it is not possible to write a debt issue valuable enough to meet the funding requirement. The solution will be second-best optimal when the corporate tax rate is zero. This is because the model has a basic value preservation property so that the value of equity equals the asset value of the firm less the value of debt less the value of expected future liquidation costs plus the value of future tax shields. Thus absent taxes, the problem reduces to minimising liquidation costs subject to a funding constraint.

This design problem does not admit of analytical solutions in non-trivial cases, and its numerical solution is itself rather time-consuming.[14] We report here the results of two sets of design experiments. The first restricts the contracts to be non-amortising bonds and varies the payout rate, β; the liquidation costs, K; and the debt ratio, D. The second focuses on the relationship between the speed of amortisation and the remaining parameters.

The results of the first experiment are summarised in Table 3. Each row reports the optimal contract for a given combination of β, K, D and T. We describe the optimal contract by the contractual interest payment expressed as a per cent of the value of debt and report the associated value of equity.

The most striking pattern in Table 3 is that as the payout rate β increases, the contractual interest increases and the value of equity decreases. To interpret this it is important to recall that the payout rate determines the expected growth rate of the asset value of the firm: the higher β the slower growth in the asset value of the firm. Fast-growing, that is, low β, firms favour relatively low interest burdens because this reduces the chances of an early forced liquidation that would deny the owner the benefits of future growth. In doing so they forego the tax shield benefits that would be associated with higher contractual interest. As we consider progressively slower growing firms, balancing the trade-off leads to higher and higher interest payments. The cash cow (the high β firm with poor growth prospects) chooses a relatively higher interest burden because the value of the near-term tax shield is substantial and the prospect of a near-term illiquidity default is less daunting because cash is relatively abundant and the future cut off by such a default is less bright. The value of equity with the optimal contract is a decreasing function of β because most of the benefits of a high growth rate accrue to equity.

The effect of leverage on contract design is seen by comparing the cases of D = 0.25 and D = 0.5, holding other parameters constant. In most cases the firm with the higher leverage selects a lower level of contractual interest. The reason is that higher leverage tends to make forced liquidation in advance of maturity more likely. The firm can offset this by decreasing contractual interest payments. This appears to be the dominant effect, but as is often the case in the comparative statics of bond valuation, other effects are present that can sometimes dominate. In particular, for T = 10 and K = 0.1 higher leverage is associated with higher interest. By raising interest payments the firm is able to increase the tax shield benefits which are substantial for the highly levered firm. This plus the reduced probability of default at maturity are sufficient to compensate the heightened prospect of an early illiquidity default.

From Table 3 we also see that the value of equity with the optimal design is decreasing in the cost of liquidation. This might at first seem surprising because our game form allowed equity holders to extract a surplus by reducing debt service by an amount determined by liquidation costs. Here we see that some of the burden of *realised* liquidation costs falls on equity. The reason is that such liquidations sometimes occur when liquidation values exceed the face value of debt so that equity holders benefit from smaller liquidation costs.

The last point to be made with reference to Table 3 is that equity is higher for T = 10 than for T = 5 for most of the cases. Here firms tend to prefer to issue longer-term debt

Table 3 Optimal design of debt contracts

$V = 1, \tau = 0.10, r = 1.05$

Debt ratio	β	Coupon (%)	Equity
Time to maturity = 5 years			
Liquidation cost = 0.1			
0.250	0.040	3.69	0.66470
0.250	0.060	5.25	0.65648
0.250	0.080	7.69	0.64992
0.250	0.100	7.80	0.64380
0.500	0.040	2.58	0.42695
0.500	0.060	4.44	0.41670
0.500	0.080	5.44	0.41174
0.500	0.100	5.60	0.40583
Liquidation cost = 0.2			
0.250	0.040	3.71	0.61797
0.250	0.060	5.67	0.60799
0.250	0.080	7.71	0.60086
0.250	0.100	9.67	0.59649
0.500	0.040	3.00	0.39859
0.500	0.060	0.43	0.38694
0.500	0.080	5.43	0.38262
0.500	0.100	7.36	0.38149
Time to maturity = 10 years			
Liquidation cost = 0.1			
0.250	0.040	2.55	0.67233
0.250	0.060	3.93	0.66082
0.250	0.080	4.73	0.65229
0.250	0.100	5.94	0.64701
0.500	0.040	4.85	0.43718
0.500	0.060	6.59	0.41544
0.500	0.080	3.17	0.40767
0.500	0.100	8.99	0.39768
Liquidation cost = 0.2			
0.250	0.040	3.32	0.64787
0.250	0.060	5.17	0.63919
0.250	0.080	7.12	0.63675
0.250	0.100	8.85	0.63431
0.500	0.040	2.09	0.41478
0.500	0.060	3.78	0.40842
0.500	0.080	5.24	0.40157
0.500	0.100	8.00	0.35585

Figure 8 Optimal design of debt contracts

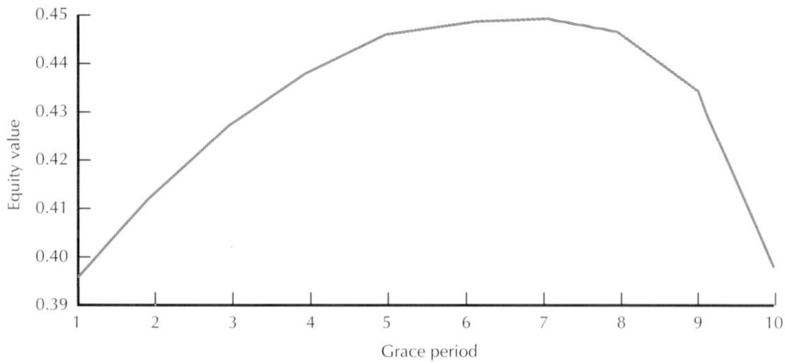

This figure examines the effect of sinking fund provisions on equity values. We assume that the debt has 10 years to maturity, beta = 0.10, and the debt level d = 0.50. The optimal grace period is 7 years.

rather than shorter-tem debt, because with non-amortising debt there is relatively high change of liquidation at maturity. The longer this is forestalled the better. Again this is not the only effect at work, as can be seen in the case of higher leverage, higher β firms with lower liquidation costs. For a sufficiently high β the firm is expected to shrink with time. If so, repayment of principal is more problematic with long-term debt than short-term debt.

In this first experiment the sensitivities with respect to several parameters reflected two considerations, namely, the implications for the tax shield and for the probability of liquidation. For example, raising coupon interest brings carry tax benefits, but its effect on default is ambiguous. By reducing the principal needed to produce a debt of given value, higher coupon reduces the chances of difficulties in repaying the loan at maturity. But it raises the chances of provoking an illiquidity default prior to maturity. This suggests that there may be some advantages to considering designs that involve some amortisation of principal or of a sinking fund. We explore this in a second design experiment in which we consider various possible values of the grace period before starting a straight-line amortisation of principal. Other parameters considered are T, β and D.

The general result that emerges from this experiment is that optimal debt contracts will typically require partial repayments of principal prior to maturity, but that it is often beneficial to delay the start of repayments of principal. This is depicted clearly in Figure 8. Here, as the grace period is increased, we see the highest value of equity that is attainable by varying coupon and principal so as to produce debt that meets the funding requirement. It can be seen that the equity increases until g = 7, beyond which it declines. The reason for this is that by setting grace too low, amortisation will start too early, reducing the tax shield that comes exclusively from the payments of interest. On the other hand, by allowing too long a period of grace, the debt service payments towards the end of the life of the bond are relatively heavy. This runs a relatively great risk of provoking a liquidation. The optimal grace period balances these two considerations.

Table 4 Optimal design of debt contracts

$\tau = 0.1$, $K = 0.1$

T	β	D	g	Contractual interest (%)	Equity
5	0.04	0.25	5	3.69	0.66470
5	0.04	0.50	5	2.58	0.42695
5	0.10	0.25	4	3.45	0.64814
5	0.10	0.50	4	0.98	0.44243
10	0.04	0.25	9	2.42	0.67808
10	0.04	0.50	8	1.77	0.45896
10	0.10	0.25	9	0.81	0.66795
10	0.10	0.50	7	3.51	0.44917

The second experiment is further reported in Table 4, which lists the optimal grace and contractual interest as well as the associated value of equity attained. Making the grace period an object of design choice of course has implications for the optimal contractual interest, and this alters significantly some of the patterns seen in Table 3. In particular, in most cases we see that the higher values of β are associated with lower values of contractual interest. This result emerges because in these cases the higher β firms select contracts with shorter grace periods. That is, slower growth firms choose to begin repaying principal earlier in order to reduce the chances of realising costly bankruptcies as the project matures. This may be desirable even if it implies that coupon levels must be kept moderate so that tax shield benefits are sacrificed.

In our view, our findings on amortising debt provide a rationale for the widely observed use of sinking funds in corporate bond issues. We see that the optimal sinking fund feature balances off tax shield benefits with potential costs of financial distress that can fall on shareholders. Furthermore, despite the fact that we have not explicitly dealt with monitored financial contracts, our model helps explain the frequent use of grace periods in bank loans. Again, borrowers delay principal payments so that they can reap the tax benefits of interest payments. However, beyond a certain point amortisation must begin in order to control the risks of costly liquidations.

CONCLUSIONS

We have presented an internally consistent, arbitrage-free framework for valuing risky debt contracts which incorporates a number of the features emphasised in recent contributions to corporate finance. The model allows for strategic debt service because the costliness of formal bankruptcy may induce creditors to accept deviations from contractual payments. When the framework is applied to the problem of valuing standard debt contracts using plausible parameters, we find yield spreads over Treasuries that correspond closely to observed levels even if costs of bankruptcy are relatively low. This suggests that the framework offers promise for the econometric modelling of spreads of corporate bonds and other risky debt contracts.

Our framework is flexible enough to allow altering assumptions about contractual features, bankruptcy procedures, or other aspects of the economic environment. We examined a model variant with corporate taxes. Other bankruptcy games seen in the static theory of corporate finance could be introduced into our dynamic framework at the cost of adding parameters and computational complexity.

The framework is used to address the question of debt contract design, that is, to find the combination of contractual features that minimise the combined tax burden and expected bankruptcy costs. Our results show that high growth (low β) firms in general tend to use a low coupon debt contract. On the other hand, low growth firms tend to use a high coupon debt contract. Although not reported here, as the tax rate increases, firms tend to increase the coupons to take advantage of the tax shield from interest payments. Highly levered firms tend to use low coupon debt. In general, sinking fund provisions are used so as to better match the cash payout from the assets: here the trade-offs are between tax shields and forced liquidations.

In our view, there is considerable opportunity for additional work along the lines of the framework we have developed here. In particular, the extension of the model to callable and convertible debt and the incorporation of a stochastic term structure would be very desirable. Second, it is clear that the modelling of the bankruptcy game could have an important impact on the results. The game form that we have adopted gives a first-mover advantage to the owner-manager with the result that he has the most scope for opportunism. Other models of bankruptcy might alter the distribution of the gains from renegotiation.[15] Changing the bankruptcy model is very likely to have important implications for optimal security design. For example, in our model strategic debt service tends to reduce the frequency of illiquidity default. Models with less scope for renegotiation may result in lower coupons or greater reliance on sinking funds. Finally, we believe that the valuation of subordinated debt and the issue of traded debt versus bank loans could be usefully studied within a framework similar to the one we have used here.

APPENDIX

This appendix shows that the martingale probability which makes the cashflow process internally consistent with the cum cashflow value process of the firm is given by $p = [(1 - \beta)r - d]/u - d$. The proof is provided for two time steps: it is easy to verify that the arguments go through for any number of future time periods.

The cum cashflow value process is specified below. The associated cashflow process is obtained by simply multiplying the cum cashflow values at each node by the factor β.

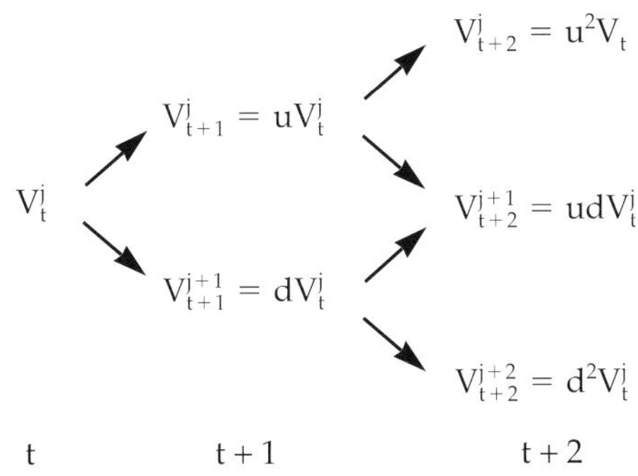

Let p be martingale probability. Then,

$$V^j_{t+1} = \frac{pV^j_{t+2} + (1 - p)V^{j+1}_{t+2}}{r} + \beta V^j_{t+1}$$

Recognising that $V^j_{t+2} = uV^j_{t+1}$ and $V^{j+1}_{t+2} = dV^j_{t+1}$ and substituting these relationships in the previous equation we get

$$r(1 - \beta) = pu + (1 - p)d$$

Solving, we get

$$p = \frac{r(1 - \beta) - d}{u - d}$$

Note that the same relationship holds for p in each step of the recursion. The martingale probability p is also state independent. It is useful to point out that $\partial p/\partial \beta = -r/(u - d)$. This implies that as the cash payout increases, the martingale probability p falls: this will have the effect of giving greater weight to cashflows in relatively "poorer" states of the world and lesser weight to "richer" states of the world in the discounting process. This occurs in our model so as to keep the cum cashflow value of the firm the same, *irrespective of the payout*. Note that this effect becomes less important as σ^2 increases: this is due to the fact that as σ^2 increases (u – d) increases as well, causing p to depend less on the payout factor β. Thus, in our model, as the volatility of the firm increases, the payout factor β becomes less important in affecting the martingale probability. These factors play a role in the valuation of alternate debt contracts.

1 Other recent papers concerned with debt include Bergman and Callen (1991), Dewatripont and Tirole (1992) and Franks and Nyborg (1993).

2 This assumption is also used by Aghion and Bolton (1992) and Hart and Moore (1991) among others to study capital structure and security design. In this context, contingent transfer of control through a debt contract is preferred to outside equity since the latter would incur an agency cost from the outset. Later, we will see that the tax deductibility of debt service would provide an additional reason for issuing debt.

3 Our approach is easily applied to more general two-state branching processes with state-dependent probabilities. For expositional clarity, we are illustrating the model with a simple binomial lattice.

4 This can be derived by solving for p such that V_t is the sum of the current cashflow plus the expected discounted value of V_{t+1}. This solution is unique. The Appendix provides a proof of these assertions.

5 See Mello and Parsons (1992) for an attempt to measure debt agency costs arising both from lack of investment commitment by shareholders and from bankruptcy costs. However, this study does not endogenise the lower reorganisation boundary of the firm, nor does it accommodate strategic behaviour in bankruptcy.

6 It should be stressed that K = 0.2 is used to reflect an upper bound on the direct and indirect costs of liquidation.

7 Below we make an explicit comparison to observed spreads with reference to coupon bearing debt.

8 Kim, Ramaswamy and Sundaresan (1993) extend the Merton model to coupon debt and allow for interest rate uncertainty. Our model with $c > 0$ and $K = 0$ is a discretised version of their model in the case of zero interest rate volatility.

9 Note that since this differs from the quasi-debt ratio d, Tables 1 and 2 are not directly comparable.

10 The volatility of equity as related to the value of the firm is given as

$$\sigma_e = \sigma E_V V/E \leq \sigma V/E$$

where E_V is the partial derivative of equity with respect to underlying value. Thus $\sigma^2 = 0.1$ and $P = 0.2$ give an upper bound for σ_e of 0.395.

11 The probability of a "down" move is $1 - p = [u - r(1 - \beta)]/u - d$ which is increasing in β.

12 Hart and Moore (1991) working in a certainty setting find multiple solutions, which leads them to introduce some restrictions on contract forms.

13 A term annuity is taken as the limiting case as principal tends toward zero and coupon is adjusted to maintain cP constant.

14 Our program written in C produces the results for Table 4 in about 72 hours on a RISC 6000 machine.

15 Interesting alternative simplified models can be found in Aghion, Hart and Moore (1992), Bergman and Callen (1991), Hart and Moore (1991) and Kahn and Huberman (1989).

BIBLIOGRAPHY

Aghion, P. and P. Bolton, 1992, "An Incomplete Contracts Approach to Financial Contracting", *Review of Economic Studies* 59, pp. 473–94.

Aghion, P., O. Hart and J. Moore, 1992, "The Economics of Bankruptcy Reform", working paper 148, LSE Financial Markets Group.

Allen, F. and D. Gale, 1988, "Optimal Security Design", *Review of Financial Studies* 1, pp. 229–63.

Asquith, P., R. Gertner and D. Scharfstein, 1991, "Anatomy of Financial Distress: An Examination of Junk Bond Issues", working paper, Massachusetts Institute of Technology.

Bergman, Y. and J. Callen, 1991, "Opportunistic Underinvestment in Debt Renegotiation and Capital Structure", *Journal of Financial Economics* 29, pp. 137–71.

Black, F. and M. Scholes, 1973, "The Pricing of Options and Corporate Liabilities", *Journal of Political Economy* 81, pp. 637–54.

Bolton, P. and D. Scharfstein, 1993, "Optimal Debt Structures with Multiple Creditors", Working Paper 32, ESF Network in Financial Markets.

Boot, A.W.A. and A.V. Thakor, 1993, "Security Design", *Journal of Finance* 48, pp. 1349–78.

Dewatripont, M. and J. Tirole, 1992, "A Theory of Debt and Equity: Creditor Diversity and Manager-Shareholder Congruence", Working Paper, ECARE.

Franks, J. and K. Nyborg, 1993, "Debt Structure and the Loss of Private Benefit: the Case of the UK Insolvency Code", Working Paper, London Business School.

Franks, J. and W. Torous, 1989, "An Empirical Investigation of Firms in Reorganization", *Journal of Finance* 44, pp. 747–79.

Franks, J. and W. Torous, 1993, "A Comparison of Financial Recontracting in Workouts and Chapter 11 Reorganizations", Working Paper 26, ESF Network in Finance Markets (forthcoming: *Journal of Financial Economics*).

Fudenberg, D. and J. Tirole, 1985, "Preemption and Rent Equalization in the Adoption of New Technology", *Review of Economic Studies* 52, pp. 383–401; reprinted as Chapter 13 of the present volume.

Harrison, J.M. and D.M. Kreps, 1978, "Martingales and Arbitrage in Multiperiod Security Markets", *Journal of Economic Theory* 20, pp. 381–408.

Hart, O. and J. Moore, 1991, "A Theory of Debt Based on the Inalienability of Human Capital", LSE Financial Markets Group Discussion Paper Series.

Jensen, M.C. and W. Meckling, 1976, "The Theory of the Firm: Managerial Behavior, Agency Cost, and Ownership Structure", *Journal of Financial Economics* 3, pp. 305–60.

Jones, E.P., S.P. Mason and E. Rosenfeld, 1984, "Contingent Claims Analysis of Corporate Capital Structures: An Empirical Analysis", *Journal of Finance* 39, pp. 611–25.

Kahn, G. and G. Huberman, 1989, "Default, Foreclosure, and Strategic Renegotiation", *Law and Contemporary Problems* 52, pp. 49–61.

Kim, I.J., K. Ramaswamy and S.M. Sundaresan, 1993, "Valuation of Corporate Fixed-Income Securities", *Financial Management,* special issue on financial distress, Autumn, pp. 117–31.

Leland, H., 1993, "Risky Debt, Bond Covenants and Optimal Capital Structure", Working Paper, University of California, Berkeley.

Maloney, K.J., 1992, "A Contingent Claims Model of Corporate Security Valuation using a Realistic Model of Financial Distress", Working Paper, Dartmouth University.

Mella-Barral, P. and W.R.M. Perraudin, 1993, "Strategic Debt Service", Working Paper 39, Centre for Economic Policy Research, London; reprinted as Chapter 12 of the present volume.

Mello, A. and J. Parsons, 1992, "Measuring the Agency Cost of Debt", *Journal of Finance* 47, pp. 1887–904.

Merton, R.C., 1974, "On the Pricing of Corporate Debt: The Risk Structure of Interest Rates", *Journal of Finance* 29, pp. 449–70.

Myers, S., 1977, "Determinants of Corporate Borrowing", *Journal of Financial Economics* 5, pp. 147–75.

Salomon Brothers, 1992, "Analytical Record of Yields and Yield Spreads", Salomon Brothers, New York.

Standard and Poor's, 1982, "Credit Overview", Standard and Poor's, New York.

Weiss, L., 1990, "Bankruptcy Resolution: Direct Costs and Violation of Priority of Claims", *Journal of Financial Economics* 27, pp. 285–314.

Zender, J., 1991, "Optimal Financial Instruments", *Journal of Finance* 46, pp. 1645–63.

12

*Strategic Debt Service**

Pierre Mella-Barral; William Perraudin

London School of Economics; Birkbeck College and CEPR

When firms experience financial distress, equity-holders may act strategically, forcing concessions from debt-holders and paying less than the originally-contracted interest payments. This chapter incorporates strategic debt service in a standard, continuous time asset pricing model, developing simple closed-form expressions for debt and equity values. We find that strategic debt service can account for a substantial proportion of the premium on risky corporate debt. We analyse the efficiency implications of strategic debt service, showing that it can eliminate both direct bankruptcy costs and agency costs of debt.

The contingent claims literature on corporate debt initiated by Merton (1974) interprets risky debt as a portfolio comprising the safe asset plus a short position in a put option written on the value of the firm's underlying assets. Under appropriate assumptions about asset return distributions, this interpretation implies simple, closed-form expressions for risky bond prices. However, empirical evidence suggests that risk premia on corporate debt significantly exceed those implied by Merton's model, except if one includes unrealistically high bankruptcy costs. Recent research has suggested two approaches whereby the high, observed risk premia may be rationalised.

1. Franks and Torous (1989) document the fact that, in many bankruptcy settlements, junior claimants including equity-holders receive a greater share of residual value than is consistent with absolute priority.

*This paper was first published in the *Journal of Finance* and is reprinted with the permission of Blackwell Publishers. The authors wish to thank Anthony Neuberger, Suresh Sundaresan, three anonymous referees and the editor for valuable comments. Any errors remain the authors' responsibility. Financial support from the ESRC, the Newton Trust and INQUIRE (UK) is gratefully acknowledged.

Longstaff and Schwartz (1995) incorporate such departures from absolute priority in a contingent claims model, taking the allocation of the bankrupt firm's residual value among stake-holders as exogenously given. They show that risk premia prior to bankruptcy may be significantly boosted by expectations of deviations from absolute priority.

2. An alternative approach takes as its starting point the observation that, in many cases, debt-holders in distressed firms are persuaded by equity-holders to accept concessions prior to formal bankruptcy proceedings. Bergman and Callen (1991) study the extraction of such concessions by equity-holders in a static model of capital structure determination. Anderson and Sundaresan (1996) include such concessions in a binomial pricing model and examine the design of debt contracts.

In this chapter, we incorporate strategic debt service by equity-holders in a standard, contingent claims asset pricing model. Employing continuous time methods, we obtain closed-form analytic expressions for the values of debt and equity when the debt issued is perpetual. Our research yields an explanation of the large default premia observed in actual bond markets. For reasonable parameter values, we find that strategic debt service may account for 30–40% of the premium on risky debt. For debt issued by firms whose asset values have low volatility, this percentage can easily become much higher.

The fact that we obtain closed-form solutions for debt and equity values and the trigger points for state variables, enables us to analyse the efficiency implications of strategic debt service. In our model, debt imposes costs both by introducing direct bankruptcy costs and by creating agency-theoretic 'under-investment' problems of the kind discussed by Myers (1977).[1] We show that strategic debt service by equity-holders eliminates both sets of costs. This result replicates in a dynamic, continuous time pricing model the earlier insight of Bergman and Callen (1991) that renegotiation of coupon payments may eliminate bankruptcy costs.

In Bergman and Callen, the bargaining framework, based on Rubinstein's (1982) model of alternating offers, is efficient in the sense that value-reducing strategies are threatened by equity-holders but never actually implemented. However, the ex post bargaining places a limit on the maximum possible debt service payment that equity-holders can be induced to make and hence constrains the amount of debt the firm can issue *ex ante*. This prevents the firm from fully exploiting the tax advantages of debt and hence reduces the firm's *ex ante* value. While the firm does have a debt capacity in our model, this does not generate inefficiencies since we suppose that there is no tax advantage to debt. In this case, when renegotiation is possible, other costs of debt are eliminated and the Modigliani-Miller theorem holds. As we show, strategic debt service has important implications for pricing even in this case since agents know *ex ante* that equity-holders will use the threat of bankruptcy costs to reduce the coupon payments they make when the firm approaches closure.

It should be stressed that our paper fits within a broader literature that has investigated the effects on debt valuation of different assumptions about the strategies open to stake-holders in the firm. Brennan and Schwartz (1984), Mello and Parsons (1992) and Mello, Parsons and Triantis (1995) study contingent claims models in which real investment and closure decisions interact with debt valuation. Fischer, Heinkel and Zechner (1989) allow dynamic adjustments in the firm's capital structure. Mauer and Triantis (1994) analyse the interaction of leverage adjustments with dynamic investment and operating decisions. Leland (1994) looks at comparative static properties of infinite maturity coupon bond prices under different assumptions about what triggers bankruptcy. Our research and that of Anderson and Sundaresan (1996) contributes to this literature by examining the effect on valuation of expanding the strategy space open to equity-holders by introducing take-it-or-leave-it offers on debt service to the firm's creditors.

Our chapter is structured as follows. The first section presents a continuous time model of a levered firm without renegotiation of the debt contract. We obtain analytic solutions for debt and equity and examine the dependence of total firm value on the level of leverage. In the following section, we allow equity-holders to make take-it-or-leave-it offers to debt-holders regarding coupon payments. With renegotiation, the firm's service flow varies stochastically in a way that depends on the state variable, the price of the firm's output. We characterise properties of equity-holders' optimal service flow function. Finally, we describe the solution of the model when instead debt-holders make take-it-or-leave-it offers to reduce the coupon. The chapter then discusses extensions of the model and goes on to present numerical simulations that show how changes in firm parameters affect the contribution of renegotiation effects to risk premia.

DEBT AND AGENCY COSTS

Basic assumptions

Throughout our analysis, we suppose that capital markets are frictionless and free of informational asymmetries and that agents are risk neutral and can borrow and lend freely at a constant, sure interest rate, r. The assumption of risk neutrality represents little loss of generality. If agents are risk averse, the analysis may be developed under risk neutral rather than actual probabilities. (See Harrison and Kreps (1979).)

Consider a firm which produces a unit of output which it sells for a price, p_t. Let p_t be a geometric Brownian motion

$$dp_t = \mu p_t + \sigma p_t dB_t \tag{1}$$

where μ and σ are constant parameters and B_t is a standard Brownian motion. While in production, the firm incurs costs per period of $w > 0$, so its net earnings flow is

$$p_t - w \tag{2}$$

Suppose that bankruptcy impairs the firm's efficiency in that, after bankruptcy, new owners of the firm can only generate earnings of

$$\xi_1 p_t - \xi_0 w \tag{3}$$

where $\xi_1 \leq 1$ and $\xi_0 \geq 1$. At any time, whoever owns the firm can opt for liquidation. The scrapping value of the firm at liquidation is a constant γ.

The expected, discounted value of the decline in earnings that accompanies bankruptcy may be regarded as the *direct cost of bankruptcy*. If the prospect of bankruptcy distorts real investment decisions, one may refer to these as agency costs of debt or *indirect bankruptcy* costs. It will turn out in our model that the firm's financing may in certain circumstances lead to inefficiently early liquidation. Since liquidation may be regarded as a real investment decision, the model contains the possibility of both direct and indirect bankruptcy costs.

The pure equity firm

We begin by considering the firm's value under pure equity financing, W_t. Since W_t is the value of an asset with an income flow $p_t - w$, financial market equilibrium under risk neutrality requires that

$$rW_t = p_t - w + \frac{d}{d\Delta} E_t(W_{t+\Delta})\Big|_{\Delta=0} \tag{4}$$

The stationarity of the payoffs involved implies that W_t depends on time only through its dependence on the state variable, p_t, and that liquidation will occur the first time that p_t hits some constant low level p_c. If $W_t = W(p_t)$ is a twice continuously differentiable function of the state variable, p_t, by Itô's lemma, this function must satisfy the differential equation

$$rW(p) = p - w + \mu p W'(p) + \frac{\sigma^2}{2} p^2 W''(p) \tag{5}$$

The boundary conditions that $W(p)$ must obey are as follows. In the absence of arbitrage, the firm value must equal the scrapping price at closure, $W(p_c) = \gamma$. If asset prices are free of bubbles, as p_t becomes large, $W(p_t)$ must approach the expected discounted integral of future income flows:

$E_t(\int_t^\infty(p_v - w)\exp[-r(v-t)]dv) = p_t/(r-\mu) - w/r$. Under the assumption of pure equity financing, the trigger point for closure, p_c, will be chosen so as to maximize total firm value, W(p). A necessary condition for this is the smooth-pasting condition, $W'(p_c) = 0$.

An alternative way in which the firm may end up operating with pure equity financing is if it starts with equity and debt, goes bankrupt and then is run as a pure equity operation by its former debt-holders. By our earlier assumptions, after bankruptcy, the former debt-holders may either liquidate immediately or maintain it as a going concern in which case the maximum earnings flow they can generate is $\xi_1 p_t - \xi_0 w$. Let $X_t = X(p_t)$ denote the value of the firm in the hands of the new owners. By arguments similar to those given above, X(p), satisfies a differential equation like (5) except that the non-homogeneous term $p - w$ is replaced with $\xi_1 p - \xi_0 w$. This equation is subject to the boundary conditions $X(p_x) = \gamma$ (no arbitrage), $\lim_{p\to\infty} X(p) = \xi_1 p/(r-\mu) - \xi_0 w/r$ (no bubbles), and $X'(p_x) = 0$ (smooth-pasting), where the last condition implicitly defines the trigger price p_x which maximises the value of X(p).

As we show in the Appendix, solving the differential equations for W(p) and X(p), one obtains the following result.

Proposition 1. *Let* $W(p_t)$ *and* $X(p_t)$ *denote the total value of the pure equity firm in the hands (i) of its initial equity-holders and (ii) of other owners after bankruptcy.* W(p) *and* X(p) *are equal to:*

$$W(p) = \frac{p}{r-\mu} - \frac{w}{r} + \left[\gamma - \frac{p_c}{r-\mu} + \frac{w}{r}\right]\left(\frac{p}{p_c}\right)^{\lambda} \quad \text{for } p \geq p_c \tag{6}$$

$$X(p) = \frac{\xi_1 p}{r-\mu} - \frac{\xi_0 w}{r} + \left[\gamma - \frac{\xi_1 p_x}{r-\mu} + \frac{\xi_0 w}{r}\right]\left(\frac{p}{p_x}\right)^{\lambda} \quad \text{for } p \geq p_x \tag{7}$$

Here, λ *is the negative root of the quadratic equation* $\lambda(\lambda-1)\sigma^2/2 + \lambda\mu = r$, *and* p_c *and* p_x *are given by:*

$$p_c = -\frac{\lambda}{(1-\lambda)}\,\frac{w + r\gamma}{r}\,(r-\mu) \tag{8}$$

$$p_x = -\frac{\lambda}{(1-\lambda)}\,\frac{\xi_0 w + r\gamma}{\xi_1 r}\,(r-\mu) \tag{9}$$

For $p < p_c$, $W(p) = \gamma$ *and for* $p < p_x$, $X(p) = \gamma$.

Using the solutions W(p) and X(p) as building blocks, we now turn to the valuation of the levered firm.

Debt and equity valuation

Suppose that the firm has issued perpetual debt with principal b/r and a contractual coupon flow b per period of time. Let $\hat{L}_t$ and $\hat{V}_t$ denote the levered firm's debt and equity values. Financial market equilibrium requires that $\hat{L}_t$ and $\hat{V}_t$ satisfy:

$$r\hat{L}(p) = b + \mu p\hat{L}'(p) + \frac{\sigma^2}{2} p^2\hat{L}''(p) \quad \textbf{(10)}$$

$$r\hat{V}(p) = p - w - b + \mu p\hat{V}'(p) + \frac{\sigma^2}{2} p^2\hat{V}''(p) \quad \textbf{(11)}$$

Assume that equity-holders are free to cover the firm's operating losses by injecting capital and that, so long as they do this, bankruptcy cannot occur. The timing of bankruptcy is thus determined by the decision of equity-holders to cease injecting capital. In adopting this assumption, we abstract from bond covenants that trigger bankruptcy when the firm's net worth deteriorates sufficiently far. Our approach is broadly similar to that of Black and Cox (1976) and Leland (1994) and may be contrasted with that of Merton (1974) which supposes that bankruptcy occurs when net worth becomes negative and implicit debt covenants are triggered.[2]

Again, the stationary nature of the payoffs involved implies that bankruptcy occurs when the output price, p_t, first hits some constant low level, p_b. In the absence of arbitrage and assuming that strict seniority of claims is respected, $\hat{L}$ and $\hat{V}$ must satisfy

$$\hat{L}(p_b) = \min\{X(p_b), b/r\} \quad \text{and} \quad \hat{V}(p_b) = \max\{0, W(p_b) - b/r\} \quad \textbf{(12)}$$

at the bankruptcy point, p_b. Since equity-holders decide the timing of bankruptcy, the bankruptcy trigger, p_b, will be selected so as to maximise the value of equity. It is therefore the root of the smooth-pasting condition, $\hat{V}'(p_b) = 0$. Solving the differential equations, (10) and (11), subject to the boundary conditions just described together with a no bubbles condition, yields the following result.

Proposition 2. *Assume that the firm has issued perpetual debt with coupon,* b, *and principal* b/r. *Suppose that the terms of the debt cannot be renegotiated. If* $\hat{V}_t$ *and* $\hat{L}_t$ *denote the total values of the firm's equity and debt under these assumptions, then, if* $\gamma < b/r$, *the debt is risky and*

$$\hat{V}(p) = \frac{p}{r-\mu} - \frac{w+b}{r} - \left[\frac{p_b}{r-\mu} - \frac{w+b}{r}\right]\left(\frac{p}{p_b}\right)^{\lambda} \quad \textbf{(13)}$$

$$\hat{L}(p) = \frac{b}{r} + \left[X(p_b) - \frac{b}{r}\right]\left(\frac{p}{p_b}\right)^{\lambda} \quad \textbf{(14)}$$

where p_b *is given by*

$$p_b = -\frac{\lambda}{(1-\lambda)}\frac{w+b}{r}(r-\mu) \quad \textbf{(15)}$$

Otherwise, if $\gamma \geq b/r$, *the debt is riskless and*

$$\hat{L}(p) = \frac{b}{r} \quad \text{and} \quad \hat{V}(p) = W(p) - \frac{b}{r} \quad \textbf{(16)}$$

The solutions, $W(p)$, $\hat{V}(p)$, and $\hat{L}(p)$ from Propositions 1 and 2 are illustrated in Figure 1. The assumption in Proposition 2 that the principal value is b/r is adopted for expositional simplicity. Otherwise, the value of debt depends on the bankruptcy trigger, p_b, even when it is riskless.[3]

Figure 1 Security valuation with no renegotiation

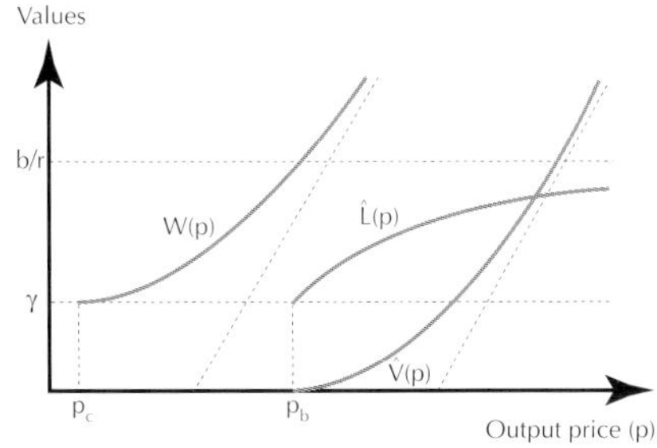

The values of the firm as a whole under efficient closure at p_c (W(p)), and of equity and debt under inefficient closure at p_b ($\hat{V}$(p) and $\hat{L}$(p)), are shown as functions of the output price, p. Note that W(p) > $\hat{V}$(p) + $\hat{L}$(p) reflecting our assumptions of both direct and indirect bankruptcy costs. W(p) smooth pastes to the liquidation value of the firm's assets, γ, at p_c. $\hat{V}$(p) and $\hat{L}$(p) value match to zero and γ respectively at p_b. The asymptotic values to which W, $\hat{V}$ and $\hat{L}$ tend as $p \to \infty$ in the absence of bubbles are shown as dashed lines.

To interpret the expressions in Proposition 2, note that if $\gamma > b/r$, bankruptcy will never occur since there will always be sufficient assets to meet the firm's liabilities. In this case, when closure occurs, the equity-holders are the residual claimants on the firm's assets and hence have an incentive to liquidate the firm at the trigger level, p_c, which maximises total firm value. If $\gamma < b/r$, the debt is risky and bankruptcy takes place when p_t first hits the level p_b which maximises the value of the levered firm's equity.

Figure 2 Total firm value and leverage

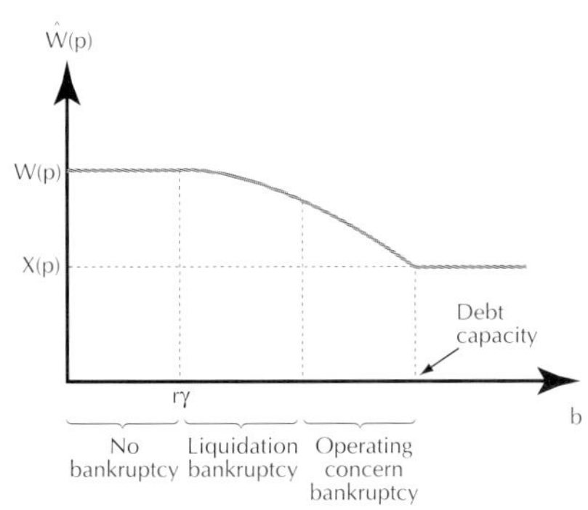

The total value of the levered firm, $\hat{W}$(p), is depicted as a function of the contracted coupon flow, b. For low b, existing equity-holders are residual claimants and no inefficiency occurs. For b slightly above rγ (where r is the interest rate and γ is the liquidation value of the firm's assets) when bankruptcy occurs, liquidation takes place instantaneously and is *ex ante* inefficient. For large b, when bankruptcy occurs, former bond-holders maintain the firm in operation but earnings are reduced due to direct bankruptcy costs.

Firm value and leverage

To understand the inefficiencies that result from the presence of debt, consider how the total value of the levered firm, $\hat{W}(p) \equiv \hat{V}(p) + \hat{L}(p)$ depends on the contracted coupon flow, b. This dependency is illustrated in Figure 2. When the principal of the debt, b/r, is less than the scrapping value γ, equity-holders act as residual claimants even at closure and the firm is operated and liquidated efficiently. Leverage imposes no efficiency losses, and the value of the firm is then the same as under pure equity financing

$$\hat{W}(p) = W(p) \qquad \textbf{(17)}$$

When the debt principal, b/r, is greater than the scrapping value γ, bond-holders are the residual claimants and the debt is risky. The value of the firm is

$$\hat{W}(p) = \frac{p}{r-\mu} - \frac{w}{r} + \left[X(p_b) - \frac{p_b}{r-\mu} + \frac{w}{r}\right]\left(\frac{p}{p_b}\right)^{\lambda} \qquad \textbf{(18)}$$

Given our assumptions about direct bankruptcy costs (that $\xi_1 < 1$ and $\xi_0 > 1$), to the right of $r\gamma$, there exists an interval for b such that $p_b < p_x$, and bond-holders who take over at bankruptcy will prefer to liquidate the firm instantly. In this case, $X(p_b) = \gamma$. For b in this range, leverage is costly because it results in liquidation at p_b rather than at the trigger, p_c. The firm is liable to such a "Liquidation Bankruptcy" in the future if

$$r\gamma < b < \frac{\xi_0 w + r\gamma}{\xi_1} - w \qquad (19)$$

If b exceeds $(\xi_0 w + r\gamma)/\xi_1$ (but $p \geq p_b$ so the levered firm is still operating), then when future bankruptcy occurs, former bond-holders will prefer to maintain the firm as a going concern. In the case of such an "Operating Concern Bankruptcy", $X(p_b)$ is given by equation (7). Here, leverage generates efficiency losses from an *ex ante* point of view (i) because of the direct bankruptcy costs it entails and (ii) because liquidation ultimately occurs at p_x rather than at p_c.

One should note that, when b goes to zero, the firm is pure equity financed and our model becomes very similar to the real options model of irreversible investment studied by Dixit (1989). In fact, when costs of entry are zero and b = 0, the two are identical. Another model that, restricted in various ways, closely resembles the model of this section is the one investigated by Mello, Parsons and Triantis (1995).

PRICING WITH RENEGOTIATION

Service flows with equity-holder offers

Now, consider how the value of the firm's securities is affected if debt- and equity-holders can renegotiate coupon payments. Initially, we suppose that equity-holders can make take-it-or-leave-it offers to bond-holders. We shall assume that possible strategies for equity-holders consist of piecewise right-continuous service flow functions of the state variable, p_t.[4] Let s(p) denote the equity-holders' optimal service flow function, and let V_t and L_t denote the value of the firm's equity and debt when equity-holders adopt s(p).

Now, what characteristics might one expect the optimal service to have? If the output price falls sufficiently far, one may presume that equity-holders will eventually wish to wind up the firm. Let p_c^* denote the trigger price at which this occurs. Consider some price slightly higher than p_c^*. Suppose that equity-holders service the debt as originally intended by paying a flow of coupon payments, b, to debt-holders. If the value of the firm's debt exceeds $X(p_t)$ then equity-holders can extract a surplus by offering debt-holders a service flow less than b. Bond-holders will not wish to declare bankruptcy since the value of the firm's assets will be less than the current value of the debt. As p_t increases, equity-holders will eventually have an incentive to pay the full, initially contracted service flow since otherwise the liquidation value will be so large that debt-holders will definitely wish to close the firm if given the opportunity. In these circumstances, there exists a price, p_s, which is the lowest price at which equity-holders pay the full coupon, b.

The structure of the argument in this and the next two subsections will be as follows. We shall (i) formalise the discussion of the last paragraph as a hypothesis about the optimal service flow function, s(p), (ii) show that this hypothesis uniquely determines s(p), and (iii) verify that the hypothesis is satisfied by an optimal service flow function. As the first stage of this argument, we suppose the following.

Hypothesis 1. *If equity-holders can make take-it-or-leave-it offers to bond-holders regarding debt service then there exist trigger levels,* p_s *and* p_c^*, *such that*

1. *bankruptcy occurs when* p_t *first hits* p_c^*,
2. *for all* $p < p_s$, $s(p) < b$ *and* $L(p) = X(p)$ *(ie, when debt service is less than the contracted coupon, the value of debt equals that of debt-holders' outside option,* $X(p)$*),*
3. *for all* $p \geq p_s$, $s(p) = b$ *(ie, the contracted coupon is paid).*

Now, consider the restrictions placed on s(p) by the above hypothesis. In equilibrium under risk neutrality, L(p) must satisfy an equation like (10) except with b replaced with s(p). If $L(p) = X(p)$ for all $p \in [p_c^*, p_s]$, then we may substitute X(p) and its derivatives for the level and corresponding derivatives of L(p) in this equation. Hence, s(p) must satisfy the equation

$$rX(p) = s(p) + \mu pX'(p) + \frac{\sigma^2}{2} p^2 X''(p) \tag{20}$$

Since X(p) is known exogenously from the analysis of basic assumptions in the first section and since $s(p) = b$ for all $p > p_s$, it follows that, for given p_s

and p_c^*, equation (20) uniquely determines s(p). Substituting for X(p) and its derivatives in equation (20), we obtain a simple explicit form for s(p):

$$s(p) = \begin{cases} r\gamma & \text{for} \quad p \in [p_c^*, p_x] \\ \xi_1 p - \xi_0 w & \text{for} \quad p \in [p_x, p_s) \\ b & \text{for} \quad p \in [p_s, \infty) \end{cases} \tag{21}$$

An intuitive explanation of the service flows in equation (21) is that equity-holders must provide debt-holders with an income flow whose capitalised value is sufficient to dissuade them from forcing bankruptcy to obtain their 'outside option'. Thus, for example, when $p \in [p_c^*, p_x)$, debt-holders require a service flow of $r\gamma$ to dissuade them from declaring bankruptcy to obtain the outside option, γ.

Debt and equity values with equity-holder offers

To complete the derivation of the service flow function, s(p), we need to solve for the two triggers, p_c^* and p_s. To accomplish this requires that one obtain the equity and debt values, V_t and L_t. Equilibrium with risk neutral agents implies that V(p) and L(p) satisfy

$$rL(p) = s(p) + \mu pL'(p) + \frac{\sigma^2}{2} p^2 L''(p) \tag{22}$$

$$rV(p) = p - w - s(p) + \mu pV'(p) + \frac{\sigma^2}{2} p^2 V''(p) \tag{23}$$

The absence of arbitrage implies that $V(p_c^*) = 0$ and $L(p_c^*) = X(p_c^*)$. In fact, the construction of the service flow implies that $L(p) = X(p)$ for all $p \in [p_c^*, p_s]$. No bubbles conditions include $\lim_{p\to\infty} L(p) = b/r$ and $\lim_{p\to\infty} V(p) = p/(r - \mu) - (w + b)/r$. Since equity-holders are effectively selecting p_c^* by their choice of service flows, p_c^* must be such as to maximise the value of the firm's equity, and hence, $V'(p_c^*) = 0$. Finally, the absence of arbitrage implies that in the range, (p_c^*, ∞), V(p) and L(p) must be continuous both in levels and in first derivatives.

Solving the differential equations in equations (22) and (23) subject to the above boundary conditions, we obtain the following result.

Proposition 3. *Under Hypothesis 1, equity-holders adopt the service flow function,* s(p), *defined in equation (21), and the values of equity,* V(p), *and debt,* L(p), *are as follows.*

$$V(p) = W(p) - L(p) \tag{24}$$

where, if $\gamma \geq b/r$, *the debt is riskless and* $L(p) = b/r$. *If* $\gamma < b/r$, *the debt is risky and*

$$L(p) = \begin{cases} b/r + [X(p_s) - b/r](p/p_s)^{\lambda} & \text{if} \quad p > p_s \\ X(p_s) & \text{if} \quad p \leq p_s \end{cases} \tag{25}$$

where p_s *is given by*

$$p_s = -\frac{\lambda}{1-\lambda} \frac{b + \xi_0 w}{\xi_1 r} (r - \mu) \tag{26}$$

The firm is liquidated the first time that p_t *hits* p_c, *where* p_c *is defined as in equation (8).*

The last sentence of the proposition makes the point that liquidation occurs at the trigger price which maximises the value of the pure equity firm. Since it is also true that prior to liquidation the firm is operated by the original equity-holders, it is clear that renegotiation generates the fully efficient outcome. To stress this finding, we state it in the form of a corollary.

Corollary 1. *If equity-holders adopt the service flow function,* s(p), *leverage does not reduce the ex ante value of the firm.*

The mathematical explanation for the result in Corollary 1 may be seen directly from equations (24) and (25). With renegotiation, V(p) equals W(p) – L(p). From equation (25), it is clear that L(p) is independent of the liquidation trigger, p_c^*. Hence, the same liquidation trigger maximises both V(p) and W(p). It follows that $p_c^* = p_c$ where p_c is defined as in equation (8).

The economic intuition for Corollary 1 is that when equity-holders can make take-it-or-leave-it offers to bond-holders, they can extract any surplus in the value of bonds over and above the bond-holders' exogenously-given outside option, X(p) (see Figure 3).

Figure 3 Security valuation with equity-holder offers

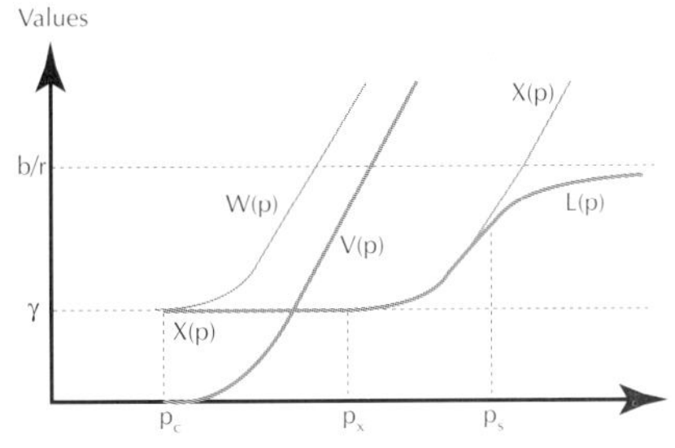

The values of equity (V(p)) and debt (L(p)) with equity-holder take-it-or-leave-it offers are shown as functions of the output price, p. L(p) and V(p) smooth paste respectively to the liquidation value of the firm's assets, γ, and to zero at the efficient liquidation price, p_c. L(p) value matches to the bond-holders' 'outside offer', X(p), for all prices less than p_s, the trigger price for renegotiation.

Equity-holders thus become the residual claimants to firm value in all states of the world and have an incentive to maximise the total value of the firm's assets.

A second important corollary of Proposition 3 which follows from its proof, is

Corollary 2. *If* $\gamma < b/r$, *the trigger price for efficient liquidation,* p_c, *is less than* p_s. *Hence, if debt is risky and equity-holders employ the service flow function,* s(p), *renegotiation always occurs for some interval of output prices.*

This result shows that *whenever debt is risky*, there is scope for renegotiation. The intuition here is that if debt becomes risky, equity-holders are no longer the residual claimants on the firm's income stream in states of the world involving bankruptcy. Hence, the decision to trigger bankruptcy will be made inefficiently. The loss in value associated with this inefficiency creates scope for renegotiation.

The optimality of s(p)

In Hypothesis 1, we claim that the optimal service flow function for equity-holders who can make take-it-or-leave-it offers consists of the function s(p) such that $L(p) = X(p)$ and $s(p) < b$ in some interval $[p_c^*, ps]$, and that $s(p) = b$ for all $p \geq p_s$. This subsection provides a formal proof of this claim. While this result is technical it deserves emphasis which is why we present it in the main body of the text.

Let $\mathcal{A}$ denote the set of pairs consisting of a liquidation price, $p_c^r \in \mathbf{R}^+$ and a piecewise right-continuous, bounded service flow function, $s^r(.)$, defined on $[p_c^r,\infty)$ such that

$$\forall p \geq p_c^r,\ s^r(p) < b \Rightarrow L^r(p,p_c^r) \geq \min\{X(p),b/r\} \tag{27}$$

where r indexes members of $\mathcal{A}$. The set $\mathcal{A}$ consists of those bankruptcy price and service flow pairs such that coupon payments fall short of the contracted level only when the value of debt exceeds the post-bankruptcy value of the firm.

Let $V(p,p_c)$ and $L(p,p_c)$ denote the equity and debt values derived in the previous subsection on equity holders.[5] The main result of this subsection is

Proposition 4. *For any* $(p_c^r, s^r(.)) \in A$, $V(p, p_c) \geq V(p, p_c^r)$ *for all* $p \geq p_c$. *Hence, Hypothesis 1 is true.*

This result is quite strong since it shows that, provided equity-holders can make take-it-or-leave-it offers to debt-holders, for any output price, p_t, the pair $(p_c, s(\cdot))$ derived above will generate a higher equity value than any other feasible $(p_c^r, s^r(.))$ pair.

Properties of s(p)

Let us study properties of the service flow function, s(p), depicted in Figure 4. Clearly, s(p) is at and continuous for prices below p_x and above p_s, while it is linearly increasing with slope ξ_1 for $p \in (p_x, p_s)$. At p_s and p_x, s(p) exhibits discontinuities in that

$$\lim_{p \downarrow p_x} s(p) - \lim_{p \uparrow p_x} s(p) = -(r\gamma + \xi_0 w)(r - \mu)/(r(1-\lambda)) < 0$$

$$\lim_{p \downarrow p_s} s(p) - \lim_{p \uparrow p_s} s(p) = (b + \xi_0 w)(r - \mu)/(r(1-\lambda)) > 0 \tag{28}$$

To understand why these discontinuities occur, note first that L(p) satisfies the differential equation (22) in the neighbourhoods of p_x and p_s. But, in the absence of arbitrage, L(p) and L′(p) must both be continuous at p_x and p_s. From inspection of the equation, it follows that s(p) is discontinuous at a given point *if and only if* the second derivative, L″(p) exhibits discontinuities at the same points. What then is the intuition for the discontinuities in the second derivative, L″(p), at p_x and p_s?

p_x and p_s are reversible switch points in that the change which occurs when the state variable crosses these levels is reversed if p_t moves back to its original position. It is well known that, in general, the second derivatives of claim values are not continuous at reversible switch points (see Dumas (1991) for a discussion). Continuity *is* implied if the switch point is chosen optimally so as to maximise the value

Figure 4 Service flow and equity earnings under equity-holder offers

(a)

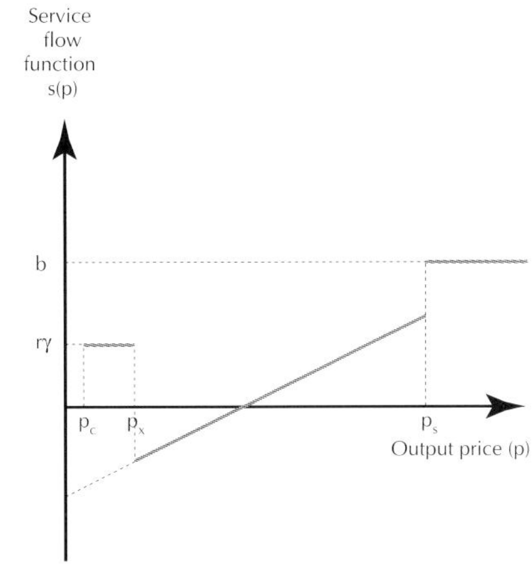

(b)

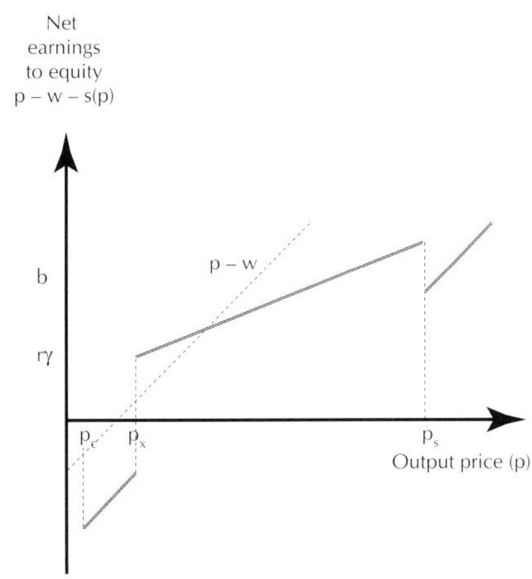

(a) shows the service flow under equity-holder take-it-or-leave-it offers, s(p), as a function of the output price, p. For p greater than p_s (the trigger price for renegotiation) the service flow equals the contracted coupon, b. For $p \in [p_x, p_s]$, (where p_x is the highest price at which liquidation bankruptcies occur) the service flow is linearly increasing. For $p \in [p_c, p_x]$, (where p_c is the trigger price for firm closure), the service flow equals the product, rγ, of the interest rate, r, and the liquidation value of the firm's assets, γ. (b) Shows equity earnings, p – w – s(p), where w is the firm's flow costs.

of the claim.[6] In our application, the switch points are not freely chosen by equity-holders to maximise their claim values, but are implied by the form of the outside option function, X(p). Hence, there is no reason to expect continuity of the second derivative or of the service flow function, s(p).

Now, consider the sign of the s(p) function. An interesting possibility that arises for certain parameter values is that s(p) actually turns negative for some output prices. In effect, bond-holders are then injecting cash into the firm. In fact, one may show that s(p) is negative for some interval of p if and only if $\xi_0 > \rho r\gamma/w$ where the constant ρ is defined as $\rho \equiv 1/[(\lambda - 1)/\lambda(r/(r - \mu))^2 - 1] > 0$. Since a larger ξ_0 corresponds to higher bankruptcy costs, the intuition here is that the greater the cost of bankruptcy to debt-holders, the more their reluctance to take over the firm can be exploited by equity-holders. Lastly, note that our assumptions that $\xi_0 \geq 1$ and $0 < \xi_1 \leq 1$ imply that for $p \in [p_x, p_s]$, $s(p) = \xi_1 p - \xi_0 w \leq p - w$. Hence, whenever bond-holders inject cash, $p - w - s(p) \geq 0$ and equity-holders are actually receiving non-negative dividends.

The above results on the sign of s(p) very much underline the importance of the bankruptcy cost parameters, ξ_0 and ξ_1, in determining how much equity-holders can extract from the firm. It is instructive in this regard to consider what happens as $\xi_1 \uparrow 1$ and $\xi_0 \downarrow 1$, ie, direct bankruptcy costs shrink to zero. Inspection of equations (8), (9), (15), and (26), reveals that $p_s \downarrow p_b$ and $p_x \downarrow p_c$. The limiting solution involves the bond-holders receiving the full, contracted coupon flow, b until p_t first hits p_b and then paying $p - w$, the firm's operating losses, until liquidation takes place at p_c.

Indeed, evaluation of equations (24) and (25) for the limiting values of ξ_1 and ξ_0 reveals the following result:

Proposition 5. *As $\xi_1 \uparrow 1$ and $\xi_0 \downarrow 1$ (so direct bankruptcy costs are eliminated), the service flow function becomes:*

$$s(p)\big|_{\xi_1 = \xi_0 = 1} = \begin{cases} p - w & \text{for} \quad p \in [p_c, p_b) \\ b & \text{for} \quad p \geq p_b \end{cases} \tag{29}$$

$V(p) \to \hat{V}(p)$ *and* $L(p) \to \hat{L}(p)$.

The implication of this proposition is that, when $\xi_1 = \xi_0 = 1$, any surplus to be extracted from bond-holders disappears and equity-holders can do no better than in the absence of renegotiation, even when they have maximum bargaining power. What drives this result is our assumption that bond-holders can exploit the scrapping value of the firm just as well as equity-holders. In this case, threatening bond-holders with firm closure does not enable equity-holders to extract concessions when direct bankruptcy costs are zero since, for all p, equity- and debt-holders are equally able to exploit the firm's productive potential.

Bond-holder offers

The above analysis presumes that equity-holders have maximum bargaining power and hence can make take-it-or-leave-it offers to bond-holders. Consider the alternative polar case in which bond-holders choose a service flow function to maximise the value of the firm's debt. Let this optimal service flow function be denoted q(p). One may think of q(p) as embodying an optimal package of concessions by lenders. In making these concessions, debt-holders will increase the value of their claims compared to a situation with no renegotiation.

The logic of our analysis of equity-holder offers suggests that we look for a q(p) that maximises the total value of the firm while minimising the value of the equity. If bond-holders can make take-it-or-leave-it offers in renegotiating coupon payments, we would expect them to behave as residual claimants, maximising firm value subject to the constraint placed upon them by the `outside option' of equity-holders. In this context, the outside option of equity-holders is supplied by limited liability. That is, unless the equity value is non-negative, equity-holders will abandon the firm, precipitating bankruptcy.

Pursuing an analysis parallel to that applied above in the case of equity-holder offers, yields the following proposition.

Proposition 6. *When bond-holders can make a take-it-or-leave-it offer of concessions to equity-holders, their optimal service flow function,* q(p)

Figure 5 Service flow and equity earnings under bond-holder offers

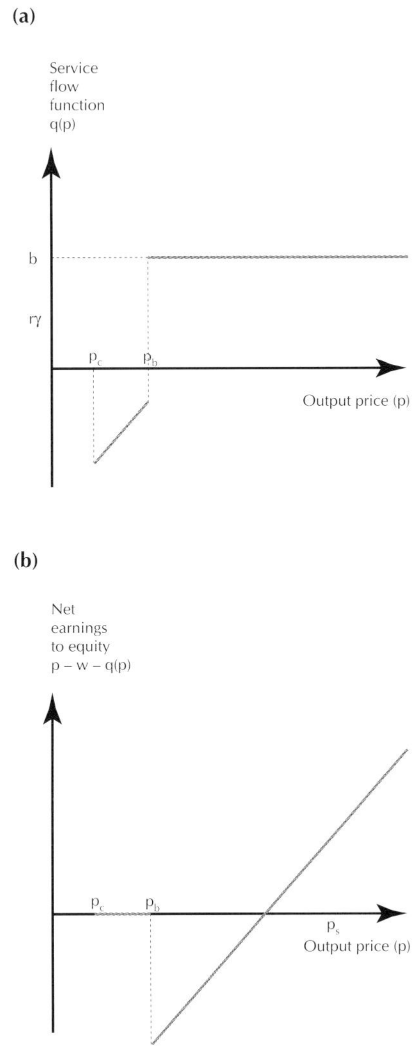

(a) Shows the service flow function under bond-holder offers, q(p), as a function of the output price, p. Bond-holders demand the contracted service flow, b, so long as p exceeds the bankruptcy trigger price without renegotiation, p_b. Thereafter, bondholders meet the operating losses of the firm until p hits the efficient closure trigger, p_c. The income flow to equity shown in (b) equals the flow equity-holders receive in the absence of renegotiation.

equals

$$q(p) = \begin{cases} p - w & \text{for} \quad p \in [p_c, p_b) \\ b & \text{for} \quad p \geq p_b \end{cases} \tag{30}$$

The value of equity is the same as in the case of no renegotiation, $\hat{V}(p)$, *and the value of the firm's debt is* $W(p) - \hat{V}(p)$.

In other words, the best package of concessions that bond-holders can make is to insist that equity-holders pay the full contracted coupon, b, so long as the output price, p_t, exceeds the trigger for bankruptcy in the case without renegotiation, p_b. For $p < p_b$, bond-holders should inject cash to cover the firm's operating losses. (It is easy to show that $p < p_b$ implies $p - w < 0$, so the firm is making operating losses at such output prices.) It is interesting to observe that, in the case of bond-holder take-it-or-leave-it offers, the cash flows to debt- and equity-holders and hence the claim values are *exactly the same* as under equity-holder offers when $\xi_0 = \xi_1 = 1$.

Another instructive fact is that, for p just greater than p_c, while on the one hand the *value* of debt is *higher* when bond holders have maximum bargaining power, *debt service* is *lower*, ie, $q(p) < s(p)$. This finding strongly underscores the fact that what matters to agents in this kind of model is the value of their claim and not simply their current income flow. An agent's bargaining power directly affects the claim value he can obtain but the implications for instantaneous flows can be quite complicated. In the models described above, bond-holders with maximum bargaining power inject more for p_t just greater than p_c because this permits them to extract a higher service payment from equity-holders when p_t is somewhat higher.

EXTENSIONS OF THE MODEL

Our analysis so far has assumed zero costs of renegotiation and no tax advantages for leverage. The net result, as we have seen, is that firms are indifferent between debt and equity financing. If one wished to develop a model of optimal capital structure, it would clearly be necessary to modify these assumptions. The most obvious approach might be to formulate the model so as to create a trade-off between renegotiation costs and tax advantages of debt. In this section, we discuss how the model may be generalised to include these two features.

Renegotiation costs

So far, we have presumed that renegotiation is costless. A natural alternative is to suppose that renegotiation of coupon payments imposes a cost δ per unit of time when the coupon is not being fully serviced. To analyse this extension of the model, define $C(p_t \mid \underline{p}, \bar{p})$ to be the value of a claim that, until p_t first hits a level $\underline{p}$, pays an income flow δ when $p_t \in (\underline{p}, \bar{p})$ and zero when $p_t > \bar{p}$. Here, $\underline{p}$ and $\bar{p}$ are arbitrary constants satisfying $0 < \underline{p} < \bar{p}$. We show in the Appendix that $C(p \mid \underline{p},\bar{p})$ is equal to

$$C(p|\underline{p},\bar{p}) = \begin{cases} \delta/r\,\varphi(p) & \text{for} \quad p \in [\underline{p},\bar{p}) \\ \delta/r\,\varphi(\bar{p})\,(p/\bar{p})^{\lambda_1} & \text{for} \quad p \in [\bar{p},\infty) \end{cases} \tag{31}$$

where

$$\varphi(p) \equiv 1 - (p/\underline{p})^{\lambda_1} + \frac{\lambda_1}{\lambda_2 - \lambda_1}\left[(p/\bar{p})^{\lambda_2} - (\underline{p}/\bar{p})^{\lambda_2}(p/\underline{p})^{\lambda_1}\right] \tag{32}$$

Here, λ_1 and λ_2 are the negative and positive roots respectively of the quadratic equation $\lambda(\lambda - 1)\sigma^2/2 + \lambda\mu = r$ already encountered.

Renegotiation costs alter the solutions for debt and equity obtained in earlier sections in a simple fashion. Consider first the case in which equity-holders make the offers. Once again, debt values are largely determined by bond-holder's outside offers. Let $L(p \mid \underline{p})$ and $W(p \mid \underline{p})$ denote the values given in equations (25) and (6) ($L(p)$ and $W(p)$, respectively) except that firm closure occurs at some arbitrary level $\underline{p} > 0$ instead of at p_c. The value of the firm's debt and equity with renegotiation costs and equity-holder offers will then equal: $L(p \mid p_{c1})$ and $V(p \mid p_{c1}) \equiv W(p \mid p_{c1}) - C(p \mid p_{c1}, p_s) - L(p \mid p_{c1})$ where p_{c1} is the root of the smooth-pasting condition $V'(p_{c1} \mid p_{c1}) = 0$

Since debt levels with equity-holder offers are again independent of the closure trigger,[7] it follows that the trigger price that maximises $V(p \mid p_{c1})$ is the same as that which maximises $W(p \mid p_{c1}) - C(p \mid p_{c1}, p_s)$. In other words, the new closure point is *constrained Pareto optimal,* ie, efficient subject to the inclusion of renegotiation costs. The fact that debt values are completely unaffected by the inclusion of renegotiation costs under equity-holder offers also implies that the entire incidence of the renegotiation cost is on equity-holders.

It is not hard to see what the debt and equity values look like with debt-holder offers. Again, the incidence will be entirely on the residual claimants, in this case the bond-holders. Thus, equity values will equal those obtained in the model with no renegotiation and the value of debt will be $W(p \mid p_{c2}) - C(p \mid p_{c2}, p_b) - \hat{V}(p)$ where p_{c2} is the root of $W'(p_{c2} \mid p_{c2}) - C'(p_{c2} \mid p_{c2}, p_b) = 0$.

Corporate taxes

The above analysis is developed under the assumption that there is no tax advantage to leverage. It is straightforward to incorporate such a tax advantage by supposing, for example, that the income flow to equity is taxed at a rate τ but that interest payments are tax deductible. The firm's debt and equity may be valued in the same way as in previous sections except that the equity value must be multiplied by $1 - \tau$. The deductibility of coupon payments implies that debt values are precisely the same as in the absence of taxes.

The introduction of a tax advantage for debt does have one important implication stressed by Bergman and Callen (1991), however. The fact that agents know coupon payments will be renegotiated imposes an absolute ceiling on the value of debt that the firm can issue. If leverage has a tax advantage, it is optimal to issue as much debt as possible so this ceiling or `debt capacity' will act as a binding constraint. Renegotiation thus limits the degree to which the firm may exploit the tax deductibility of interest payments. In this case, the maximum firm value will be achieved if renegotiation is possible but bond-holders have maximum bargaining power. Increasing bond-holder bargaining power raises the debt capacity and hence permits better use of tax deductibility.

Debt principal

In the first section, we adopted the assumption that the debt principal equals b/r. If the firm's debt is risky, its value at flotation will be strictly less than b/r, implying that the debt is sold at a discount. The motivation for this assumption is as follows. So far in our discussion, we have only considered renegotiation when the firm's finances are relatively weak, ie, in an interval above the closure point. However, if the debt is sold at par, then there will be opportunities for renegotiation when the firm is doing well.[8] In fact, for any $p_t > p_e$ where p_e is the output price prevailing when the firm is set up, $\hat{L}(p_t)$ will exceed the par value if the latter is $\hat{L}(p_e)$ and hence there will be surplus for equity-holders to extract.

In reality, call provisions may reduce the scope for renegotiation when the firm is doing well. Technically, it would be possible to incorporate some types of call provisions in our model but this would not add much to the analysis. We, therefore, adopt the simplifying assumption that the debt principal is no less than b/r. Since, for all p_t, $L(p_t) \leq b/r$, this rules out renegotiation at high levels of the state variable.

RENEGOTIATION AND RISK PREMIA

Decomposing the risk premium

The fraction of corporate debt risk premia attributable to renegotiation may be measured by the ratio $R \equiv [\hat{L}(p_t) - L(p_t0)]/[b/r - L(p_t)]$. Note the interesting fact that, for any $p_t \in [p_s,\infty)$, $R = 1 - (p_s/\hat{p}_b)^{\lambda}[b/r - X(\hat{p}_b)]/[b/r - X(p_s)]$ is independent of p_t. In other words, the proportional contribution of equity-holder take-it-or-leave-it offers to corporate risk premia is constant for all prices to the right of the renegotiation region.

Figure 6 shows plots of R against parameter values ranging from half to twice the baseline levels. Our choices of baseline parameters are justified in the Appendix. For the baseline parameter values (see the Appendix for details), R equals just over 0.3. The ratio is highly sensitive to volatility but

Figure 6 The contribution of strategic debt service to default premia

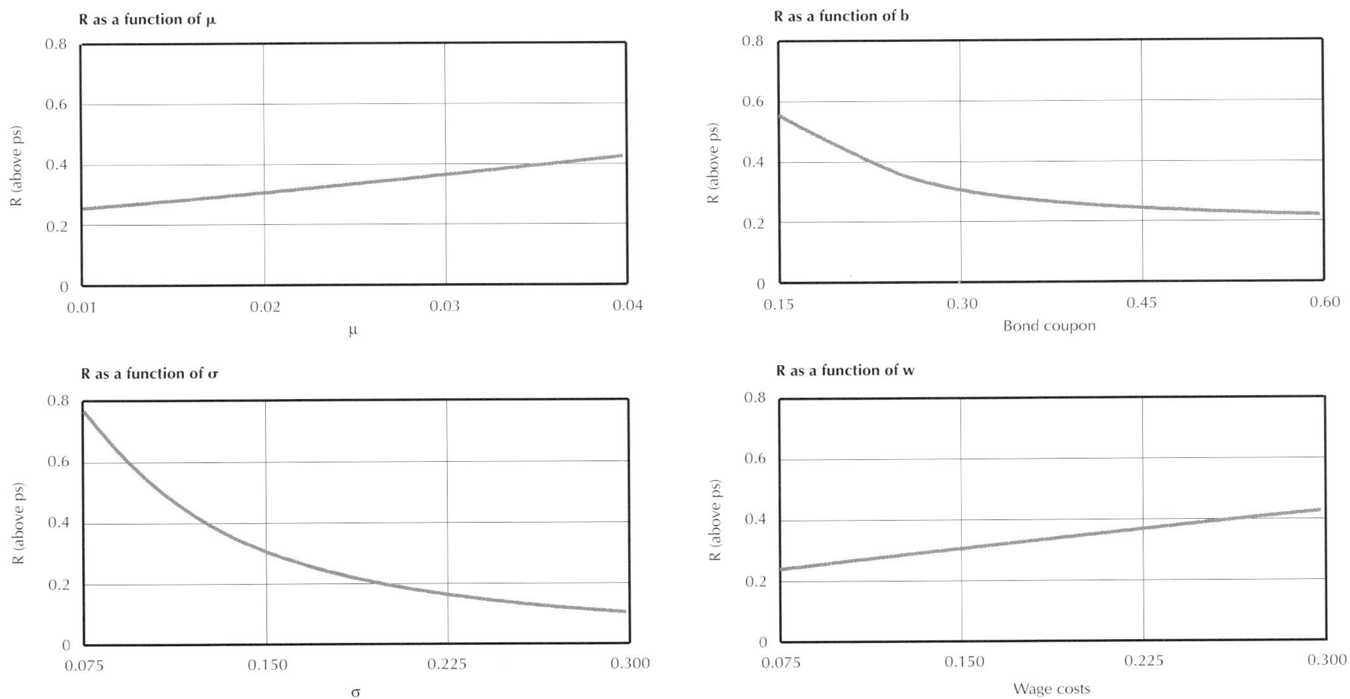

R is the fraction of the total premium over the safe return attributable to strategic debt service when equity-holders can make take-it-or-leave-it offers. R is constant for output prices, p, exceeding the trigger for renegotiation, p_s. The plots show R for $p > p_s$.

relatively impervious to changes in other parameters. One may interpret the sensitivity of R to the various parameters as follows.

Volatility effects

For a fixed residual value, $X(p_t)$, higher σ reduces the relative contribution of renegotiation due to three effects. First, one may think of the corporate debt premium as embodying option values associated with coupon cuts and bankruptcy. The latter is further 'out of the money' if the firm is at a p_t greater than p_s. Raising volatility increases the impact of the more 'out-of-the-money' component.

Second, as σ increases, the coupon cut that share-holders can extract from debt-holders declines since the continuation value of the debt is reduced. Third, increasing σ shifts up the residual, post-bankruptcy value of the firm, X(p). This increases the effective bargaining power of debt-holders, allowing them better to resist equity-holders' demands for coupon reductions, and hence reduces the contribution of strategic debt service to bond risk premia.

Together these three effects mean that the impact of renegotiation shrinks to little more than zero for very high volatility levels. Since all three effects operate in the same direction, it is not surprising that R is highly sensitive to volatility. Even though the baseline R is 0.3, it is very easy to generate much higher values of R by focusing on firms with relatively low asset value volatility.

Other comparative statics

Another comparative static depicted in Figure 6 is the impact of changes in the coupon flow parameter, b. Increasing b has two effects on the value of the debt without renegotiation. For high p, it raises the debt value for obvious reasons. For low p, a debt Laffer curve effect operates in that coupon increases accelerate bankruptcy and hence reduces debt value. These opposing effects mean that the continuation value of the debt may rise or fall when b increases. For the current parameterisation, raising b boosts the continuation value of the debt. Both the size of the renegotiation range and the cut in coupons that equity-holders increase and hence the impact of renegotiation on bond risk premia rises.

Lastly, higher wage costs reduce the post-bankruptcy value of the firm which should tend to increase the scope for strategic debt service. Higher wage costs also increase the closure price, p_c, which may partially offset this effect. For the current parameterisation, the first effect dominates as one may see in the plot.

CONCLUSION

Our study shows that, if equity-holders are able to make debt-holders take-it-or-leave-it offers of coupon payments, then when the firm is in financial distress, debt service payments will vary stochastically and debt values will approximate the firm's liquidation value[9] This situation will persist until the firm is either liquidated or recovers sufficiently to force equity-holders to pay the full, initially-contracted interest payment.

It is important to stress that our analysis does *not* imply that strategic debt service only occurs when the firm's finances are weak. If bankruptcy costs are substantial and scrapping values low, equity-holders can extract concessions from creditors even when the firm is making significant profits. In the limit, if bond-holders are unable to run the firm profitably without the equity-holders' cooperation (say because the direct bankruptcy parameter $\xi_1 \to 0$), then there is no link between the firm's ability to pay and actual debt service.[10] Whatever the level of contracted coupon payments, b, equity-holders will pay $r\gamma$, the income flow for which the capitalised value equals the scrapping price of the firm. In this case, the firm's debt capacity is simply the firm's scrapping value, γ. This finding is in conformity with the observation in Myers (1977) that debt capacity is determined by the book value of a firm's assets since the latter is a marker for the liquidation value.

Other interesting findings that emerge from our analysis include the following.

- Renegotiation of interest payments may eliminate both direct bankruptcy costs and agency costs of debt. This result is demonstrated by Bergman and Callen (1991) in a static model of optimal capital structure.
- In our model, strategic debt service exists for some level of the state variable, p_t, whenever debt is risky and there are non-zero direct costs of bankruptcy.
- Bankruptcy costs in our model are never actually incurred if renegotiation is possible. They nevertheless can have an important impact on debt pricing since they determine the size of the concessions equity-holders can extract from bond-holders.
- When a firm is not renegotiating its debt service, the proportion of the debt premium attributable to the possibility of strategic debt service is a constant, independent of the state variable. For typical parameter values this proportion is sizeable, being around 30–40%.
- For firms with relatively low asset value volatility, this percentage increases substantially. This squares with the fact that it is the relatively large default premia on high quality corporate bond yields which standard models find particularly hard to replicate.
- When direct bankruptcy costs go to zero, equity-holders can do no better than in the absence of renegotiation even if they are able to make take-it-or-leave-it offers to bond-holders and hence have maximum bargaining power.
- The optimal debt service paid by equity-holders who can make take-it-or-leave-it offers, $s(p_t)$, is a piecewise linear function of the output price.

$s(p_t)$ exhibits discontinuities at p_s, the price at which renegotiation commences, and at points at which the second derivative of bond-holders' outside option changes.

- ❑ $s(p_t)$ may go negative when the firm is close to liquidation, in which case debt-holders are injecting cash into the firm. Whenever this occurs, equity-holders receive a non-negative income flow.
- ❑ For p just greater than the efficient liquidation point, p_c, debt service is higher if bond-holders rather than equity-holders can make take-it-or-leave-it offers (ie, $q(p) > s(p)$ for p just larger than p_c). On the other hand, debt values are clearly higher when bond-holders have the greater bargaining power. This result underlines the fact that agents care about values (ie, discounted expected future income flows) and that extracting the highest value may entail accepting a relatively low income flow for some levels of the state variable, p_t.

APPENDIX

Proof of Proposition 1

One may easily verify that the general solution of equation (5) is

$$W(p) = \frac{p}{r - \mu} - \frac{w}{r} + A_1 p^{\lambda_1} + A_2 p^{\lambda_2} \qquad \textbf{(A1)}$$

where λ_1 and λ_2 are the negative and positive roots, respectively, of the quadratic equation, $\lambda(\lambda - 1)\sigma^2/2 + \lambda\mu = r$ and A_1 and A_2 are constants to be determined from the boundary conditions. The absence of bubbles in asset prices implies that $\lim_{p\to\infty} W(p) = p/(r - \mu) - w/r$. But as $p \to \infty$, p^{λ_2} is explosive, hence A_2 must equal zero. Suppose that liquidation occurs when p_t hits some low level p_c. In the absence of arbitrage (which precludes jumps in asset values), $W(p_c) = \gamma$. Hence, $A_1 = p_c^{-\lambda_1} [p_c/(r - \mu) - (w + r\gamma)/r]$ and for given p_c, W(p) is uniquely determined, and equals the expression in equation (6). If p_c is chosen to maximise W(p), the usual optimality condition implies that the derivative of W(p) at p_c must equal that of the function to which it switches. Since, the scrapping value γ is constant, $W'(p_c) = 0$, which then implies the value for p_c given in equation (8). Applying identical arguments to X(p) and p_x yields the expressions in equations (7) and (9).

Proof of Proposition 2

If $\gamma > b/r$, bankruptcy can never occur as the residual value of the firm's assets always exceeds the face value of the debt. In this case, the debt is riskless so $\hat{L}(p) = b/r$. Since equity-holders are the residual claimants in all states of the world, the total firm value is W(p). By the additivity of value in frictionless, perfectly competitive asset markets, $\hat{V}(p) = W(p) - \hat{L}(p) = W(p) - b/r$.

Now, consider the valuation of debt and equity when $\gamma < b/r$. Clearly, the firm will go bankrupt if p_t falls sufficiently far since liquidation will at some point be preferable to continued operation and at this point the residual value of the firm will be insufficient to pay off bond-holders. To derive debt and equity values, consider equations (11) and (10). The general solutions to these equations are:

$$\hat{L}(p) = \frac{b}{r} + C_1 p^{\lambda_1} + C_2 p^{\lambda_2} \tag{A2}$$

$$\hat{V}(p) = \frac{p}{r - \mu} - \frac{w + b}{r} + D_1 p^{\lambda_1} + D_2 p^{\lambda_2} \tag{A3}$$

The asymptotic conditions, $\lim_{p\to\infty} \hat{L}(p) = b/r$, and $\lim_{p\to\infty} \hat{V}(p) = p/(r - \mu) - (w + b)/r$ imply that C_2 and D_2 are zero. Determining C_1 and D_1 from the value-matching conditions at bankruptcy, $\hat{L}(p_b) = X(p_b)$ and $\hat{V}(p_b) = 0$ yields the expressions in equations (13), (14) and (16). Since p_b is chosen to maximise the value of the levered firm's equity, it must satisfy $\hat{V}'(p_b) = 0$. Solving this equation for p_b yields the expression in equation (15).

Proof of Proposition 3

Again, there are two cases. If $\gamma > b/r$, debt is riskless and the firm is liquidated efficiently by equity-holders when p_t first equals p_c. When $\gamma < b/r$, the analysis is more interesting. Suppose that Hypothesis 1 applies and that p_c is less than p_s where p_s is the trigger for renegotiation assumed in the Hypothesis. For $p < p_s$, $L(p) = X(p)$. For $p \geq p_s$, $L(p)$ satisfies equation (22) with $s(p)$ set equal to b. The general solution to (22) is that given in equation (A2). Again, a no bubbles condition implies that $C_2 = 0$. C_1 and p_s are determined by the no arbitrage conditions $L(p_s) = X(p_s)$ and $L'(p_s) = X'(p_s)$. (Note that continuity of $L'(p)$ at p_s has nothing to do with optimality. Claim values must have continuous derivatives at reversible switch points since kinks permit arbitrage gains.) Simple derivations based on these conditions yields the expression for $L(p)$ and p_s given in equations (25) and (26).

The additivity of claim values implies that $V(p) = W(p \mid p_c^*) - L(p)$ where $W(p \mid p_c^*)$ is defined to be the same as $W(p)$ given in equation (6) but with p_c replaced with the liquidation trigger p_c^*. Maximising V (p) with respect to p_c^* is the same as maximising $W(p \mid p_c^*)$ with respect to p_c^* since $L(p)$ is independent of the closure trigger. This implies that $p_c^* = p_c$. Finally, to ensure that the solution makes sense, one must check the ordering of p_c and p_s. When $b/r < \gamma$, inspection of equations (8) and (26) shows that $p_c < p_s$ as required.

Proof of Proposition 4

To prove Proposition 4, we first need to establish a lemma.

Lemma 1. *Consider two assets whose prices,* $Y_1(p)$ *and* $Y_2(p)$ *are functions of a stochastic process,* p_t, *with almost surely continuous sample paths. Asset-holders receive income flows* $f_1(p)$ *and* $f_2(p)$. *Suppose there exist scalars* p_1 *and* p_2 *where* $p_1 < p_2$ *such that* $Y_1(p_1) = Y_2(p_1)$ *and* $Y_1(p_2) = Y_2(p_2)$ *and* $f_1(p) \geq f_2(p)$ *for all* $p \in [p_1, p_2]$. *If agents are risk neutral, then* $Y_1(p) > Y_2(p)$ *for all* $p \in [p_1, p_2]$.

Proof of the Lemma

For all $p_t \in [p_1, p_2]$

$$Y_1(p_t) - Y_2(p_t) = E_t\left[\int_t^\tau (f_1(p_s) - f_2(p_s)) \exp[-r(s-t)]ds\right] \qquad \textbf{(A4)}$$

where τ is the stopping time at which p_t first hits either p_1 or p_2. The result follows from the non-negativity of the integrand.[11]

Proof of the Proposition

Let $L^r(p)$ and $V^r(p)$ denote the values of debt and equity under risk neutrality for some $(p_b^r, s^r(.)) \in \mathcal{A}$. Let $L(p,p_b^r)$ be the value of debt when the service flow is $s(p)$ but liquidation occurs at p_b^r rather than p_b. Note (i) that $L(p_b^r; p_b^r) = L^r(p_b^r,p_b^r) = \min\{X(p_b^r), b/r\}$, and (ii) that as $p \to \infty$, $X(p) \to \infty$, so that $\lim_{p\to\infty} L^r(p,p_b^r) \geq b/r = \lim_{p\to\infty} L(p, p_b^r)$. (i) and (ii), together with continuity, imply that if there exists p such that $L^r(p,p_b^r) < L(p,p_b^r)$, then there exist p_1,p_2 where $p_1 < p_2$ (and where p_2 is possibly ∞), such that for all $p \in [p_1,p_2)$, $L^r(p,p_b^r) < L(p,p_b^r)$ and $L^r(p_1,p_b^r) = L(p_1,p_b^r)$ and $\lim_{p\to p_2} [L^r(p_1,p_b^r) - L(p,p_b^r)] = 0$. But, if $L^r(p,p_b^r) < L(p,p_b^r)$ for $p \in [p_1,p_2]$, then $s^r(p) = b$ in this range since otherwise debt-holders would trigger liquidation. But Lemma 1 implies $L^r(p,p_b^r) \geq L(p,p_b^r)$ for all $p \in [p_1,p_2]$ which is a contradiction. Hence, $L^r(p,p_b^r) \geq L(p,p_b^r)$ for all $p \geq p_b^r$, which implies that $V^r(p,p_b^r) \leq V(p,p_b^r)$ for all $p \geq p_b^r$. But, $V(p,p_b) \geq V(p,p_b^r)$ for all $p \geq p_b$ by the optimality of p_b.

The renegotiation cost function

$C(p \mid \underline{p},\bar{p})$ has the general solution

$$C(p|\underline{p},\bar{p}) = \begin{cases} \delta/r + C_{11}p^{\lambda_1} + C_{12}p^{\lambda_2} & \text{for} \quad p \in [\underline{p},\bar{p}) \\ C_{21}p^{\lambda_1} + C_{22}p^{\lambda_2} & \text{for} \quad p \in [\bar{p},\infty) \end{cases} \qquad \textbf{(A5)}$$

Boundary conditions $\lim_{p\uparrow\bar{p}} C(p \mid \underline{p},\bar{p}) = C(\bar{p} \mid \underline{p},\bar{p})$, $\lim_{p\uparrow\bar{p}} C'(p \mid \underline{p},\bar{p}) = C'(\bar{p} \mid \underline{p},\bar{p})$, $C(\underline{p} \mid \underline{p},\bar{p}) = 0$, and $\lim_{p\to\infty} C(p \mid \underline{p},\bar{p}) = 0$, determine the constants C_{11}, C_{12}, C_{21}, and C_{22}.

Justification of baseline parameters for simulations

The parameter σ is chosen so that the volatility of the equity price when $p_t = p_s$ is approximately 15%. This figure is consistent with data on typical annualised equity return standard deviations. Lastly, in selecting w, it is important to recognise that the interpretation of w in the chapter as irrecoverable, flow costs is made simply to aid exposition. In fact, w measures the deviation from log-normality of the distribution of changes in our earnings latent variable (ie, if $w = 0$, earnings, $p_t = w$, are lognormal). Hence, it is not really appropriate to select w by examining data on costs (as, for example, in Dixit (1989)). We experiment with various values but report a simulation for w equal to 0.15, ie, for a latent variable reasonably close to log-normality.

We employ a scrapping value, γ, of 2 and suppose that the direct costs of bankruptcy are 20% (so $\xi_0 = 0.8$ and $\xi_1 = 1.25$). Note that when renegotiation occurs, these costs are never realised but instead simply allow equity-holders to extract concessions by threatening bankruptcy. Estimates of realised direct bankruptcy costs are typically much lower for large industrial and commercial companies but this is quite consistent with our model. Lastly, we choose a safe interest rate of 6%, approximately equaling the current US prime rate. The share of the bond risk premium associated with strategic debt service, R, is relatively insensitive to the growth rate in the output price, μ, and the coupon flow, b. We therefore simply set μ at a reasonable number of 2% and choose a value of b of 0.3.

1 In our case, 'under-investment' amounts to the fact that, if debt contracts cannot be renegotiated, the firm is liquidated inefficiently early. Titman (1984) has examined the efficiency of liquidation decisions from a somewhat similar perspective although in his case the firm's customers suffer agency costs from early liquidation.

2 A third approach, followed by Kim, Ramaswamy and Sundaresan (1993) and Anderson and Sundaresan (1996), is to suppose that bankruptcy is triggered by illiquidity, ie, when net earnings after interest fall below zero.

3 Furthermore, if the principal exceeds b/r, the value of debt will be decreasing in the state variable, p, in some cases. If the b/r exceeds the principal, for very high p, equity-holders may be able to extract concessions from bond-holders as we discuss below in the third section.

4 We assume piecewise continuity in order to avoid technical problems with strategies that involve cutting coupon payments below contracted levels on zero- measure sets. Such cuts in the coupon would allow debt holders to trigger bankruptcy but would not affect claim values. Right-continuity is assumed in order to ensure uniqueness of the optimal service flow function.

5 This notation serves to stress the dependence on the liquidation trigger price, p_c.

6 Termed by Dumas (1991) a super-contact condition, such continuity of the second derivative is the equivalent for reversible switch-points of the smooth-pasting conditions for permanent switches. The latter are encountered, for example, in the optimal exercise of an American option.

7 Our notation, $L(p \mid p_{c1})$ suggests that $L(. \mid .)$ does depend on p_{c1}. However, p_{c1} just determines the lower end of the domain on which $L(. \mid .)$ is defined and $\partial L(p \mid p_{c1})/\partial p_{c1} = 0$.

8 We thank Anthony Neuberger for pointing this out.

9 Renegotiation can arise even when the firm is quite healthy if the residual value of the firm after bankruptcy is sufficiently low. So long as direct bankrupcy costs are not too high, renegotiation only occurs when the firm is in financial distress.

10 In this, our model resembles typical analyses of sovereign debt. In sovereign defaults, lenders do not even have access to a liquidation value and borrowers can only be obliged to repay the capitalised value of the costs of default that lenders can inflict upon them.

11 Note that the proof of the lemma relies on a dynamic form of first order stochastic dominance.

BIBILOGRAPHY

Anderson, Ronald W. and Suresh Sundaresan, 1996, "Design and Valuation of Debt Contracts", *Review of Financial Studies* 9, pp. 37–68.

Bartolini, Leonardo and Avinash Dixit, 1991, "Market Valuation of Illiquid Debt and Implications for Conflict among Creditors", *IMF Staff Papers* 38, pp. 828–49.

Bergman, Yaacov Z. and Jeffrey L. Callen, 1991, "Opportunistic Underinvestment in Debt Renegotiation and Capital Structure", *Journal of Financial Economics* 29, pp. 137–71.

Black, Fischer and John C. Cox, 1976, "Valuing Corporate Securities: Some Effects of Bond Indenture Provisions", *Journal of Finance* 31, pp. 351–67.

Brennan, Michael J. and Eduardo S. Schwartz, 1984, "Valuation of Corporate Claims", *Journal of Finance* 39, pp. 593–607.

Dixit, Avinash, 1989, "Entry and Exit Decisions under Uncertainty", *Journal of Political Economy* 97, pp. 620–38.

Dumas, Bernard, 1991, "Super Contact and Related Optimality Conditions", *Journal of Economic Dynamics and Control* 15, pp. 675–85.

Franks, Julian R. and Walter N. Torous, 1989, "An Empirical Investigation of US Firms in Renegotiation", *Journal of Finance* 44, pp. 747–69.

Harrison, Michael J. and David M. Kreps, 1979, "Martingales and Arbitrage in Multiperiod Securities Markets", *Journal of Economic Theory* 20, pp. 381–408.

Kim, In Joon, Krishna Ramaswamy, and Suresh Sundaresan, 1993, "Valuation of Corporate Fixed-Income Securities", *Financial Management*, Special Issue on Financial Distress

Leland, Hayne E., 1994, "Risky Debt, Bond Covenants and Optimal Capital Structure", *Journal of Finance* 49, pp. 1213–52.

Longstaff, Francis A. and Eduardo S. Schwartz, 1995, "A Simple Approach to Valuing Risky and Floating Rate Debt", *Journal of Finance* 50, pp. 789–819.

Mauer, David C. and Alexander J. Triantis, 1994, "Interactions of Corporate Financing and Investment Decisions: A Dynamic Framework", *Journal of Finance* 49, pp. 1253–77.

Mello, Antonio S. and John E. Parsons, 1992, "The Agency Costs of Debt", *Journal of Finance* 47, pp. 1887–1904.

Mello, Antonio S., John E. Parsons and Alexander J. Triantis, 1995, "An Integrated Model of Multinational Flexibility and Financial Hedging", *Journal of International Economics* 39, pp. 27–51.

Merton, Robert C., 1974, "On the Pricing of Corporate Debt: The Risk Structure of Interest Rates", *Journal of Finance* 29, pp. 449–70.

Myers, Stuart C., 1977, "Determinants of Corporate Borrowing", *Journal of Financial Economics* 5, pp. 147–75.

Rubinstein, Ariel, 1982, "Perfect Equilibrium in a Bargaining Model", *Econometrica* 50, pp. 97–109.

Titman, Sheridan, 1984, "The Effect of Capital Structure on a Firm's Liquidation Decision", *Journal of Financial Economics* 13, pp. 137–51.

Part V

Foundations

13

*Pre-emption and Rent Equalisation in the Adoption of New Technology**

Drew Fudenberg and Jean Tirole

University of California, Berkeley; Massachusetts Institute of Technology

We study the adoption of a technology to illustrate the effects of pre-emption in games of timing. We show that the threat of pre-emption equalises rents in a duopoly, but that this result does not extend to the general oligopoly game. If the gain to pre-emption is sufficiently small, then the optimal symmetric outcome, which involves "late" adoption, is an equilibrium. This contrasts with Reinganum's result that in precommitment equilibria there must be "diffusion". We develop a new and richer formulism for modelling games of timing, which permits a continuous-time representation of the limit of discrete-time mixed-strategy equilibria.

Research alone is not sufficient for technical progress: technological innovations yield no benefits until they are employed. Therefore social policy should consider the incentives for adopting innovations as well as incentives for their discovery. This chapter analyses a market in which firms are deciding when to adopt a new technology. Adopting the innovation increases a firm's flow of profits, but adopting it sooner is more expensive than adopting it later. A monopolist would not adopt the innovation immediately, but would wait until the increase in flow profit just equalled the marginal cost of adopting sooner. The market is not monopolised, however, which has two related implications.

First, a part of each individual firm's return to adoption comes at the expense of its rivals, so the individual return exceeds the industry return. Thus we would expect the adoption times in non-cooperative equilibria to be sooner than would be optimal for the industry as a whole.

*This paper was first published in the *Review of Economic Studies* 52, pp. 383–401 (1985). The authors would like to thank Bernard Caillaud, Richard Gilbert, David Kreps, Eric Maskin and Leo Simon for helpful conversations. Financial support from NSF Grant SES 82–07928 is gratefully acknowledged.

Second, if the information lags are short, so that firms can observe and respond to their rivals' actions, firms have an incentive for "pre-emptive adoption". By this we mean that firms will adopt sooner than they would choose to, were their rivals' adoption dates fixed. Firms adopt pre-emptively to prevent or delay adoption by their opponents. One might expect the threat of pre-emption to ensure that identical firms have the same equilibrium payoffs, because were one firm to adopt at time t and do better than a rival who adopts later, the rival would pre-empt the first firm by adopting just before t. We show that this intuition is correct in a duopoly, but that with more than two firms equilibrium payoffs need not be equal.

Two previous authors have studied the timing of the adoption of new technology. Scherer (1967) analysed the introduction of a product innovation in duopoly. He concluded that if the firms were required to pre-commit themselves to introduction times, then the equilibrium outcome would be for both firms to move simultaneously, at an earlier time than would maximise joint profits. Reinganum (1981a,b) analysed the adoption process using a different specification of the payoffs, again requiring pre-commitment to adoption dates. She found that there must be a "diffusion", in that the firms adopt at different dates, even though the firms are *ex ante* identical and there is no uncertainty.

When firms pre-commit themselves to adoption dates, pre-emption is ruled out by assumption. Pre-commitment strategies are equivalent to infinite information lags. To allow for pre-emption, we consider the alternative extreme case in which firms can respond immediately to their rivals' decisions; that is, adoption is perfectly observable and is instantaneous. Our model also covers the case of an adoption process that takes a fixed amount of time, if firms can observe their rivals' actions. The equilibrium concept we use is that of Perfect Equilibrium, which rules out empty threats.

In the second section we briefly review Reinganum's model of technology diffusion. We show why in pre-commitment equilibria the early adopter performs better than the later one. While there are as many equilibria as there are permutations of the firms, at any time, each firm always prefers an equilibrium in which it is the next to adopt.

In the third section we consider the duopoly case (Reinganum (1981a)). We show that if firms cannot threaten to maintain their planned adoption times regardless of what happens before, then identical firms receive equal payoffs in equilibrium. The intuition is simple: were one firm to plan on adopting first at a time, t, which gave it a higher payoff than the second adopter, the second firm could increase its payoff by pre-empting just before t. Faced with the pre-emption, the first firm could then do no better than to adopt when the second firm would have.

If the best response to early adoption is to adopt much later, the gain from pre-emption is higher. The (unique) equilibrium times of adoption are then unequal. In this "diffusion" equilibrium, which prevails for example in a new market, both firms receive lower payoffs than with pre-commitment. If, on the other hand, the best response to early adoption is to quickly follow suit, several equilibria exist, and these equilibria are Pareto-comparable. In particular, the optimal symmetric adoption time is sustainable as an equilibrium. This time is later than all other equilibrium outcomes. Intuitively, adopting the innovation yields two types of benefits – it increases total industry profits and increases the adopter's share. Thus the industry's incentive to adopt is less than any individual competitor's, and so "late" adoption maximises total profits. This outcome is enforced by the quick response to adoption, which provides a credible threat against the temptation of pre-emption.

The third section gives the economic intuition and the main results for the duopoly case. Traditional continuous-time strategy spaces are not adequate to capture and verify these intuitions. We therefore develop in the following section, a new and richer formalism for modelling games of timing that permits a continuous-time representation of the limit of discrete-time mixed-strategy equilibria. This formalism may be of independent interest to technically-minded readers, since it can be used in more general games of timing. All the economics of the problem however, are presented in the third section. The fifth section considers the model with more than two firms, and shows that rent equalisation need not occur. The subsequent section discusses some related work, while the last section concludes.

THE MODEL

There are n identical firms, indexed by $i = 1, \ldots, n$, in the industry. At time 0, a cost reducing innovation is announced. $\pi_0(m)$ is the net cashflow of firm i when m firm(s) have adopted the innovation, but firm i has not. $\pi_1(m)$ is firm i's net cashflow if m firm(s) including i have adopted. T_i is firm i's adoption date and $c(t)$ is the present value of the cost of bringing the innovation on line by time t. Let T_{-i} denote the vector of dates of adoption by firm i's competitors. $V^i(T_i, T_{-i})$ is firm i's payoff. Suppose w.l.o.g. that firm i is the i-th to adopt. Then:

$$V^i(T_i, T_{-i}) = \Sigma_{m=0}^{i-1} \int_{T_m}^{T_{m+1}} \pi_0(m)e^{-rt}dt + \Sigma_{m=i}^{n} \int_{T_m}^{T_{m+1}} \pi_1(m)e^{-rt}dt - c(T_i)$$

where $T_0 \equiv 0$, $T_{n+1} \equiv +\infty$, and r is the common interest rate.

We will make Reinganum's assumptions (with minor changes):

Assumption 1. For all $m \leqq n$, (i) $\pi_0(m)$ and $\pi_1(m)$ are positive and decrease within; and (ii) $\pi_1(m-1) - \pi_0(m-2) > \pi_1(m) - \pi_0(m-1)$.

Assumption 2. (i) $\pi_1(1) - \pi_0(n-1) \leqq -c'(0)$. (ii) $\inf_t\{c(t)e^{rt}\} < [\pi_1(n) - \pi_0(n - 1)]/r$. (iii) $\forall t$, $\{c(t)e^{rt}\}' < 0$, and $\{c(t)e^{rt}\}'' > 0$.

Assumption 1(ii) states that there are decreasing returns in the rank of adoption. The increase in profits due to adopting (m – 1)-th exceeds that due to adopting m-th. Thus from the revenue point of view firms have an incentive to adopt earlier than their competitors. Assumption 2(i) rules out immediate adoption, even for purposes of pre-emption. Adoption at date 0 gives the firm a cashflow not exceeding $\pi_1(1)$ whereas not adopting gives it a cashflow at least equal to $\pi_0(n-1)$. Assumption 2(ii) requires that all firms adopt in finite time. It implies that if t is sufficiently large it becomes a dominant strategy to adopt and thus non-adoption is not an alternative. Assumption 2(iii) says that the "current cost", $e^{rt}c(t)$, is decreasing, but at a decreasing rate. Both conditions are satisfied by $c(t) = e^{-(r+\alpha)t}$, so that the current cost falls at rate α due to technical progress. The first part of the assumption is natural; the second ensures that the payoffs are quasi-concave. Reinganum makes somewhat stronger assumptions on c'' to obtain concavity.

Reinganum shows that for this model, there are adoption times $\mathcal{G}^* = (T_1^*, \ldots, T_n^*)$ that form a pre-commitment equilibrium, that every pure-strategy pre-commitment equilibrium is given by a permutation of $\mathcal{G}^*$, and that $T_i^* \neq T_j^*$ for $i \neq j$.

Proposition 1. *In a pre-commitment equilibrium the firms' equilibrium payoffs decline monotonically with the order of adoption.*

Proof. Let $\mathcal{G}^*$ be a pre-commitment equilibrium with $T_i^* < T_j^*$ for $i < j$. Then:

$$\begin{aligned} V^i(T_i^*, T_{-i}^*) &\geqq V^i(T_1^*, \ldots, T_{i-1}^*, T_j^*, T_{i+1}^*, \ldots, T_j^*, \ldots, T_n^*) \\ &= V^j(T_1^*, \ldots, T_{i-1}^*, T_j^*, T_{i+1}^*, \ldots, T_j^*, \ldots, T_n^*) \\ &> V^j(T_j^*, T_{-j}^*) \end{aligned}$$

where the first inequality follows from T_i^* being a best response, the equality from the symmetry of the payoff functions, and the second inequality from the monotonicity of V^j relative to the other firms' adoption times.

It is easy to see that the adoption dates $(T_1^* < T_2^* < \cdots < T_n^*)$ are given by the following marginal conditions:

$$[\pi_1(m) - \pi_0(m-1)]e^{-rT_m^*} + c'(T_m^*) = 0 \qquad \textbf{(1)}$$

While pre-commitment equilibria in general fail to capture what is meant by "strategic behaviour", in this instance, they seem especially suspect. The firm which is able to pre-commit itself to adopt first does best, yet any firm can adopt first in equilibrium. The strategic interactions suppressed by pre-commitment would resurface in a competition to be the first to pre-commit.

STRATEGIC ADOPTION IN A DUOPOLY: THE HEURISTICS AND THE MAIN RESULTS

We now study the two-firm case in order to highlight some important features of strategic adoption. The general oligopoly case is considered below.

The first thing to examine is what happens when one of the two firms has adopted at some time t. This continuation game is a simple one-player decision problem. The payoff of the remaining firm is the maximum over t′ greater than or equal to t of V(t′, t). From the quasi-concavity of the objective function of the remaining firm, its optimal time of adoption is either T_2^*, or, if t exceeds T_2^*, t itself.

Having pinned down what happened in the continuation game, let:

L(t) = (the leader's) payoff for the firm that succeeds in pre-empting the other firm at time t

$$= \begin{cases} V(t, T_2^*) & \text{if } t < T_2^* \\ V(t, t) & \text{if } t \geqq T_2^* \end{cases}$$

F(t) = (the follower's) payoff for the firm that is pre-empted at time t

$$= \begin{cases} V(T_2^*, t) & \text{if } t < T_2^* \\ V(t, t) & \text{if } t \geqq T_2^* \end{cases}$$

M(t) = payoff of both firms if they adopt together at time t

$$= V(t, t)$$

The functions L, F and M are pictured in Figures 1 and 2. Let us check that we have drawn them correctly. First consider $t < T_2^*$. It is clear that $L(t) > M(t)$ since a firm's payoff increases with its rival's date of adoption. Second, $F(t) > M(t)$ since the best response to t is T_2^*. Next, using Assumption 2(iii) we can show that

$$(L(t) - F(t))' = 0$$

implies

$$(L(t) - F(t))'' = re^{-rt}(\pi_1(1) - \pi_0(1)) - c''(t) < 0$$

so L – F is strictly quasi-concave.

We know that $L(0) < F(0)$ (from Assumption 2(i)); that $L(T_2^*) = F(T_2^*)$; and that $L(T_1^*) > F(T_1^*)$ (from Proposition 1). Thus there exists a unique T_1 in $(0, T_1^*)$ such that $L(T_1) = F(T_1)$.

Last, let $\hat{T}_2$ denote the optimal date for *simultaneous* adoption, ie, $\hat{T}_2$ maximises M(t). Let us show that $\hat{T}_2$ is unique and exceeds T_2^*. First, M(t) can be shown to be strictly quasi-concave. Second,

$$\left.\frac{\partial V(t, T_2^*)}{\partial t}\right|_{t=T_2^*} = 0 \Rightarrow M'(T_2^*) = \left.\frac{dV(t,t)}{dt}\right|_{t=T_2^*} = \left.\frac{\partial V(T_2^*, t)}{\partial t}\right|_{t=T_2^*} > 0$$

Therefore, $\hat{T}_2 > T_2^*$.

We will distinguish two cases:

Case A. $L(T_1^*) > M(\hat{T}_2)$

Case B. $L(T_1^*) \leqq M(\hat{T}_2)$.

Case A is pictured in Figure 1 and Case B in Figure 2.

Figures 1 and 2 suffice to give the economic intuition. The following reasoning is extremely loose and further analysis is conducted in the fourth section.

Consider Case A. Neither firm can obtain more than $L(T_1^*)$. Therefore, it is in each firm's interest to adopt at time T_1^* if the other firm does not adopt at exactly the same time. But if a firm knows that the other will adopt at time T_1^*, it wants to pre-empt at $(T_1^* - \varepsilon)$. Reasoning backwards, at any t between T_1 and T_1^*, firms want to pre-empt to avoid being pre-empted later on. Thus, as we show in the fourth section, there is adoption at time T_1 (and the follower adopts at time T_2^*). Before T_1 it is a dominant strategy not to adopt because the follower's payoff exceeds the leader's payoff and the joint-adoption payoff. We see that the equilibrium must involve *diffusion*, with adoption times T_1 and T_2^*.

Figure 1 Case A

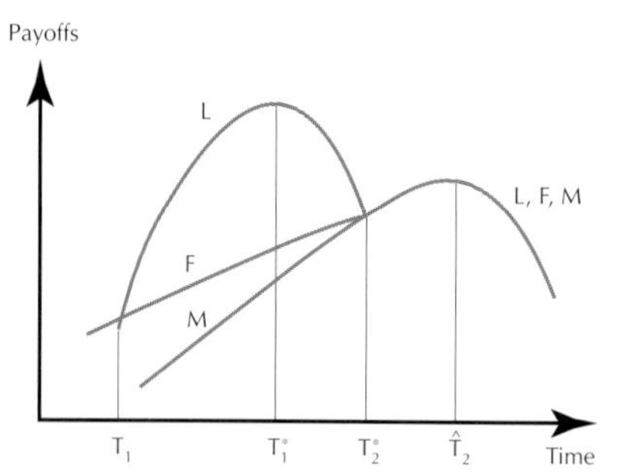

Figure 2 Case B

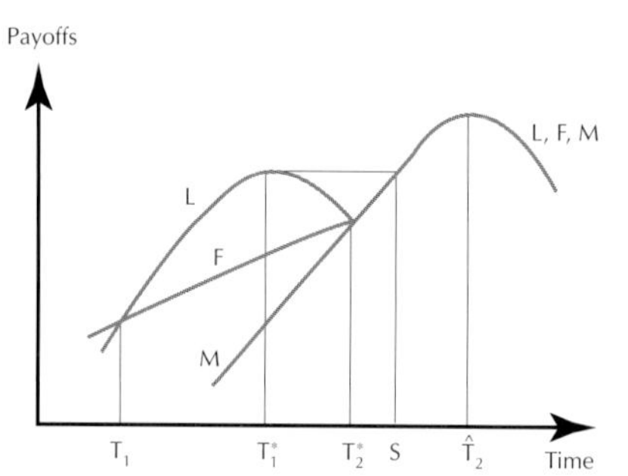

Next consider Case B. Let S be defined by $S < \hat{T}_2$ and $L(T_1^*) = M(S)$. Note that past $\hat{T}_2$ it is a dominant strategy to adopt. We can first exhibit a class of equilibrium adoption times: choose $t \in [S, \hat{T}_2]$ and let the firms wait until t, and adopt in every sub-game from t on. Clearly a firm cannot do better by pre-empting before t. Thus we have a continuum of joint-adoption equilibria indexed by the date of adoption $t \in [S, \hat{T}_2]$. These late adoption equilibria exhibit the dynamics predicted by Scherer (1980, p. 429), with "each industry member holding back initiating its R & D effort in the fear that rapid imitation by others will be encouraged, more than wiping out its innovative profits". Are there other equilibria? Imagine that firms adopt before S. Then the situation is as in Case A since a firm cannot obtain more than $L(T_1^*)$. By the same reasoning, equilibrium must involve diffusion (at dates T_1 and T_2^*). The fourth section will prove the following proposition:

Proposition 2. (A) *If* $L(T_1^*) > M(\hat{T}_2)$ *there exists a unique equilibrium distribution over outcomes. With probability one-half, firm one adopts at* T_1 *and firm two adopts at* T_2^**, and with probability one-half the roles of the firms are reversed.*[1] *Thus the equilibrium exhibits diffusion; and with probability one the adoption dates are* T_1 *and* T_2^*.

(B) *If* $L(T_1^*) \leqq M(\hat{T}_2)$ *two classes of equilibria exist. The first class is the* (T_1, T_2^*) *diffusion equilibrium. The second class is a continuum of joint-adoption outcomes indexed by the date of adoption* $t \in [S, \hat{T}_2]$.

Thus, depending on the parameters, there will be a unique equilibrium outcome (Case A) or multiple equilibrium outcomes (Case B). Note that in Case A the equilibrium is symmetric: each firm has an equal chance of adopting first. Despite this, there is zero probability of the firms' making a "mistake" and both adopting at T_1. In the fourth section we will explain how the correlation involved here arises naturally in the continuous time limit of shorter and shorter time periods. Before introducing the formal analysis we examine the properties of these equilibria. We have already "demonstrated":

Proposition 3. *In a perfect equilibrium the payoffs of the two firms are equated. The perfect equilibria are Pareto-comparable to the pre-commitment equilibria (from the firms' perspective).*

Proposition 4. *All joint-adoption equilibria, if they exist, Pareto-dominate the pre-commitment outcomes, which Pareto-dominate the diffusion equilibrium.*

Proof. The first assertion results from the fact that if $t \in [S, \hat{T}_2]$, $M(t) \geqq L(T_1^*) > F(T_1^*)$ ($L(T_1^*)$ and $F(T_1^*)$ are the firms' payoffs in pre-commitment equilibrium). To prove the second assertion notice that $L(T_1^*) > F(T_1^*) > F(T_1) = L(T_1)$.

More interesting is the fact that when there are multiple equilibria (Case B), all equilibria are Pareto-comparable.

Proposition 5. *Joint-adoption equilibria are Pareto-ranked by their date of adoption, later adoption being more efficient from the firms' perspective.*

Proof. This is a consequence of the monotonicity of M up to $\hat{T}_2$.

Thus in Case B the late joint-adoption equilibrium at $\hat{T}_2$ Pareto-dominates any other Perfect Equilibrium. As argued by Harsanyi (1964) and Fudenberg and Tirole (1983), if one equilibrium Pareto-dominates all others, it is the most reasonable outcome to expect. Consider for example the diffusion equilibrium in which firms adopt only to avoid pre-emption. This situation is comparable to that of two strangers attacking each other *solely* for self-defence (see Fudenberg and Tirole (1983) for another example). Such complete lack of trust is easily avoided by coordination on the Pareto-dominant equilibrium, which is the natural "focal point" of the game. The case against the equilibria with simultaneous adoption before $\hat{T}_2$ is even stronger. In such equilibria each firm adopts only because the other does. No pre-emption is involved, because after T_2^* there is immediate response to adoption, so coordination seems particularly easy. We stress that such coordination requires neither binding contracts (as the equilibrium is by its nature self-enforcing) nor explicit communication. We will call the joint-adoption equilibrium at $\hat{T}_2$ the Superior Perfect Equilibrium.

Last, we discuss the parameter values for which each case occurs. The joint-adoption equilibria require $M(\hat{T}_2) \geqq L(T_1^*)$, or

$$D = \int_{T_1^*}^{T_2^*} (\pi_1(1) - \pi_0(0))e^{-rt}dt + \int_{T_2^*}^{\hat{T}_2} (\pi_1(2) - \pi_0(0))e^{-rt}dt - (c(T_1^*) - c(\hat{T}_2)) \leqq 0$$

Relatively large $\pi_1(1)$ correspond to markets with a large first-mover advantage: as $\pi_1(1)$ increases, T_1^* decreases and D goes up. In such markets the highly competitive diffusion equilibrium will prevail. In particular, we get the diffusion equilibrium in a new market, with $\pi_1(1)$ the monopoly profit and $\pi_0(0) = 0$. In this case the follower's payoff is unaffected by the time at which the leader adopts, so that the optimal joint adoption time $\hat{T}_2$ is the same as the optimal reaction time T_2^* so there cannot be a joint-adoption equilibrium.

If the innovation is only advantageous when the follower has not yet adopted, so that $\pi_1(2)$ is near $\pi_0(0)$, and $(\pi_1(1) - \pi_0(0))$ is near $(\pi_1(2) - \pi_0(1))$, then the equilibrium is the (Superior) joint-adoption one. This will be the case, for example, if adopting first simply transfers profits to the adopter from the other firm, but the latter can regain an equal footing by following suit. Here the gain from pre-emption is low, as T_2^* is near T_1^*, while the joint optimum $\hat{T}_2$ is quite late, for the innovation has little joint value. Thus the firms can postpone adoption without explicit collusion because of the credible threat of quick response. In this case, joint adoption at $\hat{T}_2$ is greatly preferable to the diffusion equilibrium.[2]

STRATEGIC ADOPTION IN A DUOPOLY: A FORMAL ANALYSIS

The issues

To verify the intuitions of the last section, we must develop a new formalisation of strategy spaces and payoffs for continuous-time games of timing, in which a player's only decision is to choose a (single) time to move. We shall see that the traditional formalisation is not adequate, because it entails a "loss of information" in passing from discrete time with short periods to the continuous time limit. This loss of information prevents a continuous-time representation of the limits of discrete-time mixed-strategy equilibria. We therefore introduce a richer formalisation which does allow a continuous-time representation of these limits.

The usual way to describe a player's strategy is to choose a function $G_i(t)$ which is non-decreasing, right-continuous and has range in [0, 1] (see for example, Pitchik (1982)). The interpretation of G_i is that the cumulative probability that player i has moved (adopted in our framework) by time t (conditionally on the other player(s) not having moved before) is $G_i(t)$. Then the player's payoff, for example, would be:

$$V^i(G_1, G_2) = \int_0^\infty [L(s)(1 - G_j(s))dG_i(s) + F(s)(1 - G_i(s))dG_j(s)] + \Sigma a_i(s)a_j(s)M(s) \quad \textbf{(2)}$$

where $a_i(s) = \lim_{\varepsilon\to 0} [G_i(s) - G_i(s - |\varepsilon|)]$ is the size of the jump in G_i at s.[3] This formalisation allows mixed strategies, but it is not adequate for our purposes for two reasons. First, it does not specify what happens in all the possible subgames. For example, consider that for some player i and for some t, $G_i(t) = 1$. Then the strategies do not specify what happens if players deviate and at time $(t + \varepsilon)(\varepsilon > 0)$ no one has yet moved. The Perfect Equilibrium concept, in order to rule out empty threats, requires that players optimise whatever the history (ie, even for histories which are not reached on the equilibrium path). In particular strategies must be specified for every contingency. Of course the remedy to this problem is to "roll over" the cumulative distribution by defining $G_i^t(s)$ as the cumulative probability that player i moved by time $s \geqq t$ conditional on no one moving before t and requiring that the family of functions G_i^t satisfy an intertemporal consistency condition (Bayes rule).

The second drawback of this formalisation is that it is not quite the continuous-time analogue of discrete-time games of timing. To see this, consider the following game, dubbed the "grab the dollar" game.[4]

Time is discrete: t = 0, 1, 2, … . There are two players, with common discount factor $\delta < 1$. At each period t players can try to grab a dollar that lies between them. If at time t only one of the players tries, he gets δ^t and the other player gets zero. If both try, each loses δ^t. In both cases the dollar is gone. If nobody tries the dollar can still be grabbed in the following period. Mixed strategies are allowed.

In this game there is a unique symmetric equilibrium. At each period, if no player has tried to grab the dollar yet, each player tries to grab it with probability $\frac{1}{2}$. In this equilibrium both players have a zero payoff. If a player tries to grab the dollar at t, he wins and gets δ^t with probability $\frac{1}{2}$, and with probability $\frac{1}{2}$, there is a "mistake": both try at the same time and lose δ^t. The fact that there is a positive probability of mistake is an attractive feature of this equilibrium, because with information lags, coordination is difficult to achieve.

The formalism of equation (2) is inadequate to represent this equilibrium in continuous time. As the period length converges to zero, the probability that by any positive time a player has tried to grab the dollar converges to one, and so the strategies should be represented by a unit mass at time zero, that is, $G_i^0(s) = 1 \ \forall s \geqq 0$. Using (2) the probability of mistake would also be equal to one; however, the limiting value of the probability of mistake is $\frac{1}{3}$. The problem is that many different strategies converge to a unit mass at time zero, including, for example, the strategies that specify "adopt with probability one in the first period". There has been a loss of information in passing to the continuous-time limit; the payoffs cannot be computed using the $G_i(\cdot)$ alone.

We will enlarge the strategy space to avoid this loss of information. In the grab the dollar game, if the probability that player i moves in each period is constant and equal to α_i, we would like the probability that player i wins to be $\{\alpha_i(1 - \alpha_j)/[\alpha_i + \alpha_j - \alpha_i\alpha_j]\}$ and the probability of a mistake to be $\{\alpha_i\alpha_j/[\alpha_i + \alpha_j - \alpha_i\alpha_j]\}$. We will do this by adding an extra function, $\alpha_i(t)$, to each player's strategy space, to represent the "intensity" with which players move at times "just after" $G_i(t)$ jumps to one. That is, $\alpha_i(t)$ will represent the "intensity" of an "interval of consecutive atoms". We will define the payoffs to equal the limits of the payoffs of the corresponding discrete-time games. The reader familiar with the idea of generalised curves in optimal control theory will notice the similarity between that construction and our own: both are devices meant to allow continuous time models to more accurately represent discrete limits and both are defined through their influence on payoffs.

In our adoption game, we will be particularly interested in what happens as the α's converge to zero with the length of the time-period. Notice that if each α_i is of order (Δt), then the probability of mistake is of order $(\Delta t)^2$. In the limit we have the Poisson-like result that there is zero probability of both moving simultaneously.[5]

Before proceeding, we note that the grab the dollar game has asymmetric equilibria. For example, "player one always grabs the dollar (if they have not yet done so), player two never grabs".[6] There are also equilibria in which players alternate grabbing, for example, "player one grabs at odd-numbered periods, player two grabs at even-numbered ones". These "alternating" equilibria have no continuous-time limit due to precisely the "chattering problem" to which generalised curves are a response. We have not attempted to extend our model to allow these chattering equilibria, because the (extreme!) coordination they require seems highly unreasonable.

Faced with the inadequacy of the usual continuous-time formulation, we chose to enlarge the strategy spaces. An alternative would have been to work directly with the discrete-time equilibria. We chose our approach because we view "continuous time" as a convenient mathematical construction intended to represent the notion of very short time periods. Thus we do not view our enlarged strategy spaces as more artificial than the usual ones. On the contrary, while the usual continuous-time strategy spaces are useful for some games, in games of pre-emption they are not appropriate.

We should point out that using continuous rather than discrete time makes a difference that is unrelated to randomisation or to our enlarged strategy spaces. Continuous-time models have "more" pure strategy equilibria than discrete-time models do. More precisely, the set of continuous-time (pure strategy) equilibria is not the set of limit points of discrete-time equilibria, but rather the set of limit points of discrete-time ε-equilibria with $\varepsilon \to 0$ along with the period length. (Fudenberg and Levine (1986). In an ε-equilibria each player's chosen strategy yields within ε of his optimal payoff.) For example, cooperation can be sustained in a finite-horizon continuous-time prisoner's dilemma, but not in a discrete-time one. Similarly, the joint adoption outcomes we described in the third section are not discrete-time equilibria. To see this, let the time interval between periods be denoted Δ, and assume that the joint optimum $\hat{T}_2$ belongs to the time grid. Assume both firms waited until $\hat{T}_2$ to adopt. Then it is in the interest of any firm to adopt at $(\hat{T}_2 - \Delta)$, whatever its opponent's play at that time, because adopting at $(\hat{T}_2 - \Delta)$ has no effect on one's opponent's date of adoption. By backwards induction it is clear that from T_1^* on, both firms adopt with probability one, so that only the diffusion equilibrium remains. This is an extreme implication of perfectness, reminiscent of the finitely-repeated prisoner's dilemma. In both games a large gain from cooperation is destroyed by a small payoff to cheating in the last period.

One way to reconcile perfectness with cooperation in the discrete-time versions of these games is to allow players to ε-optimise, as we remarked above. Another way is to introduce a small amount of incomplete information, which has been shown to allow finite-horizon resolutions on the prisoner's dilemma (Kreps *et al.* (1982)) and the chain store paradox. (Kreps and Wilson (1982a), Milgrom and Roberts (1982).) Assume that with probability $(1 - \eta)$, the payoffs are as before, but with probability η the firm is "incompetent". By this we mean that the firm has to observe the other firm's adoption for at least one period before adopting itself. Then we can show (see Fudenberg and Tirole (1984)) that for arbitrarily small $\eta > 0$, if $M(\hat{T}_2) > L(T_1^*)$ so that we are in Case B, then in the limit of shorter time periods the adoption date converges to $\hat{T}_2$. The incomplete information not only restores joint adoption as an equilibrium but selects it as the *only* one. Of course, other kinds of incomplete information would yield other results. Fudenberg and Maskin (1986) have shown that any equilibrium outcome of an infinitely repeated game without discounting is an equilibrium outcome of a sufficiently long, finitely repeated game. We conjecture that there is an analogous result relating continuous- and discrete-time equilibria. This might seem to weaken the case for joint adoption, but we think not. Our discussion of ε-optimisation and incomplete information is not intended as the main argument for joint adoption. Rather, we find joint adoption plausible *a priori* and have argued against ruling them out on purely technical grounds.

Strategy spaces and payoff functions

We now introduce strategy spaces that take into account the two issues we have just discussed. We first define "simple" continuous time strategies that permit us to distinguish between "types" of atoms, and define payoffs and Nash equilibrium. We then simply extend these strategies to "closed-loop" strategies and define Perfect Equilibrium.

Definition 1. A simple strategy for player i in the game starting at t is a pair of real-valued functions (G_i, α_i): $[t, \infty) \times [t, \infty) \rightarrow [0, 1] \times [0, 1]$ satisfying:

(a) G_i is non-decreasing and right-continuous.

(b) $\alpha_i(t) > 0 \Rightarrow G_i(t) = 1$.

(c) α_i is right-differentiable.

(d) If $\alpha_i(t) = 0$ and $t = \inf(s \geqq t \mid \alpha_i(s) > 0)$, then $\alpha_i(\cdot)$ has positive right derivative at t.[7]

Condition (a) ensures that G_i is a cumulative distribution function. $\alpha_i(t)$ will measure the "intensity" of atoms in the interval $[t, t + dt]$; thus condition (b) requires that if $\alpha_i(t)$ is positive then player i is sure to adopt by t. Conditions (c) and (d) are imposed for technical convenience.

To define the payoffs resulting to a pair of simple strategies, we need a bit more notation. Let

$$\tau_i(t) = \begin{cases} \infty & \text{if } \alpha_i(s) = 0 \ \forall \ s \geqq t \\ \inf(s \geqq t | \alpha_i(s) > 0) & \text{otherwise} \end{cases}$$

$\tau_i(t)$ is the time of the first "interval of atoms" in player i's strategy. Let $\tau(t) = \min(\tau_1(t), \tau_2(t))$. In the subgame starting at t, there is sure to be an adoption by time $\tau(t)$. Finally let G_i^- be the left-hand limit of $G_i(\cdot)$ at t, that is $G_i^-(t) = \lim_{\tau \to t^-} G_i(t)$.[8] The game begins at time $t \geqq 0$; so impose $G_i^-(t) = 0$, $i = 1,2$. The payoffs are then

$$V^i(t, (G_1,\alpha_1), (G_2,\alpha_2))$$

$$= \left[\int_t^{\tau(t)^-} (L(s)(1 - G_j(s))dG_i(s) + F(s)(1 - G_i(s))dG_j(s)) + \Sigma_{s < \tau(t)} a_i(s)a_j(s)M(s)\right]$$

$$+ (1 - G_i^-(\tau(t)))(1 - G_j^-(\tau(t)))W^i(\tau(t), (G_i, \alpha_i), (G_j, \alpha_j))$$

where

$$W^i(\tau, (G_i, \alpha_i), (G_j, \alpha_j)) = \left[\frac{G_j(\tau) - G_j^-(\tau)}{1 - G_j^-(\tau)}\right][(1 - \alpha_i(\tau))F(\tau) + \alpha_i(\tau)M(\tau)] + \left[\frac{1 - G_j(\tau)}{1 - G_j^-(\tau)}\right]L(\tau)$$

if $\tau_j(t) > \tau_i(t)$

$$= \left[\frac{G_i(\tau) - G_i^-(\tau)}{1 - G_i^-(\tau)}\right][(1 - \alpha_j(\tau))L(\tau) + \alpha_j(\tau)M(\tau)] + \left[\frac{1 - G_i(\tau)}{1 - G_i^-(\tau)}\right]F(\tau)$$

if $\tau_i(t) > \tau_j(t)$, while if $\tau_1(t) = \tau_2(t)$

$$W^i(\tau, (G_1, \alpha_1), (G_2, \alpha_2))$$

$$= \begin{cases} M(\tau) \quad \text{if } \alpha_i(\tau) = \alpha_j(\tau) = 1 \\ [(\alpha_i(\tau)(1 - \alpha_j(\tau))L(\tau)) + (\alpha_j(\tau)(1 - \alpha_i(\tau))F(\tau)) + (\alpha_i(\tau)\alpha_j(\tau)M(\tau))]/ \\ (\alpha_i(\tau) + \alpha_j(\tau) - \alpha_i(\tau)\alpha_j(\tau)) \quad \text{if } 2 > \alpha_i(\tau) + \alpha_j(\tau) > 0 \\ \left[\dfrac{\alpha_i'(\tau)L(\tau) + \alpha_j'(\tau)F(\tau)}{\alpha_i'(\tau) + \alpha_j'(\tau)}\right] \quad \text{if } \alpha_i(\tau) = \alpha_j(\tau) = 0 \end{cases}$$

These payoffs are the same as those of equation (2) up to $\tau(t)$. With probability $(1 - G_i^-(\tau(t))(1 - G_j^-(\tau(t))$, none of the players has moved by $\tau(t)$. At least one of the cumulative distributions $G_i(\cdot)$ then jumps to one. If $\tau_j(t) > \tau_i(t) = \tau$, then the payoffs are computed as the limits of discrete time payoffs when firm i adopts with probability $\alpha_i(\tau)$ at each period and firm j adopts with probability $[(G_j(\tau) - G_j^-(\tau))/(1 - G_j^-(\tau))]$ at the first instant and with probability zero thereafter. This corresponds to a situation in which firm j plays an isolated jump (or a density) at τ, and firm i adopts "continuously" with intensity $\alpha_i(\tau)$ (we know that firm j does *not* have an interval of atoms at τ because $\tau_j(t) > \tau$). If $\tau_1(t) = \tau_2(t) = \tau$, the probabilities of getting L, F and M are computed from discrete-time limits with constant probabilities of moves $\alpha_i(\tau)$ and $\alpha_j(\tau)$. If $\alpha_i(\tau) = \alpha_j(\tau) = 0$, the payoffs are computed by a first-order Taylor expansion.[9]

Definition 2. A pair of simple strategies (G_1, α_1) and (G_2, α_2) is a Nash equilibrium of the game starting at time t (with neither firm having yet moved) if each player's strategy maximises his payoff $V^1(t, ., .)$ holding the other player's strategy fixed.

Definition 3. A closed-loop strategy for players is a collection of simple strategies $(G_i^t(\cdot), \alpha_i^t(\cdot))_{t \geq 0}$ satisfying the intertemporal consistency conditions:

(e) $G_i^t(t+\upsilon) = G_i^t(t+u) + (1 - G_i^t(t+u))G_i^{t+u}(t+\upsilon)$ for $t \leqq u \leqq \upsilon$
(f) $\alpha_i^t(t+\upsilon) = \alpha_i^{t+u}(t+\upsilon) = \alpha_i(t+\upsilon)$ for $t \leqq u \leqq \upsilon$.

The reason why we need a family of strategies (G_i^t) is, as we explained, that to test for perfectness we must define strategies even conditional on zero-probability events. Condition (e) ensures that the family of strategies is consistent between non-zero probability events; that is, if time t is reached, and $G_i^t(\cdot)$ puts positive weight on times from $(t+u)$ on, then $G_i^t(s)$ should be consistent between t and $(t+u)$.[10] Condition (f) is a similar consistency condition.

Definition 4. A pair of closed-loop strategies $\{(G_1^t(\cdot), \alpha_1^t(\cdot))\}_{t\geqq 0}$ and $\{(G_2^t(\cdot), \alpha_2^t(\cdot))\}_{t\geqq 0}$ is a Perfect Equilibrium if for every t, the simple strategies $(G_1^t(\cdot), \alpha_1^t(\cdot))$, $(G_2^t(\cdot), \alpha_2^t(\cdot))$, are a Nash equilibrium.

Proof of the existence of the equilibria of Proposition 2

Now that we have developed an adequate formalism we can prove Proposition 2. First let us prove the existence of the equilibria. Recall the two cases: if $L(T_1^*) > M(\hat{T}_2)$ the unique equilibrium outcome is diffusion at (T_1, T_2^*); if $L(T_1^*) \leqq M(\hat{T}_2)$ then there is also a continuum of joint-adoption outcomes. We will exhibit strategies that support these two types of equilibria because the nature of the strategies is instructive. The proof that there are no other equilibrium outcomes is more technical; it can be found in Appendix 1.

We begin with the diffusion outcome:

Lemma 1. *The following symmetric strategies are a perfect equilibrium:*

$$G^t(s) = \begin{cases} 0 & s < T_1 \\ 1 & s \geqq T_1 \end{cases}$$

$$\alpha(s) = \begin{cases} 0 & s < T_1 \\ \dfrac{L(s) - F(s)}{L(s) - M(s)} & T_1 \leqq s < T_2^* \\ 1 & s \geqq T_2^* \end{cases}$$

Proof. We will check that these strategies yield a Nash equilibrium starting at any $t \in [T_1, T_2^*]$. First consider a subgame starting at t, $T_1 < t \leqq T_2^*$. If $G_i^t(t) = 0$ then firm i receives F(t). If $G_i^t(t) = \lambda$, $0 < \lambda \leqq 1$, and firm i plays an isolated jump, firm i receives $\lambda[\alpha(t)M(t) + (1 - \alpha(t))L(t)] + (1 - \lambda)F(t) = F(t)$. If $G_i^t(t) = 1$, and $\alpha_i(t) = a$, firm i receives

$$\frac{a\left[(1 - \alpha(t))L(t) + \alpha(t)M(t)\right] + (1 - a)\alpha(t)F(t)}{a + \alpha(t) - a\alpha(t)} = F(t)$$

Thus in each subgame in $[T_1, T_2^*]$, firms are indifferent between all possible choices, and we may specify whatever strategies we please. At $t = T_1$ the payoff to the given strategies is again F(t) (here we use the formula for payoffs when $\alpha_i = \alpha_j = 0$), as is the payoff to all other choices. Finally we check that $\alpha'(T_1) > 0$ and that the strategies are intertemporally consistent.

Notice that in equilibrium, one firm adopts at T_1 and the other at T_2^*. Even though the firms play mixed strategies at T_1, the probability of mistake is zero. This should seem intuitive, because the gain to pre-empting at T_1 is zero, and so neither firm would adopt at T_1 were there any possibility of mistake.

The coordination involved here may remind some readers of Aumann's (1974) notion of a correlated equilibrium. We should therefore stress that we have not introduced a jointly-observable coordinating device. The zero probability of mistake arises naturally in continuous time. In discrete time with short periods, the gain to pre-empting in the first period after T_1 is of order (dt), so in a mixed-strategy equilibrium each player must adopt with probability of order (dt) so that his opponent is willing to randomise. Then, as noted above, the probability of a mistake converges to zero with the period length.

Lemma 2. *If* $L(T_2^*) = M(S) \leqq M(\hat{T}_2)$, *the following symmetric strategies are a perfect equilibrium for any* $t_1 \in [S, \hat{T}_2]$:

$$\begin{cases} G^t(s) = \begin{cases} 0 & s < t_1 \\ 1 & s \geqq t_1 \end{cases} \\ \alpha(s) = \begin{cases} 0 & s < t_1 \\ 1 & s \geqq t_1 \end{cases} \end{cases}$$

Proof. The given strategies yield each firm $M(t_1)$, while any deviation yields at most max $(L(T_1^*), F(s))$.

In this case each firm's credible threat to immediately adopt if the other does (at times after T_2^*) can enforce the Pareto-superior joint adoption equilibrium.

THE OLIGOPOLY CASE

We now show that the possibility of pre-emption need not enforce rent equalisation when there are more than two firms. For this purpose we need only consider the three-firm case. Equilibrium payoffs need not be equal with more than two firms for two basic reasons. First, the payoffs in the "continuation game" after one firm has already adopted need not be equal, because with two firms payoffs were only equated along the equilibrium path, and not in all subgames. With unequal continuation payoffs, different firms have different gains to moving first, and so the threat of pre-emption need not equate the equilibrium payoffs. Asymmetric continuation payoffs can arise in two ways: first, if multiple continuation equilibria exist, one can specify different continuations depending on the name of the pre-empting firm. This kind of asymmetry may in some subgames be dismissed by the requirement that the firms play the Superior Perfect Equilibrium in every continuation game, if one exists. Unfortunately, not all of the two-firm continuation games have a Superior Perfect Equilibrium. Some subgames contain equilibria which cannot be Pareto-ranked. (See Appendix 2 for an example.)

The second reason that equilibrium payoffs need not be equated with more than two firms is that the duopoly case is special because the follower's response is continuous in the leader's move. This continuity disappears with more than two firms, even if the continuation outcome is symmetric and independent of the name of the first adopter. To see this, assume that we are in Case A, and let $\bar{T} > T_1^*$ be such that $L(\bar{T}) = M(\hat{T}_2)$, where L and M are the payoffs of the remaining two firms after one firm has adopted. Thus $\bar{T}$ is the first date at which late joint adoption is a continuation equilibrium. Let us specify that if the first adoption, t, is before $\bar{T}$, the remaining two firms play the "diffusion" equilibrium, while for $t > \bar{T}$ the continuation outcome is joint adoption at $\hat{T}_2$. Then at $\bar{T}$, the payoff to pre-emption jumps up. If a firm adopts first at $\bar{T}$, the others wait until $\hat{T}_2$, while adopting first at $\bar{T} - \varepsilon$ prompts immediate adoption by one of the others.

Now even though the two-firm continuation equilibrium is symmetric, and superior when possible, there is scope for specifying asymmetric three-firm Perfect Equilibria. For example, "firm three adopts with probability one from $\bar{T}$ on and the other firms wait". Firm three makes more profit than its competitors. Nevertheless, firm one, say, cannot impose rent equalisation by pre-empting firm three, since if it adopted at $(\bar{T} - \varepsilon)$ it would trigger immediate adoption by firms two and three. The discontinuity in the payoffs as a function of time gives scope for rents to be unequal.

RELATED WORK

We have already mentioned the Reinganum and Scherer work on adoption with pre-commitment. Scherer himself realised that the pre-commitment assumption "is vulnerable to the criticism of myopia in ignoring subsequent rival reactions to one's own decision". This criticism is addressed in the work of Rao and Rutenberg (1979) on the timing of capacity expansion. They use a discrete-time sequential-move model, and (implicitly) employ the Perfect Equilibrium solution concept. Because of their complicated specification of payoffs, (quite similar to that of Scherer) they were forced to use numerical techniques and no general conclusions emerged. Gilbert and Harris (1984) analysed pre-emptive capacity expansion on a continuous-time model. They concluded that with two or more identical firms, not only would payoffs be equalised but that all new plants would earn zero profits. The zero-profit equilibrium arises because, in contrast to the other models cited, each firm may build many plants, so that there is no opportunity cost to new investment.[11]

To avoid considering mixed strategies. Gilbert and Harris divide each time t into two "moments", t^- and t^+. Thus their game, like Rao and Rutenberg's, is one with sequential moves – firms are *unable* to move simultaneously. Our formalism makes this restriction unnecessary. We feel this is an improvement because imperfect information and the accompanying possibility of "mistakes" are an important aspect of game timing. While it is true that on the path of the diffusion equilibrium the probability of a mistake is zero, off the equilibrium path mistakes do occur. In contrast, with the Gilbert and Harris formalism, in no subgame would both firms adopt before T_2^*.

CONCLUSION

While we have termed our model one of the adoption of new technology, the main insights extend to more general games of timing. For example, we have seen that the threat of pre-emption equalises rents in a duopoly, but need not do so if there are more than two firms. The duopoly case is special because under reasonable conditions, the follower's response is continuous in the leader's move. This continuity disappears with more than two firms. The same phenomenon will naturally arise in other games of timing.

We have also seen that when the gain to pre-emption is sufficiently low, as for an innovation which simply captures a rival's market share, the optimal symmetric outcome is to be expected; while if moving first is very profitable, as in a new market, the equilibrium will involve early adoption and "diffusion". Again, such conclusions should emerge in more general settings.

One result which might not hold with a different payoff structure is the existence of a diffusion equilibrium Pareto-inferior to the equilibrium with pre-commitment. This may depend on the existence of pre-commitment equilibria with diffusion.

The two key assumptions underlying the pre-emption argument are that information lags are short and that the payoffs are common knowledge. We have already observed that relaxing the first assumption makes late adoption less plausible. Relaxing the assumption that the payoffs are common knowledge means that a firm is no longer sure that it will be pre-empted if it waits past T_1 to adopt. Because the rival's (distributional) strategy is now "smoothed out", the incentive to play aggressively is lessened. We conjecture that in this case the time of first adoption would be later than with complete information; this is an interesting topic for future research.

This paper considered only the case of identical firms. In our survey, Fudenberg and Tirole (1986), we extend the analysis to consider the case of an incumbent and an entrant, each of whom can adopt the new technology. We find that if the products are perfect substitutes and the flow profits are determined by Bertrand competition then, in equilibrium the incumbent adopts first at the time where the entrant is just indifferent between adopting first and adopting second. This strong result is the dynamic version of Gilbert and Newberry's work on the persistence of monopoly. It heavily relies on the absence of incomplete and imperfect information, and on the assumption of (zero-profit) Bertrand competition as discussed in the survey. If the incumbent either is unsure about the entrant's adoption cost (or more generally profit function) or observes the entrant's entry decision only with a lag, it may err and let the entrant adopt. Similarly, changing the nature of competition in the product market can reverse the persistence result.

APPENDIX 1

Proof of Proposition 2

In the text we exhibited equilibrium strategies. We now show that there are no other equilibrium outcomes. Let $\tau_i(t)$ denote firm i's first date of "intervals of atoms", in the subgame starting at time t, ie, $\tau_i(t) = \inf\{s \geqq t|\alpha_i(s) > 0\}$, and let $\eta_i(t) = \inf\{s \geqq t|G_i^s(s) > 0\}$. Note that if $\eta_i(0) < \tau_i(0)$, $\eta_i(0)$ is the first time of an isolated jump. And let $\tau(0) = \min\{\tau_1(0), \tau_2(0)\}$ and $\eta(0) = \min\{\eta_1(0), \eta_2(0)\}$.

Let us first assume that $\tau(0) \leqq \eta(0)$.

Lemma 2. $\tau(0) \leqq \hat{T}_2$.

Proof. Imagine that $\tau(0) > \hat{T}_2$, and consider the subgame starting at $\hat{T}_2$ when none of the firms has adopted. Since there is no jump in G_i at $\hat{T}_2$ $(i = 1, 2)$, there is a positive probability of adoption strictly after $\hat{T}_2$. But it is a dominant strategy for both firms to adopt at $\hat{T}_2$ and thereafter (both L(t) and M(t) decrease). Therefore the firms would not be optimising. (Note that the same lemma holds if $\tau(0) > \eta(0)$: then $\eta(0) \leqq \hat{T}_2$.)

Lemma 3. $\forall i, G_i^0(T_1^-) = 0$.

Proof. Let us show that a firm does not want to adopt at $t < T_1$. If it did, its payoff would be either M(t) or L(t) depending on whether the other firm adopts or not. But each firm can guarantee itself at least F(t) by waiting until T_2^*, say. Since $F(t) > \max\{M(t), L(t)\}$, a firm indeed does not want to adopt before T_1. Note that this lemma holds whether or not $\tau(0) \leqq \eta(0)$.

Lemma 4. *Let* $t \in [T_1, T_1^*)$. *If* $\tau(t) \leqq T_1^*$, *then* $\tau(t) = t$.

Proof. Assume that $t < \tau(t) \leqq T_1^*$. Since neither firm's strategy contains an atom before $\tau(t)$, adopting at $(\tau(t) - \varepsilon)$ gives either firm a payoff $L(\tau(t) - \varepsilon)$. Thus, if $V^i(s)$ denotes firm i's valuation at time s, $\min\{V^1(\tau(t) - \varepsilon), V^2(\tau(t) - \varepsilon)\} \geqq L(\tau(t) - \varepsilon)$ and $V^1(\tau(t) - \varepsilon) + V^2(\tau(t) - \varepsilon) \geqq 2L(\tau(t) - \varepsilon)$. Now there is adoption with probability one at $\tau(t)$. Therefore, using the facts that F and L are increasing, and both exceed M, we have:

$$V^1(\tau(t) - \varepsilon) + V^2(\tau(t) - \varepsilon) \leqq F(\tau(t)) + L(\tau(t))$$

Lastly $F(\tau(t)) < L(\tau(t))$ and therefore the last two inequalities are in contradiction for ε small enough.

Let us now consider the two cases of Proposition 2.

(A) $L(T_1^*) > M(\hat{T}_2)$. Let us show that $\tau(0) \leqq T_1^*$. Imagine that $\tau(0) > T_1^*$. Then in the subgame starting at T_1^* each firm can obtain payoff $L(T_1^*)$ since no firm has an atom at T_1^*. Since both firms cannot guarantee themselves a payoff $L(T_1^*)$, $\tau(0)$ must be at most T_1^*. Thus from Lemma 4, $\tau(0) = T_1$: the only possibility is the diffusion outcome (T_1, T_2^*) unless there is a positive probability of mistake at T_1. But since $F(T_1) = L(T_1) > M(T_1)$, this probability must be zero (if $\tau_i(0) = T_1$, say, j would be better off not adopting at T_1 if there is a probability of mistake).

To complete our treatment of Case A we must now show that Prob{firm one moves first} = Prob{firm two moves first} = $\frac{1}{2}$. From Lemma 4 and the right-continuity of the α's we know that firm one, say, has $\alpha_1(t) > 0$ in some open interval beginning at T_1. We claim that firm two also has $\alpha_2(t) > 0$ in some open interval beginning at T_1. If $\tau_2(T_1) > T_1$, then $G_2^{T_1}(T_1) = 0$, as adopting at T_1 is weakly dominated. As $G_2^{T_1}$ is right-continuous, for any $\delta > 0$, there exists $\varepsilon > 0$ such that $G_2^{T_1}(T_1 + \varepsilon) < \delta$. By waiting until $(T_1 + \varepsilon)$, firm one increases its profit $(L(T_1) = F(T_1))$ by at least $\{(L(T_1 + \varepsilon) - L(T_1)(1 - \delta)\}$ (since both F and L are increasing). So $\tau_2(T_1) = T_1$, which means that $\alpha_2(t) > 0$ in some open interval beginning at T_1. On some such interval we thus have $\alpha_1(t) > 0$ and $\alpha_2(t) > 0$. From the definition of payoffs, this implies that $\alpha_1(t) = \alpha_2(t) = [L(t) - F(t)]/[L(t) - M(t)]$. So starting at t slightly bigger than T_1, the two firms have equal probabilities of adopting first. This also holds at T_1 as the α's are right continuous.

(B) $L(T_1^*) = M(S) \leqq M(\hat{T}_2)$. We know from Lemma 2 that if $\tau(0) \leqq T_1^*$, the equilibrium must be diffusion at (T_1, T_2^*). Let us assume that $T_1^* < \tau(0) < S$. Consider the subgame starting at T_1^*. Since there is no atom at T_1^*, both firms can guarantee themselves $L(T_1^*)$. Therefore

$$V^1(T_1^*) + V^2(T_1^*) \geqq 2L(T_1^*)$$

But since the game ends with probability one at $\tau(0) < S$, and $L \geqq M$ and $F \geqq M$,

$$V^1(T_1^*) + V^2(T_1^*) \leqq \max_{t\in[T_1^*,\tau(0)]}(L(t) + F(t)) < 2L(T_1^*)$$

a contradiction. Next assume that $S < \tau(0) \leqq \hat{T}_2$. Let us assume that no firm ever wants to adopt between $\tau(0)$ and $\hat{T}_2$. Indeed since no firm adopts with probability one at any instant before $\tau(0)$ and M is increasing, it is in the interest of any firm to wait. Let $u \leqq T_2^*$ denote the supremum of the times such that at least one firm is willing to adopt at these times. There are two possible cases. Either for some ε and for some i, $G_i^{u-\varepsilon}(u) < 1$ or for all ε and all i, $G_i^{u-\varepsilon}(u) = 1$. Let us consider the former case. Then $\lim_{\varepsilon\to 0} G_i^{u-\varepsilon}(u) = 0$ since there is no atom before u. Thus i's probability of adoption between $(u - \varepsilon)$ and u converges to zero, and j's payoff to not adopting between $(u - \varepsilon)$ and $\tau(0)$ converges to $M(\tau(0)) > L(T_1^*) \geqq L(t)\ \forall t$. This contradicts the definition of u: close to u, firm j would not want to adopt. Next assume that for all ε, all i, $G_i^{u-\varepsilon}(u) = 1$. Then for any ε, the probability of adoption by time u is one. Therefore

$$V^1(u - \varepsilon) + V^2(u - \varepsilon) \leqq \max_{t\in[u-\varepsilon,u]}(L(t) + F(t)) \tag{3}$$

But since there is no atom at $(u - \varepsilon)$, each firm can guarantee itself $L(u - \varepsilon)$ by adopting at $(u - \varepsilon)$ in the subgame starting at $(u - \varepsilon)$. Therefore:

$$V^1(u - \varepsilon) + V^2(u - \varepsilon) \leqq 2L(u - \varepsilon) \tag{4}$$

Clearly (3) and (4) are in contradiction for ε small if $u < T_2^*$. If $u = T_2^*$, the reader will check that the left-hand derivative of $(L + F)$ at T_2^* is negative.

Therefore:

$$\max_{t\in[u-\varepsilon,u]}\{L(t) + F(t)) = L(u - \varepsilon) + F(u - \varepsilon) < 2L(u - \varepsilon), \text{ a contradiction.}$$

Therefore no firm adopts before $\tau(0)$. If $\tau(0) = S$, then the same reasoning holds except that a firm is indifferent between adopting at T_1^* and waiting until $\tau(0)$. Since $T_1^* < \tau(0)$, almost surely no firm adopts at T_1^*. Therefore, almost surely the equilibrium involves joint adoption at S. Because of the "almost surely" statement we ignored the possible outcome (T_1^*, S) in the theorem.

Second, assume $\eta(0) < \tau(0)$.

Lemma 5. *There are no isolated jumps in* (T_1, T_1^*).

Proof. Consider that $\eta_1(0) = \eta(0) < T_1^*$. Note that firm two does not adopt at time $\eta(0)$ if the subgame starting at $\eta(0)$ is reached: it can avoid a strictly positive probability of mistake by waiting a bit. There are two cases. Either there exists an interval $(\eta(0), \eta(0) + \varepsilon]$ in which firm two does not adopt, or for any $\varepsilon > 0$, firm two adopts with some probability in $(\eta(0), \eta(0) + \varepsilon]$. In the former case, firm one does not optimise when it adopts at $\eta(0)$: it could wait without fearing pre-emption, and adopt later; it would do better because L increases. Therefore consider the latter case. Then $\forall \varepsilon > 0$,

$$V^2(\eta(0)) \leqq a_1 F(\eta(0)) + (1 - a_1) L(\eta(0) + \varepsilon)$$

where $a_1 > 0$ is the conditional jump in the distribution G_1 at time $\eta(0)$. By taking the limit when ε tends to zero:

$$V^2(\eta(0)) \leqq a_1 F(\eta(0)) + (1 - a_1) L(\eta(0))$$

Since firm two can guarantee itself $L(\eta(0) - \varepsilon')$ in any subgame starting at $(\eta(0) - \varepsilon')$, then firm two adopts with probability one before $\eta(0)$ (otherwise:

$$L(\eta(0) - \varepsilon') \leqq V^2(\eta(0) - \varepsilon') \leqq F(\eta(0)) + (1 - a_1) L(\eta(0))$$

a contradiction for ε' small). We are back to the proof of Lemma 4:

$$2L(\eta(0) - \varepsilon') \leqq V^1(\eta(0) - \varepsilon') + V^2(\eta(0) - \varepsilon') \leqq L(\eta(0)) + F(\eta(0))$$

a contradiction. We have already noticed that Lemma 3 always holds. Therefore $\eta(0) = T_1$.

Lemma 6. *There cannot exist an isolated jump in* $[T_1^*, T_2^*]$.

Proof. First assume that $T_1^* < \eta(0) \leqq T_2^*$. Let $\eta_1(0) = \eta(0)$, say. Since firm one is willing to adopt at $\eta(0)$ if it is not pre-empted before, its valuation at T_1^* is strictly less than $L(T_1^*)$. But it can guarantee itself $L(T_1^*)$ by adopting immediately in the subgame starting at T_1^*, a contradiction. Second, assume that $\eta(0) = T_1^*$. Firm two's valuation at T_1^* is at most:

$$V^2(T_1^*) \leqq a_1 F(T_1^*) + (1 - a_1) L(T_1^*)$$

if $L(T_1^*) > M(\hat{T}_2)$ (Case A). Since firm two can guarantee itself $L(T_1^* - \varepsilon)$ in the subgame starting at $(T_1^* - \varepsilon)$ and $L(T_1^* - \varepsilon) > V^2(T_1^*)$ for ε sufficiently small, it must be the case that firm two adopts with probability one before T_1^* in the subgame starting at $(T_1^* - \varepsilon)$. Again we are back to the proof of

Lemma 4. Since the probability of adoption before T_1^* is one, there is:

$$2L(T_1^* - \varepsilon) \leqq V^1(T_1^* - \varepsilon) + V^2(T_1^* - \varepsilon) \leqq L(T_1^*) + F(T_1^*)$$

a contradiction.

Next consider Case B: $L(T_1^*) = M(S) \leqq M(\hat{T}_2)$. Let t denote the first time after T_1^* such that firm two is willing to adopt at t conditionally on none of the firms having adopted yet. We claim that $t > S$. If not

$$V^2(T_1^*) \leqq a_1F(T_1^*) + (1 - a_1)L(T_1^*)$$

and the previous reasoning obtains. Let us therefore assume that firm two does not adopt in $[T_1^*, t)$ for $t > S$. Then firm one could do better than $L(T_1^*)$ in the subgame starting at T_1^*: it could wait until slightly after S to adopt without being pre-empted. Therefore in the subgame starting at T_1^* firm one does not optimise.

Last, consider the case $\hat{T}_2 \geqq \eta(0) > T_2^*$. Let $\eta(0) = \eta_1(0)$, say. Notice that $G_1^{\eta(0)}(\eta(0)) = 1$. Since M is increasing before $\hat{T}_2$, firm two would not adopt in a neighbourhood of $\eta(0)$. But this in turn implies that firm one itself should not adopt at $\eta(0)$. Similarly $G_2^{\eta(0)}(\eta(0)) = 1$: both firms must adopt with probability one at $\eta(0)$. For the rest of the proof of Proposition 2, it does not matter whether certain adoption by both firms at a given time corresponds to isolated or non-isolated jumps. This concludes the proof of Proposition 2.

APPENDIX 2

This appendix shows that in some continuation games there is no Superior Perfect Equilibrium. Let L, F and M be the payoffs of the remaining two firms: let U satisfy $T_1^* < U < T_2^*$ and $L(U) > M(\hat{T}_2)$, so we are in Case A.

There are several undominated equilibria at U: let $\bar{T} > U$ be such that $L(\bar{T}) = M(\hat{T}_2)$, and let one and two be the remaining firms.

Let

$$U \leqq \upsilon \leqq s < \bar{T} \quad \begin{cases} G_1^{\upsilon}(s) = 1 \\ \alpha_1(s) = 1 \end{cases} \quad \begin{cases} G_2^{\upsilon}(s) = 0 \\ \alpha_2(s) = 0 \end{cases}$$

and after $\bar{T}$ the equilibrium strategies correspond to late joint-adoption at $\hat{T}_2$ (see the fourth section). Then firm one makes more profit than firm two if firm three pre-empts at U. It can be checked that these strategies are indeed an equilibrium of the continuation game and that the outcome is undominated (firm one gets L(U), which is the maximum it can get in the continuation game). Naturally, the strategies of firms one and two can be reversed, yielding a second undominated equilibrium.

1 The equal probability property is proved assuming that firms do not play weakly dominated strategies. Nothing else in the chapter relies on this assumption.
2 Scherer (1967), points out that the pre-commitment equilibrium may involve adoption even when never adopting would maximise joint profits.
3 Pitchik used the equivalent formula

$$V^i = \int_0^{\infty} \left[\int_0^t F(s)dG_j(s) + L(t)(1 - G_j(t)) + \alpha_j(t)M(t) \right] dG_i(s)$$

4 Richard Gilbert suggested this game as a stylised model of pre-emption. See Gilbert and Stiglitz (1979).
5 Strictly speaking, Poisson processes do not exhibit discontinuous cumulative distribution functions. However, consider two independent stationary Poisson processes each with intensity λ. For any λ there is no probability of mistake, and in the limit ($\lambda \to \infty$) both processes have an atom at time zero yet there is still zero probability of both events occurring simultaneously.
6 The "always grab" equilibria do not exist if $\delta > 1$.
7 This condition is not necessary. We make it to simplify the definition of the payoffs.
8 This limit exists for all i because $G_i(\cdot)$ is monotone.
9 More generally, we could recover the limiting probability distribution from the nth derivative of α_1 at τ if the first (n – 1) were zero.
10 This is analogous to condition (5.3) of Kreps–Wilson (1982b).
11 While Gilbert–Harris do not examine the possibility of other equilibria in which new plants do earn a profit, they conjecture that this cannot occur.

BIBLIOGRAPHY

Aumann, R., 1974, "Subjectivity and Correlation in Randomized Strategies", *Journal of Mathematical Economics*, pp. 67–96.

Fudenberg, D. and D Levine, 1986, "Limit Games and Limit Equilibria", *Journal of Economic Theory* 38, pp. 261–79.

Fudenberg, D. and E. Maskin, 1986, "Folk Theorems for Repeated Games with Discounting or with Incomplete Information", *Econometrica* 54, pp. 533–56.

Fudenberg, D. and J. Tirole, 1983, "Capital as a Commitment: Strategic Investment to Deter Mobility", *Journal of Economic Theory* 31, pp. 227–50.

Fudenberg, D. and J. Tirole, 1984, "Pre-emption and Rent Equalization in the Adoption of a New Technology", Discussion Paper 29, (CERAS).

Fudenberg, D. and J. Tirole, 1986, "Dynamic Models of Oligopoly", in J. Lesourne and H. Sonnenschein (eds), in *Encyclopaedia of Economics*.

Gilbert, R. and R. Harris, 1984, "Competition and Mergers with Lumpy Investments", *Rand Journal of Economics* 15, pp. 197–212.

Gilbert, R. and J. Stiglitz, 1979, "Entry, Equilibrium, and Welfare", presented at the NBER Conference on Theoretical Industrial Organisation (Montreal).

Harsanyi, J., 1964, "A General Solution for Finite Noncooperative Games, Based on Risk Dominance", in M. Dreshner *et al.* (eds.), *Advances in Game Theory, Annals of Mathematics, Study* 52, pp. 627–50 (Princeton).

Kreps, D., P. Milgrom, J. Roberts and R. Wilson, 1982, "Rational Cooperation in the Finitely-Repeated Prisoner's Dilemma", *Journal of Economic Theory* 27, pp. 245–52.

Kreps, D. and R. Wilson, 1982a, "Reputation and Imperfect Information", *Journal of Economic Theory* 27, pp. 253–79.

Kreps, D. and R. Wilson, 1982b, "Sequential Equilibria", *Econometrica* 50, pp. 863–94.

Milgrom, P. and J. Roberts, 1982, "Predation, Reputation and Entry Deterrence", *Journal of Economic Theory* 27, pp. 280–312.

Pitchik, C., 1982, "Equilibria of a Two-Person Non-Zerosum Noisy Game of Timing", *International Journal of Game Theory* 10, pp. 207–21.

Prescott, E. and M. Visscher, 1977, "Sequential Location Among Firms with Foresight", *Bell Journal of Economics* 8, pp. 378–93.

Rao, R. and D. Rutenberg, 1979, "Preempting an Alert Rival: Strategic Timing of the First Plant by an Analysis of Sophisticated Rivalry", *Bell Journal of Economics* 10, pp. 412–28.

Reinganum, J., 1981a, "Diffusion of New Technology: A Game-Theoretic Approach", *Review of Economic Studies* 153, pp. 395–406.

Reinganum, J., 1981b, "Market Structure and the Diffusion of New Technology", *Bell Journal of Economics* 12(2), pp. 618–24.

Scherer, F.M., 1967, "Research and Development Resource Allocation Under Rivalry", *Quarterly Journal of Economics* 81, pp. 359–94.

Scherer, F.M., 1980, *Industrial Market Structure and Economic Performance,* Second Edition (Chicago: Rand MacNally).

Young, L.C., 1980, *Lectures on the Calculus of Variations and Optimal Control Theory,* (New York: Chelsea Publishing Co.).

14

(s, S) *Equilibria in Stochastic Games**

Prajit K. Dutta and Aldo Rustichini

Columbia University; Boston University

We study a class of two-player continuous time stochastic games in which agents can make (costly) discrete or discontinuous changes in the variables that affect their payoffs. It is shown that in these games there are Markov-perfect equilibria of the two-sided (s, S) *rule type. In such equilibria at a critical low state (respectively high state), player 1 (respectively 2) effects a discrete change in the environment. In some of these equilibria either or both players may be passive. On account of the presence of fixed costs (to discrete changes), the payoffs are non-convex and hence standard existence arguments fail. We prove that the best response map satisfies a strong monotonicity condition and use this to establish the existence of Markov-perfect equilibria. The first-best solution is also a two-sided* (s, S) *rule but the symmetric first-best solution has a wider* (s, S) *band than the symmetric Markov equilibrium.*

In many economic applications, agents can make discrete or discontinuous changes in variables that affect their payoffs. Such changes may complement continuous changes: for example, a firm can decide to introduce an entirely new model of a product that it has on the market rather than make minor improvements or modifications in it. A similar problem is encountered in the decision between learning more about a currently used technology or discovering a new and totally unknown one, as in Jovanovic

*This chapter was first published in *Journal of Economic Theory* 67, 1995, pp. 1–39 and is reprinted with the permission of Academic Press Inc. We have benefited from the comments of seminar participants at Chicago, Columbia, McGill, NYU, Pennsylvania, Stanford and Western Ontario, and especially Ariel Rubinstein, Ricardo Cabellero, Andrew Caplin, Bentley McLeod, Joseph Greenberg, Charles Wilson and Roy Radner. Several anonymous referees contributed to the improvement of the original paper, both in the form and in the substance. The published version was largely prepared while the first author was visiting the University of Rochester, and he thanks them for their support.

and Rob (1990). Similarly it is important to consider discontinuous adjustments through devaluations in foreign exchange markets as an alternative to open market interventions. Yet another example of discrete change is the entry and exit decision of a firm (Dixit, 1989a). A final example is the choice faced by an agent between making discrete price changes and letting relative prices change continuously on account of inflation (Caplin and Sheshinski, 1988). Unfortunately the distinction between continuous and discrete changes is necessarily absent from discrete time models[1] while this difference can be seen immediately and crisply in a continuous time formulation. In the latter framework one can think of continuous changes as determining the rate of change of the economic environment, whereas discrete changes determine "jumps" in such an environment. In this chapter we study a continuous time stochastic model whose outstanding characteristic is that agents can make discontinuous changes in the environment.

In many applications in which a consideration of discrete change is important, such changes are made in a strategic setting. For instance each of the examples cited above is naturally embedded in models in which there are only a few economic agents who behave strategically.[2] This suggests the importance of a game theoretic formulation for "discontinuous action" models. Since the basic problem involves changes in the game environment, a purely repeated game model (in which the game environment is unchanging) obviously cannot be employed. The appropriate framework is that of a stochastic (or in other usage, dynamic or Markovian) game. We analyse a general stochastic game (see Parthasarathy 1973 or Fudenberg and Tirole, 1992 for a description of such games) with the further specification that players' actions have discrete effects on the state variable of the game.

It is well known that in continuous time games in which players are allowed to make instantaneous moves, problems of consistency in the descriptions of outcomes arise (see Stinchcombe, 1992 for a number of examples). We analyse here a game in which discrete changes are costly (in a sense which will be made precise later). We then introduce a framework for a continuous time game which is flexible enough to allow players to react instantaneously to their opponents' moves. This framework may be of more general interest in the consistent analysis of continuous time games.

Description of the game and results

A real-valued state variable X(t) (the game environment) evolves according to some exogenous stochastic process. Either of two players can at any time of their choosing change the state by a discrete and arbitrary amount. There is a fixed cost to making any such change. Players get per period returns which have two components: there is a direct payoff which depends on the state and from this is subtracted the costs of the discrete action undertaken. The critical assumption made on the direct payoffs is that they are monotone: player 1's payoff increases in the state, while that of player 2 decreases. A strategy of a player specifies a set of stopping times and associated jump sizes. We say that a strategy is Markovian if it depends only on the payoff relevant variable X. Note incidentally that on account of the fixed cost the (gross of cost) return function of a player is necessarily non-concave (and this often implies non-existence of Nash equilibria, of course).

Our main theorems (Theorem 5.4 and 5.5) state that there is a class of simple Markov-perfect equilibria in the stochastic game which are characterised by two-sided (s, S) rules. In such an equilibrium the strategies of the two players are completely described by four parameters L, $L+\theta$ (for player 1) and U, $U-\mu$ (for player 2). If the state is ever at or below L, player 1 jumps it up to $L+\theta$ whereas if it ever gets up to U or above, player 2 jumps it down to $U-\mu$. Of course $L \leqslant L+\theta < U$ and $L < U-\mu \leqslant U$. The parameters are determined by the primitives of the game (the fixed costs and period payoffs): by way of illustration we explicitly compute these parameters in some special cases (in which the flow payoffs are linear). We further show that it is not essential that both players be "active" in equilibrium. There are equilibria in which one or more players never take any discrete actions, ie it may be the case for instance that $\theta = 0$ and $L = \underline{E}$, where $\underline{E}$ is the minimum value of the state.

There are two results preliminary to the main theorems which are of independent interest. The first (Theorem 4.1) establishes that the best response to a one-sided (s, S) rule is a one-sided (s, S) rule. An equivalent way of stating that result is that the optimal policy of a single agent, whose payoffs are any monotone increasing function and who is faced with exogenous two-sided uncertainty and an impulse barrier, is an (s, S) rule. This is a result new to a optimal control literature (the previous results were for the special case of linear flow payoffs (see Harrison, Selke and Taylor, 1983 and Constantanides and Richard, 1978)). The second result is a monotonicity lemma. Recall that a player's best response problem is non-convex. Hence in games with discrete actions one cannot appeal to standard fixed point arguments. We define instead a partial order on the set of Markovian strategies of the players and the monotonicity lemma establishes that the best response is monotonic in this order. Equilibrium can then be derived by appealing to Tarski's fixed point theorem for monotone maps. These constructions may be of more general interest.[3]

We further show that the first-best solution is also a two-sided (s, S) rule. We compare the symmetric first-best solution with the symmetric Markov-perfect equilibrium. *A priori* there is no reason to expect a systematic comparison between these two solutions. The objectives are different in the two cases (the first-best internalises all returns and costs) and the optimisation problems are different too (the first-best is unconstrained, whereas the equilibrium best response is against an exogenous impulse barrier). Consider the symmetric first-best solution (in which a discrete action is taken whenever the state goes down to L_c, where $L_c < 0$, or up to $|L_c|$, and the action is to move the state to 0). The corresponding symmetric Markov-perfect equilibria is one in which the state is jumped whenever it gets below a value L_n or above $|L_n|$ and then it is moved to 0.[4] Theorem 6.2 establishes that $L_c < L_n$, ie, that the (s, S) band is wider in the first-best than in equilibrium.

Turning to a general characterisation of MPE, we show in Theorem 7.1 that any MPE that satisfies a mild restriction[5] is a simple generalisation of the two-sided (s, S) equilibrium. In particular, any such equilibrium has to be of the form that there are at most countably many closed intervals $\{[\underline{L}_n, \overline{L}_n], n = 1, 2, \ldots\}$ for player 1 (and respectively $\{[\underline{U}_n, \overline{U}_n], n = 1, 2, \ldots\}$ for player 2 such that player 1 effects a change whenever the state is in some $[\underline{L}_n, \overline{L}_n]$ (and, respectively, player 2 affects a change if the state is in some $[\underline{U}_n, \overline{U}_n]$. Furthermore, the players alternate in their activity regions, ie, between any two intervals in which player 1 moves there must be an interval where player 2 moves, and conversely. Note that a two-sided (s, S) equilibrium is a special case where there is a single interval for each of the two players.

A final contribution of this chapter is the introduction of a general framework for continuous time games with discrete instantaneous actions. Our formalisation involves the consideration of the play of an extensive form repeated game "within every instant". Such a device allows a consistent modelling of instantaneous reactions to other players' moves. Moreover the fact that moves are costly allows us to assign payoffs in a natural way to inconsistent outcomes. A detailed discussion of these issues is contained in Appendix 1.

The next section sets up the model and the following section contains illustrative examples. The chapter then analyses the best-response and equilibrium issues. We consider the first-best problem and then go on to further analyse the MPE we have found.

THE MODEL

A description of the game

Let $E \subseteq \mathbb{R}$. E will be the state space. We shall consider either of two specifications for E. In the first specification we take $E = \mathbb{R}$ and refer to this as the non-compact domain case. Alternatively we shall take $E = [\underline{E}, \overline{E}]$, $-\infty < \underline{E} < \overline{E} < \infty$, and refer to this as the compact domain case.[6] The index i will refer to a generic player. In all statements pertaining to i, the index j will refer to the other player. Time is continuous and the horizon is infinite.

Assume until further notice that $E = \mathbb{R}$. A state variable, the environment of the game, evolves as a joint consequence of the actions of both players and an underlying Markov process. Suppose that, in the absence of intervention by either player, the environment evolves according to an exogenous stochastic process $\{Z(t) : t \geqslant 0\}$. Additionally, at any instant, either player can jump the state by any finite amount of his choosing. The environment of the game at time t, X(t), is then determined by adding to the exogenous environment Z(t) the total displacement effected by the two players up to time t.[7]

We make the following assumption on the exogenous stochastic process:

(A0) The stochastic process $\{Z(t) : t \geqslant 0\}$ is a *Brownian motion* with drift and variance (m, v).

A convenient restriction to consider, although inessential for us, is that $m = 0$, ie exogenous factors on average leave the game environment unchanged.

The *direct flow payoffs* of each player depend on the state variable and will be denoted $r_i(x)$. Purely for notational purposes we make throughout a symmetry assumption on this payoff function. It should be understood (and the reader can easily check this) that none of the analysis that follows is contingent on this assumption:

(A1) *Symmetry.* $r_i(x) = r_j(-x)$ for all $x \in \mathbb{R}$. Let us denote the returns of players 1 and 2 as r(x) and r(– x) respectively. The only critical assumption on returns will be:

(A2) *Monotonicity.* r is strictly increasing in x and absolutely continuous. When the domain of the state variable is $\mathbb{R}$ we shall employ the following[8]:

(A3) *Boundary conditions*: (i) $\lim_{x \uparrow \infty} r(x) = \bar{r} < \infty$; (ii) $\lim_{x \downarrow -\infty} r'(x) = -\infty$.

A principal characteristic of discrete changes (in the game environment) is the fact that even small changes require a minimal outlay of resources. The most natural way to model such a phenomenon would be to assume that there is a fixed cost to any intervention. Let $c(\xi)$ denote the *cost to a discrete change* of amount ξ. We assume for simplicity that the only cost is a fixed cost[9]:

(A4) *Fixed Cost*: $c(\xi) = c_0$ for any ξ.

Note that in the pure fixed cost case described by (A4), player 1 can make the state arbitrarily large (which he prefers) at no extra cost (and similarly 2 can make the state arbitrarily small at no extra cost). Indeed as single decision makers this is exactly what they would attempt to do, given the assumption on the direct flow payoff.

Histories and strategies

The continuous time formulation of the stochastic game, although natural and simplifying from several viewpoints, presents substantial technical problems in specifying strategies and outcomes.

In particular, in continuous time models there is no "next instant after time t". This is problematical when player i moves at instant t and j wishes to react to i's action "as soon as possible". To allow for instantaneous reactions we will consider an extended definition of "time", which admits "reactions within an instant". So we assume that, at every time t, within that instant, players can act countably many times and in essence play an infinitely repeated game. Time then is a two variable index: the first component of it is the continuous time component, and the second the index number of the move in the extensive form game which takes place at that instant. The notation will be $\tau = (\tau_1, \tau_2)$; the continuous time component is τ_1 and will also be denoted by t and takes values in the positive real line. The second component, τ_2, takes values in the set of positive integers and counts the number of moves, taking place at each instant, in the infinitely repeated game. In Appendix 1 we develop formally this interpretation of time. So the change in the state at instant t is defined as

$$X(t) - X(t -) = \sum_{\{\tau:\tau_1 = t\}} \xi_1(\tau) + \sum_{\{\tau:\tau_1 = t\}} \xi_2(\tau) \qquad \textbf{(1)}$$

Of course given that there is a fixed cost to be paid whenever a player moves, in a best response only a finite number of potential action opportunities within the instant are actually used, and therefore the summation in equation (1) is well defined. In this setup a partial history at time τ is a description of activity for all $s \leq \tau_1$ and moves within the instant τ_1 up to τ_2. Consider the subset of histories at τ which have the property that the two players have intervened only a finite number of times; call this set the set of finite histories at τ.

A strategy σ_i prescribes an action for player i after every finite history up to every instant τ (and further satisfies certain measurability requirements detailed in Appendix 1). A strategy pair (σ_i, σ_j) is said to be *consistent* if it generates almost surely a finite history for all τ. An inconsistent strategy pair generates an outcome under which, with positive probability, there is an infinite number of moves by some finite time. Along every such sample path the game terminates (and the players are assessed a lifetime payoff of minus infinity).

We will analyse consistent strategies. Indeed since our focus in this chapter is on Markov strategies, we present in the next subsection a sufficient condition under which a pair of such strategies is consistent (see Appendix 1 for the general discussion of histories, strategies and outcomes).

The lifetime payoffs for a pair of strategies (σ_i, σ_j) is given by

$$W_i(\sigma_i,\sigma_j)(x) = E_x\left\{\delta\int_0^T r_i(x_s)e^{-\delta s}ds - \sum_{\tau_i} c(\xi_i(\tau))e^{-\delta\tau_i}\right\} \tag{2}$$

where T is the termination date (possibly ∞) and τ_i are the dates of player i's discrete actions; both of these random variables have distributions generated by the strategies. Further, x is the initial state, and E_x denotes the expectation conditional on x being the initial state.

Markovian strategies and Markov-perfect equilibrium

In a Markov strategy, a player only considers the state at any instant to decide his action. Formally, a Markov strategy for player i is given by a pair (O_i, g_i), $O_i \subseteq E$, $g_i : E \to E$, Borel measurable function, where player i moves if and only if $x(\tau) \in O_i$, and $g_i(x) = x$ if $x \notin O_i$. If $x \in O_i$, $g_i(x)$ indicates the value to which player i moves the state variable.

A particularly simple example of Markov strategies is an (s, S) strategy of the type used, for instance, in inventory control problems:

$$O_1 = (-\infty, L],\ g_1(x) = L + \theta \qquad \text{for} \quad x \leqslant L$$

for player 1 and

$$O_1 = [U, +\infty),\ g_2(x) = U - \mu \qquad \text{for} \quad x \geqslant U$$

for player 2. We shall refer to these strategies as (L, θ) and (U, μ) strategies respectively.

If $L + \theta < U$, and $U - \mu > L$, then each time the state variable hits the value L it is displaced by the first player to $L + \theta$, and the second player does not react; similarly when U is reached the state is moved to $U - \mu$ by the second player.

Consider, however the case $L + \theta \geqslant U$, $U - \mu > L$. If $\lim_{t\uparrow t_0} x_t = L$, where $x_t > 0$ for $t > t_0$, then the outcome naturally associated with this pair of strategies involves a simultaneous move at time t_0 of both players, with the process starting again at $x_{t_0} = U - \mu$, and no further intervention by either player, till a later date at which the process hits either L or U. In the previous section we introduced the expanded definition of time which allows players to make simultaneous moves within an instant precisely to accommodate such strategy configurations.

There is however still one more possibility: $L + \theta > U$ and $U - \mu < L$. The natural outcome associated with this pair of strategies, if the state reaches U or L at t_0, seems to be an infinite number of moves at the same instant by the two players. There seems to be no apparent way to define the value of the state at t_0; indeed this seems irrelevant since the payoff to date of both players is already $-\infty$ and so, without loss of generality, we say that the game terminates at time t_0.

There are simple restrictions on a pair of Markov strategies under which the above pathology is excluded; indeed that guarantee that in any finite time interval there is almost surely a finite number of jumps, ie that the pair is consistent. A clarification may be necessary. The conditions that we define below, and denote by (M1) and (M2), do not impose restrictions on the strategies that the players are allowed to use. Rather, they describe a subset of the strategy pairs for which the sample path of the state variable is well defined and the payoff is finite (that is, different from $-\infty$) for both players. That (M1) and (M2) are sufficient to ensure this is proved in Lemma 2.3.1 below.

Consider any two Markov strategies. The first restriction simply requires that there are a finite number of moves within each instant. Call the final destination G(x) if the initial jumping point is x. The second restriction says that the distance between the set of final destinations and the set of jumping points is positive. This restriction rules out the possibility that the time intervals between moves can become arbitrarily small. Note that in this case even when outcome is well defined, the payoffs of the players are $-\infty$.[10]

Formally let $D(x) \equiv g_1(x) + g_2(x) - 2x$ be the displacement of the state after the action of the two players; (recall that $g_i(x)$ is the state to which player i moves the game). Now define

$$G^0(x) = x;\ G^{n+1}(x) = G^n(x) + D(G^n(x))$$

So $G^n(x)$ denotes the position of the state variable after n simultaneous moves of the two players. The first restriction is:

(M1) For every $x \in O_1 \cup O_2$, there exists a finite n such that

$$G^i(x) = G^{i+1}(x) \qquad \text{for all} \quad i \geqslant n$$

We denote $G(x) = \lim_{i\to\infty} G^i(x)$. The second restriction is:

(M2) The distance $d(O_1 \cup O_2, G(O_1 \cup O_2)) \geqslant a > 0$.

Here $d(A, B) = \inf_{a,b} d(a, b)$ for $a \in A$ and $b \in B$, and d is the Euclidean distance. We can now state our sufficient condition for a pair of Markov strategies to be consistent.

Lemma 2.2.1. *If a pair of Markov strategies satisfies* (M1) *and* (M2) *above, then almost surely there is a failure history until time τ produced by the two strategies, for any time τ.*

Proof. Under the conditions (M1) and (M2) above, the proof follows the lines of proof of the Theorem 1.1 in Bensoussan and Lions, 1982 (p. 632). The intuition behind this proof is the following. Fix any time index τ and consider the sample paths over which there are a number of jumps larger than any number N. Consider the distance that the original Brownian motion has to cover to make these jumps possible. The restriction M2 makes the total distance at least as large as aN, for every N. Since in finite time the Brownian motion can cover only a finite distance, with probability one, we obtain the result.

For the pair of (s, S) strategies (U, μ) and (L, θ), the two restrictions are satisfied under all possible configurations of U, μ, L, θ except two.

In the first, either $\mu = 0$ or $\theta = 0$, and in the second $L + \theta = U$ or $U - \mu = L$. In the first case condition M2 is not satisfied, while in the second, condition M1 is not. Indeed in each case there is an infinite number of jumps by any finite time, ie the strategies are inconsistent. Hence in each case the game has terminated by such a time period.

A Markov-perfect equilibrium (MPE) is a pair of Markovian strategies O_i^*, g_i^* such that

$$W_i(O_i^*, g_i^*; O_j^*, g_j^*)(x) \geqslant W_i(O_i, g_i; O_j^*, g_j^*)(x) \qquad \textbf{(3)}$$

for all x, i and Markovian g_i.

From herein we restrict attention to Markovian strategies and analyse MPE. Of course it is easy to show that in our model MPE continue to be subgame perfect equilibria when all admissible strategies are considered. Also when the set O_i is not explicitly described, we mean $O_i \equiv \{x : g_i(x) \neq x\}$.

SOME EXAMPLES

Before we proceed to the general analysis later in the chapter, let us consider three simple examples. These examples will illustrate the theorems that follow. In each of these examples we make some special simplifying assumptions that allow us to compute the equilibria of the game. They may not therefore satisfy all the conditions of Theorem 4.1. Throughout $E = \mathbb{R}$.

Example 3.1. $r(x) = x$ and $Z(t)$ is a deterministic process with $Z = 0$.

Example 3.1 analyses a trivial deterministic law of motion which could be thought of as an extremely simple Brownian motion with $m = 0$ and $v = 0$. We claim that the following strategies constitute the only Markov-perfect equilibria in the game:

Take any pair (U^*, L^*) satisfying $U^* - L^* = c_0$. Player 1 acts iff $X < L^*$ and then jumps the state to U^*. Player acts iff $X > U^*$ and then jumps the state to L^*.

Consider player 1's best-response problem. He clearly does not want to jump the state beyond U^*. If he jumps at all his best option is to move the state to U^*. A discrete move at positive cost is worthwhile only if the current state is sufficiently payoff-poor. The point of indifference is precisely L^*. The result is a two-sided (s, S) rule equilibrium. A series of simple arguments should convince the reader that the above conditions are also necessary for an equilibrium. Note that the bandwidth, $U^* - L^*$, is increasing in the fixed costs. Also note that the first-best (with flow payoffs $r(x) + r(-x)$) is never to make discrete changes.

Example 3.2. $r(x) = 1$ for $x \geqslant 0$, $r(x) = 0$ when $x < 0$ and $Z(t)$ is a deterministic process with $Z = 1$.

This example is also defined for a trivial Brownian motion with $v = 0$ and $m = 1$. The flow payoffs are only weakly monotone; this simplifies the analysis but is not crucial to the example. Indeed, at some analytical cost, one could use here the linear returns specification of Examples 3.1 and 3.3. The *only* Markov-perfect equilibria of this game are:

Case 1. $c_0 \leqslant 1$. Player 1 acts iff $X < L^*$ and then jumps the state to $X = 1$ where L^* is defined by $e^{\delta L^*} = 1 - c_0$. Player 2 never acts.

Case 2. $c_0 > 1$. Neither player acts.

That the strategies of case 1 form an equilibrium can be checked directly: suppose player 2 is a passive player. If player 1 ever jumps he obviously wants to jump the state to some positive number, say $X = 1$. The question is: will he ever act and if yes, which states will he act at? The maximised returns (or value function) must be increasing in X and hence the jumping region must be of the form $(-\infty, L)$. It is clear that an immediate jump yields the player $1 - c_0$, whereas waiting for the state to improve yields lifetime returns of $e^{\delta x}$ if the current state is $x < 0$. A comparison of the two options yields the critical value of L^*. Let us now check that it is a best response for player 2 to be passive. If he acts he would clearly like to move the state to L^*. Pick any state from which player 2 acts. His returns are clearly cyclical, each cycle yielding $1 - e^{\delta L^*} - c_0$. But by definition of L^* this is exactly zero. So player 2 is indifferent between acting and not acting at any state.

That this is the only equilibrium can be seen by noting that if there is an alternate equilibrium in which player 2 is active so must player 1 be. Else given the upward drift of the state process, player 2 would like to jump the state to its minimum and with a non-compact domain this implies that he would have no best response. For the same reason an active player 2 in equilibrium must jump the state down to the minimum point at which 1 is inactive. Let this point be denoted L. As the state progresses up from L to 0, player 2 makes $1 - e^{\delta L}$ over this cycle and this return must be worthwhile, ie it must be that $1 - e^{\delta L} \geqslant c_0$. If the inequality is strict, ie 2 makes strictly positive returns over the cycle, he initiates the cycle as soon as possible, ie at $x = 0$. But this is not possible since in that case the active player 1 makes negative returns in the game. So $1 - e^{\delta L} = c_0$, ie $L = L^*$. Finally, player 1 would initiate a jump at L^* only if 2 is passive; else 1 makes losses in equilibrium.

It is clear that if the flow payoffs over the infinite horizon do not cover fixed costs, then neither player acts, ie the strategies of case 2 are in equilibrium and only those under those cost specifications. Example 3.2 illustrates two-sided (s, S) equilibria with one or both players passive. Note that L^* is decreasing in the fixed cost. Further, the first-best policy is again "no jump".

Example 3. $r(x) = x$ and Z is a Brownian motion.

The previous two examples have easily computable equilibria but on account of the deterministic transitions, no interesting state dynamics. In this example we compute the equilibrium when the underlying state process is in fact a (non-trivial) Brownian motion.

The calculations here are more laboured and hence have been consigned to Appendix 2.

Lemma 3.1 (Best response). *Suppose that player 2's strategy is jump iff* $X \geqslant U$ *and then jump down to* $U - \mu$, *for some* $U - \mu < U < \infty$. *Then player 1's best response is given by jump iff* $X \leqslant L$ *and then jump to* $L + \theta$, *where* θ *is a constant independent of* U, μ *and the "band"* $U - L$ *is determined solely by* μ.

Lemma 3.2 (Equilibrium). *The symmetric equilibrium of the game is given by the requirement that player 1 (respectively 2) jumps iff* $X \leqslant L^*$ *(respectively* $X \geqslant |L|^*$*). The jump size is* θ^*, *where* θ^* *is unique, and is given by*

$$-\left[\frac{1 - e^{\theta}}{1 + e^{\theta}} - \frac{1 - e^{-\theta}}{1 + e^{-\theta}}\right] + c_0 - \theta = 0 \tag{4}$$

The band size can be computed from

$$e^{2L} = \theta\left(1 - \left(1 - \frac{4(e^{\theta} - 1)(1 - e^{-\theta})}{\theta^2(1 + e^{-\theta})(1 + e^{\theta})}\right)^{1/2}\right)\frac{1 + e^{-\theta}}{2(e^{\theta} - 1)} \tag{5}$$

where $\theta = \theta^*$. *Again the first-best policy is no jump.*

A BEST RESPONSE TO AN (s, S) RULE IS AN (s, S) RULE

Suppose player 2's strategy is in fact given by (U, μ) where $-\infty < U - \mu \leqslant U < \infty$. Let V denote the value function in the corresponding best-response problem of the first player:

$$V(x) = \sup_g W_1(g, (U,\mu))(x)$$

where (with some abuse of notation) $W_1(g, (U,\mu))(x)$ denotes the discounted expected return to a Markovian strategy g in response to a (U, μ) strategy, with initial state x. We now show that there is a unique Markovian best-response strategy, say g^*. Further, it is also of the class (s, S) in that it is characterised by a single point L and a single destination $L + \theta$.

Theorem 4.1. *Consider a given Markovian policy of player 2, characterised by a pair* (U, μ). *Then*

(i) *There exists a unique optimal policy of player 1, say* g^*, *in the set of admissible strategies; this optimal policy is a Markovian policy, characterised by a pair* (L, θ); *furthermore,* $\theta > 0$.

(ii) *The value function* V *is differentiable.*

Proof. The proof of Theorem 4.1 is somewhat long and will therefore be divided into several lemmas (Lemma 4.2–4.7). These lemmas follow after a brief overview. For future consideration, define $M = \sup_{x \in E} V(x)$. It is easy to see that $V(x) \geqslant M - c_0$ for all x. Further, $V(x) > M - c_0$ implies that $g^*(x) = x$.

The logic of the proof of Theorem 4.1 is as follows: a straightforward modification of the argument in Bensoussan and Lions, 1982 (Chap. 6, Theorem 1.1) establishes that there is an optimal policy which is characterised by a set of jump points E^* and associated destinations.[11] Whenever the state is at $x \in E^*$, it is jumped to a destination g(x). We next prove some properties of the value function which are then used to characterise optimal policies. We establish first that the value function in the player's best-response problem has a unique maximum, denoted $L + \theta$. This suffices to show that whenever a discrete action is undertaken, it always involves a jump to $L + \theta$, ie for all $x \in E^*$, $g(x) = L + \theta$. We then show that the value function to the left of $L + \theta$ is strictly increasing. Hence there is a unique state, denoted L, such that if $x \leqslant L + \theta$, it is optimal to jump if and only if the state is less than L. Finally we demonstrate that it is never optimal to jump when $x > L + \theta$. The implication of these arguments is that for any optimal policy $E^* = (-\infty, L]$ and, consequently, that there is a unique optimal policy.

The following operator notation is used repeatedly in the sequel. Suppose we take an initial state $x \in (a, b)$. Let us consider the discounted returns to the policy: take no action as long as the state stays in the interval (a, b). The first time the process $\{X_t : t \geqslant 0\}$ hits either a (and this time is denoted T_a) or b (denoted T_b), the decision problem is terminated with terminal rewards V(a) and V(b) respectively. This policy defines an operator $\Gamma V(x; a, b)$ as

$$\Gamma V(x;a,b) = E_x\left\{\delta\int^{T_a \wedge T_b} e^{-\delta s} r(x_s) ds \right.$$
$$\left. + e^{-\delta T_a} V(a)\, I(T_a \leqslant T_b) + e^{-\delta T_b} V(b)\, I(T_a > T_b)\right\} \tag{6}$$

where I(C) is the indicator function of a set C.

We first argue the straightforward point that the value function must be continuous. Pick any $x \leqslant U$ and suppose that it is a point of discontinuity. It cannot be that there is a sequence $\{x_n : n \geqslant 0\}$ such that $\lim_{n\to\infty} V(x_n) < V(x)$. Else there is an interval (a, b), $a < x < b$ such that no ε-optimal policy from initial state x jumps from any state in (a, b). Clearly $V(x_n) \geqslant (\Gamma V(x_n; a, b)$. Further $V(x) \leqslant \Gamma V(x; a, b) + \varepsilon$ and the operator Γ is continuous. Combining all this yields $\lim_{n\to\infty} V(x_n) \geqslant V(x) - \varepsilon$. This is true for all $\varepsilon > 0$ and hence we conclude that if there is a discontinuity it must be the case that $\lim_{n\to\infty} V(x_n) > V(x)$. In that case none of the states x_n can be a jumping state. In fact there must be an interval (a, b), $a < x < b$ such that no ε-optimal policy from any initial state x_n jumps in this interval. Hence, $\Gamma V(x_n; a, b) \geqslant V(x_n) - \varepsilon$ and $V(x) \geqslant \Gamma V(x; a, b)$. The above arguments repeat to show that the discontinuity is impossible. Finally from (A3) it follows that this continuous function cannot attain its supremum for arbitrarily small x. Hence V attains the maximum on a finite domain.

Lemma 4.2. *Suppose* $V(x^*) = M$. *Consider* $x_1 < x_2 < x^*$, *with the property that* $V(x_1) > M - c_0$. *Then,* $V(x_2) > V(x_1)$.

Proof. Let $\Delta \equiv x_2 - x_1$, and define $x_3 = \sup\{x \in (-\infty, x_1) : V(x) = M - c_0\}$. By (A3) this set is non-empty and by the continuity of the value function, $x_3 < x_1$. Similarly, define $x_4 = \inf\{x \in (x_1, x^* - \Delta): V(x) = M - c_0\}$. We have two cases to consider:

Case 1. If the relevant set is empty, then define $x_4 = x^* - \Delta$. Let T_3 and T_4 denote the first hitting time of x_3 and x_4, respectively, when the initial state is x_1. Then

$$V(x_1) = \Gamma V(x_1; x_3, x_4) \tag{7}$$

Given the continuity of sample paths of the stochastic process, Z_t, the state remains within (x_3, x_4) prior to hitting either boundary. Hence, before

$T_3 \wedge T_4$ there is no action by either player. Further, given the spatial homogeneity of the process, we know that

$$S_3 \equiv \inf\{t \geqslant 0 : X(t) = x_3 + \Delta;\ X(0) = x_2 = x_1 + \Delta\}$$

has the same distribution as T_3. Similarly,

$$S_4 \equiv \inf\{t \geqslant 0 : X(t) = x_4 + \Delta;\ X(0) = x_2 = x_1 + \Delta\}$$

has the same distribution as T_4. Hence,

$$V(x_2) \geqslant \Gamma V(x_2; x_3 + \Delta, x_4 + \Delta) > \Gamma V(x_1; x_3, x_4) = V(x_1) \qquad \textbf{(8)}$$

The first inequality in (8) follows from the fact that the proposed policy for initial state x_2 may not of course be the optimal one. The strict inequality uses the following facts: the respective stopping times have the same distributions and further, $V(x_3 + \Delta) \geqslant V(x_3) = M - c_0$ and $V(x_4 + \Delta) = V(x^*) \geqslant V(x_4)$. Add to this the fact that the flow payoffs r are strictly increasing in the state and the identity in distribution of S_3 and T_3 (and S_4 and T_4), ie

$$E_{x_2}\int_0^{S_3 \wedge S_4} e^{-\delta s} r(x_s)ds > E_{x_1}\int_0^{T_3 \wedge T_4} e^{-\delta s} r(x_s)ds$$

Combining all this, (8) is proved.

Case 2. The set in the definition of x_4 is non-empty. Compare now the lifetime returns to starting at x_1 and eventually acting at either x_3 or x_4 against starting at x_2 and eventually acting at either $x_3 + \Delta$ or $x_4 + \Delta$. The above argument repeats. The proof of Lemma 4.2 is complete.

Remark. Assumption (A3), the boundary condition, is really not required for the above proof. The only modification required, in the absence of this assumption, would come from the fact that the set defining x_3 might be empty. In such a case take $x_3 = -\infty$. The arguments then are identical.

It immediately follows that V has a unique global maximum:

Corollary 4.3. *There does not exist* $x^* \neq x^{**}$, *s.t.* $V(x^*) = V(x^{**}) = M$.

Let g^* denote an optimal policy. Since the cost of discrete actions are independent of the size of the action it follows that if a player is to act at any state he must necessarily move the state to the unique maximum of the value function, ie that $g^*(x) \neq x$ implies that $g^*(x) = x^*$. Of course, $x^* < U$. This follows immediately from Corollary 4.3. In the sequel we call the chosen x, $L + \theta$. Since $g^*(x) \neq x$ if $V(x) > M - c_0$ it should further follow that if player 1 decides to act at some $\hat{x}$ then he acts at all $x < \hat{x}$. We actually prove a somewhat weaker version of this claim now. Note that the policy of acting on an entire interval $(\infty, \hat{x})$ is equivalent, from a payoff standpoint, to one in which the player acts at almost all $x < \hat{x}$. In other words the equivalence class for such a strategy is player 1 acts at all but some isolated points in $(-\infty, \hat{x}]$. We now show that the optimal policy must be from this equivalence class:

Corollary 4.4 *If $\hat{x} < x^*$ is an "action state", then so is almost every state smaller than $\hat{x}$, ie, if $g^*(\hat{x}) \neq \hat{x}$, then there does not exist an interval (a,b), $b \leqslant \hat{x}$ such that there is no intervention by player 1 for every $x \in (a,b)$.*

Proof. By Lemma 4.2 for all $x < \hat{x}$ it must be that $V(x) = M - c_0$. The proof of the lemma immediately implies that there cannot be an interval of states on which there is inaction and at the same time the value function is constant.

Now, let $L \equiv \sup\{x < x^* : g^*(x) \neq x\}$. In the next lemma we show that $L > -\infty$.

Lemma 4.5 *The "jumping point" is finite, ie, $L > -\infty$.*

Proof. Suppose to the contrary that $g^*(x) = x$ for every x. Fix $\varepsilon > 0 > K$ and pick $\hat{x} < x^*$, such that $r(\hat{x}) < K$ for all $x \leqslant \hat{x}$. This is possible by A3(ii). Let $\tilde{x} < \hat{x}$ further satisfy $1 - Ee^{\delta \hat{T}} > 1 - \varepsilon$, where $\hat{T} = \inf\{t \geqslant 0 : X(t) = \hat{x};\ X(0) = \tilde{x}\}$. By hypothesis,

$$V(\tilde{x}) = \Gamma V(\tilde{x}; -\infty, \hat{x}) \tag{9}$$

$$< K(1 - Ee^{\delta \hat{T}} + \varepsilon V(\hat{x})) \tag{10}$$

Note that by hypothesis it is not optimal to ever jump, ie,

$$V(\hat{x}) \geqslant M - c_0 \tag{11}$$

Letting $K \to -\infty$ and $\varepsilon \downarrow 0$, equations (10) and (11) yield $M = -\infty$ and this is clearly a contradiction. The lemma is proved.

Remark. In the absence of the assumption (A3ii) one cannot rule out the case that "passive" play may be optimal, ie, that $L = -\infty$.

What we have established so far is that in any optimal policy discrete actions are taken for any state below $L > -\infty$ with a jump up to $L + \theta$. Further, for any $x \in (L, L + \theta]$, player 1 refrains from acting and $\theta > 0$. To complete the proof of our result we now show that no action is taken for $x \in (L + \theta, U)$.

Lemma 4.6 $g^*(x) = x$, *for every* $x \in (L + \theta, U)$.

Proof. Let $x_1 = \inf\{x \in (L+\theta, U) : g^*(x) \neq x\}$. If the set is empty the lemma is proved. Else let $T_2 = \inf\{t \geqslant 0 : X(t) = x_1, X(0) = L+\theta\}$ and $T_1 = \inf\{t \geqslant 0; X(t) = L, X(0) = L+\theta\}$. Then define $\hat{x} \in (L+\theta, U)$ in such a way that $x_1 + \hat{x} - (L+\theta) < U$. Denote $\Delta = \hat{x} - (L+\theta)$. Then

$$V(L+\theta) = M = \Gamma V(L+\theta; L, x_1)$$

$$V(\hat{x}) \geqslant \Gamma V(\hat{x}; L+\Delta, x_1+\Delta) > \Gamma V(L+\theta; L, x_1) = M \geqslant V(\hat{x})$$

The reasoning which results in the above inequalities is identical to that of the proof of Lemma 4.2. The strict inequality clearly contradicts the definition of M.

We have proved that any optimal policy must be of the (s, S) type.[12] Since L and $L+\theta$ are uniquely defined, we have additionally established the uniqueness of best response. We now complete the proof of the theorem by establishing the differentiability of the value function.

Lemma 4.7. *The value function is differentiable.*

Fix the strategy of player 2, characterised by the pair (U, μ), and consider for any x the supremum of the payoff to player 1 over the set of admissible strategies. Straightforward modifications of the arguments in Bensoussan and Lions, 1982 (Chap. 6) give that this is equal to the supremum over the set of Markovian strategies, ie it is given by V(x) as defined at the beginning of this section.

From Itô's lemma, it follows immediately that the payoff, for any initial point $x \in [L, U]$, of the following Markovian policy characterised by the pair $(L, L+\theta)$ is given by the solution of a differential equation with two boundary conditions:

$$\frac{\sigma^2}{2} V''(x) + mV'(x) - \delta V(x) + \delta r(x) = 0 \qquad \text{for } x \in (L, U)$$

$$V(U) = V(U - \mu) \tag{12}$$

$$V(L+\theta) - c_0 = V(L)$$

We recall that a solution to (12) is a function of the form

$$V(x) = a_1(L, \theta)\, e^{-\alpha_1 x} + a_2(L, \theta)\, e^{-\alpha_2 x} + R(x)$$

where $\alpha_1 = 1/\sigma^2[m + \sqrt{m^2 + 2\delta\sigma^2}]$, $\alpha_2 = 1/\sigma^2[-m + \sqrt{m^2 + 2\delta\sigma^2}]$, and R(x) is a special solution. For example, if $m = 0$, then R can be taken to be

$$R(x) = \frac{\alpha_2}{2} \int_0^x \left[e^{\alpha_1(x-y)} - e^{\alpha_2(x-y)}\right] r(y)\, dy$$

From the boundary conditions that follow the differential equation (12), one can straightforwardly calculate $a_1(L, \theta)$ and $a_2(L, \theta)$. It is easy to see that these depend on the special solution and c_0. Clearly the special solution is a continuously differentiable[13] function and hence so are $a_1(L, \theta)$ and $a_2(L, \theta)$. The value function corresponds to the optimal choice of L and θ and in particular an optimal choice of $a_1(L, \theta)$ and $a_2(L, \theta)$. The argument in Dixit (1989b) now can be applied directly, and it shows that the first-order conditions for such a choice imply that V is a differentiable function. This completes the proof of Theorem 4.1.

Remark. Bensoussan and Lions (1982), Constantanides and Richard (1978) and Harrison, Selke and Taylor (1983) have also analysed this problem. The Bensoussan and Lions formulation, although very general, does not directly cover the case of a boundary with impulse reflection (as it is at U in our case). Consequently their arguments on the sufficiency of Markovian policies and the existence of an optimal policy have to be appropriately modified. More importantly, in the general formulation of their problem it is not possible to establish the sharp (s, S) rule characterisation for optimal policies. Harrison, Selke and Taylor (1983) and Constantanides and Richard (1978) do establish the optimality of (s, S) policies but restrict attention to the case of linear payoffs. (Further, they consider a somewhat different behaviour at the boundary of the state and their analyses admit fixed as well as proportional costs.) Caplin and Sheshinski (1988) analyse a discrete time version of this problem, use a weaker restriction in which r need only be quasi-concave, but restrict attention to decreasing processes. Hence, it would appear that the optimality of (s, S) policies that we establish in Theorem 4.1 is a more general result, in some dimensions, than those hitherto available.

MONOTONICITY OF BEST RESPONSE AND EXISTENCE OF EQUILIBRIUM

In this section we show that the best response map is "monotone" in a sense made precise shortly. Then we use the Tarski fixed point theorem to establish that an equilibrium in two-sided (s, S) rules exists in the stochastic game. A few remarks on the two steps are in order.

The objective function in a player's optimisation problem is fundamentally non-concave. This is so since any intervention, however small, involves a positive fixed cost whereas there is no cost for inaction. Non-concavities create well-known problems for equilibrium existence. For our game these problems may be summarised as follows: when the domain of X is compact (a case we detail a little later), ie, when $E = [\underline{E}, \overline{E}]$, then the best response is a non-convex valued correspondence. For some values of (U, μ) it can be shown that player 1 is indifferent between zero intervention and $L = \underline{E}$ and $L + \theta = x^*$. Of course, given the non-concavity in the objective function no convex combination of these two policies is as good. Further, for (U, μ) in one neighbourhood, $L = \underline{E}$ and $L + \theta = x^*$ is the unique best response and in a second neighbourhood, zero intervention is the unique optimal choice. In other words, the best-response correspondence is neither convex valued nor does it admit a continuous selection. If the domain of X is the real line, then Theorem 4.1 establishes that, despite the non-concavity, the best response is unique and indeed further arguments show that the best-response function is continuous. In this case, standard fixed point arguments do not work because the domain is not compact. We now show that the best response map is however monotone once the right order is specified. This result, interesting in its own right, enables us to apply a fixed point theorem for monotone maps to guarantee existence of equilibrium.

For the next two subsections we continue to discuss the case where the domain of X is $\mathbb{R}$. Let $\Pi = \{(U, U - \mu) \in \mathbb{R}^2 : -\infty < U - \mu \leqslant U < \infty\}$. Π can be identified with the space of (s, S) policies for player 2. Denote a generic element by π. Similarly, define Σ as the set of (s, S) policies and σ as a generic policy for player 1. We define a partial order on Π and Σ as follows:

Definition. (i) $\pi \geqslant \hat{\pi}$ iff $U \leqslant \hat{U}$ and $U - \mu \leqslant \hat{U} - \hat{\mu}$;
(ii) $\sigma \geqslant \hat{\sigma}$ iff $L \geqslant \hat{L}$ and $L + \theta \geqslant \hat{L} + \hat{\theta}$.

The partial order can be given the following interpretation: $\pi \geqslant \hat{\pi}$ may be thought of as implying more "aggressive" behaviour by player 2 under π rather than $\hat{\pi}$, in that player 2 acts sooner ($U \leqslant \hat{U}$) and lowers the state further ($U - \mu \leqslant \hat{U} - \hat{\mu}$) under the former regime. We now prove the rather surprising result that although the order on the strategy space is partial, the best-response map is completely ordered. Let $B_1\pi$ denote a best response (of player 1) to π.

Proposition 5.1. *Consider* $\pi \neq \hat{\pi}$. *Then either* $B_1(\pi) \geqslant B_1(\hat{\pi})$, *or* $B_1(\hat{\pi}) \geqslant B_1(\pi)$, *or both.*

Proof. For notational convenience, suppress in the proof the subscript on B_1. So suppose $B(\pi)$ and $B(\hat{\pi})$ are not ordered. Without loss of generality, suppose that $\hat{L} < L < L + \theta < \hat{L} + \hat{\theta}$.

We use V (respectively $\hat{V}$) to refer to the value function in response to (U, μ) (respectively $(\hat{U}, \hat{\mu})$). Then for any $\varepsilon > 0$, let $x \in (L, L + \theta - \varepsilon)$. Define $T_1 = \inf\{t \geqslant 0: X(t) = L; X(0) = x\}$, and $T_2 = \inf\{t \geqslant 0: X(t) = L + \theta - \varepsilon; X(0) = x\}$. Then

$$V(x) = \Gamma V(x; L, L + \theta - \varepsilon) \tag{13}$$

Similarly, let $T_3 = \inf\{t \geqslant 0: X(t) = L + \varepsilon; X(0) = x + \varepsilon\}$ and $T_4 = \inf\{t \geqslant 0: X(t) = L + \theta; X(0) = x + \varepsilon\}$. By the spatial homogeneity of Brownian motion, the distribution of T_1 (respectively T_2) is identical to that of T_3 (respectively

T_4). Note that

$$V(x+\varepsilon) = \Gamma V(x+\varepsilon; L+\varepsilon, L+\theta) \tag{14}$$

Combining equations (13) and (14) and writing

$$\alpha(x) \equiv E_x \delta \int_0^{T_1 \wedge T_2} e^{-\delta s} r(x_s)\, ds$$

$$\alpha(x+\varepsilon) \equiv E_{x+\varepsilon} \delta \int_0^{T_3 \wedge T_4} e^{-\delta s} r(x_s)\, ds$$

from the identity in distribution of the stopping times we have

$$\begin{aligned} V(x+\varepsilon) - V(x) = {} & [\alpha(x+\varepsilon) - \alpha(a)] \\ & + [V(L+\theta) - V(L+\theta-\varepsilon)]\, E_x e^{-\delta T_1} I(T_1 \leq T_2) \\ & + [V(L+\varepsilon) - V(L)]\, E_x e^{-\delta T_2} I(T_1 > T_2) \end{aligned} \tag{15}$$

From Theorem 4.1, we know that V is differentiable. Hence, from equation (15) it follows that

$$\begin{aligned} V'(x) &= \alpha'(x) + V'(L+\theta)\, E_x e^{-\delta T_1} I(T_1 \leq T_2) + V'(L)\, E_x e^{-\delta T_2} I(T_1 > T_2) \\ &= \alpha'(x) \qquad (16) \\ &< \alpha'(x) + \hat{V}'(L+\theta)\, E_x e^{-\delta T_1} I(T_1 \leq T_2) + \hat{V}'(L)\, E_x e^{-\delta T_2} I(T_1 > T_2) \\ &= \hat{V}'(x) \qquad (17) \end{aligned}$$

Note that equation (16) follows since, $\hat{V}'(L+\theta) > V'(L+\theta) = 0$ (because $\hat{L} + \hat{\theta} > L + \theta$) and $\hat{V}'(L) > V'(L) = 0$ (because $\hat{L} < L$) (and all of this from Lemma 4.2). Equation (17) follows from a derivation identical to equations (13) and (14). But if $V'(x) < \hat{V}'(x)$, for all $x \in (L, L+\theta)$, then $c_0 = V(L+\theta) - V(L) < \hat{V}(L+\theta) - \hat{V}(L) < c_0$, a contradiction. Hence, the proposition is proved.

Proposition 5.2. *Suppose* $\pi \geq \hat{\pi}$. *Then*, $B_1(\hat{\pi}) \geq B_1(\pi)$.

Proof. By Proposition 5.1, a contradiction to this claim implies that $B(\pi) \geq B(\hat{\pi})$ but not $B(\hat{\pi}) \geq B(\pi)$. Hence, we have $\hat{L} \leq L$ and $\hat{L} + \hat{\theta} \leq L + \theta < U \leq \hat{U}$, with at least one of the first two weak inequalities strict. Now let

$$\hat{T} = \inf\{t \geq 0 : X(t) = L + \theta;\ X(0) = L\}, \text{ and}$$

$$V(L) = -c_0 + V(L+\theta) \geq \beta(L) + E e^{-\delta \hat{T}} V(L+\theta) \tag{18}$$

$$\hat{V}(L) = \hat{\beta}(L) + E e^{-\delta \hat{T}} \hat{V}(L+\theta) \geq -c_0 + \hat{V}(L+\theta) \tag{19}$$

where

$$\hat{\beta}(L) = E_L\left[\delta\int_0^{\hat{T}} e^{-\delta s} r(x_s)ds - c_0 \sum_{1 \leqslant \hat{T}} e^{-\delta r}\right]$$

for the policy $(\hat{L}, \hat{L} + \hat{\theta})$. Combining equations (18) and (19), and noting that at least one inequality is strict, we get

$$[V(L+\theta) - \hat{V}(L+\theta)] \geqslant Ee^{-\delta\hat{T}}[V(L+\theta) - \hat{V}(L+\theta)] \qquad \textbf{(20)}$$

It is straightforward to show (and five do this in Lemma 5.3) that whenever $\pi \geqslant \hat{\pi}$, $V(x) \leqslant \hat{V}(x)$. This fact and equation (20) now give a contradiction. This concludes the proof of proposition 5.2. We can now present the last step, lemma 5.3.

Lemma 5.3. *Suppose* $\pi > \hat{\pi}$. *Then* $\hat{V}(x) > V(x)$.

Proof. Suppose, to begin with, that $x \in (L, U)$. A candidate policy that could be employed by player 1 in response to $(\hat{U}, \hat{\mu})$ is precisely $B(\pi)$. Since $\hat{U} \leqslant U$ and $\hat{V} - \hat{\mu} \geqslant U - \mu$, with at least one strict inequality, the strict monotonicity or r and special homogeneity imply the lemma. But this in turn implies that the lemma holds for all x in $(-\infty, L]$ or $[U, \infty]$.

A similar set of arguments establishes monotonicity in the best response of player 2: $\sigma \geqslant \hat{\sigma}$ implies that $B_2(\hat{\sigma}) \geqslant B_2(\sigma)$. Define the composite map $B = B_2 \circ B_1$. Clearly, $\pi \geqslant \hat{\pi}$ implies that $B(\pi) \geqslant B(\hat{\pi})$. We shall use this very useful property within the context of Tarski's fixed point theorem for complete lattices.

The principal obstacle to a direct application of the theorem to the stochastic game is that the space of strategies Π (recall Definition 5.1) is a lattice but not complete in the order induced by $\geqslant$ (see a statement of Tarski's theorem in Appendix 3). To complete the lattice we need to allow for the extended real variables $U = \pm\infty$, $U - \mu \equiv \pm\infty$ (respectively $L = \pm\infty$, $L + \theta = \pm\infty$). Directly incorporating these into the space (and interpreting $U = \infty$ as "player 2 never acts") does not work, since there is no best response to such a policy. Since player 1's payoffs are increasing in the state, it is immediate that if he acts against a passive player he would like to increase the state arbitrarily. To get around this "problem at infinity", we define a best response (to passive play) as the limit of best responses, as $U \uparrow \infty$. By Proposition 5.2, this limit is well defined, and the best-response map remains monotone. Hence Tarski's fixed point can be applied. An additional argument is then required to show that the fixed point has not been created by the artificial extension of the map. We conclude by showing that the fixed point must be interior, ie that it is in fact an equilibrium in the stochastic game. The boundary arguments outlined above are stated precisely and proved in Appendix 3.

Theorem 5.4. *The stochastic game has an equilibrium, given by the two pairs* (L^*,θ^*) *and* (U^*,μ^*), *such that* $-\infty < L^* < L^* + \theta^* < U^* < \infty$ *and* $\infty < L^* < U^* - \mu^* < U^* < \infty$. *The equilibrium behaviour is therefore described by a two-sided* (s, S) *rule: whenever* $x \leqslant L^*$, *player 1 acts and pushes the state up to* $L^* + \theta^*$. *Conversely, whenever* $x \geqslant U^*$, *player 2 lowers the state down to* $U^* - \mu^*$.

Proof. Define $\tilde{\Pi} \equiv \Pi \cup \{(U, U - \mu) : U = \pm\infty, U - \mu = \pm\infty\}$. $\tilde{\Pi}$ is clearly a complete lattice under $\geqslant$. Further, the best response is monotone on $\tilde{\Pi}$ by Proposition 5.2 and the extension of the map at the boundaries (see Appendix 3), equations (A3.1)–(A3.4) for details). Take $\hat{U}_0 = (\infty,\infty)$, $U_0 = (-\infty, -\infty)$ as the largest and smallest elements of the space in order to satisfy the boundary requirements of Tarski's Theorem (see Appendix 3 for a statement). Thus, there is a fixed point such that $\pi^* = B(\pi^*)$. We shall denote $\pi^* = (U^*, U^* - \mu^*)$ and $B_1(\pi^*) = (L^*, L^* + \theta^*)$ with the further convention that $B_{11}(\pi^*) = L^*$ and $B_{12}(\pi^*) = L^* + \theta^*$ (similarly, $B_{21}(B_1(\pi^*))$ and $B_{22}(B_1(\pi^*))$. Let us show that the fixed point is interior. It cannot be that $U^* = \infty$. Else, $B_{11}(\pi^*) \leqslant \overline{L} < \infty$ by Proposition A3.1(i) and (ii) and hence $B_{21}(B_1(\pi^*)) \leqslant U(\overline{L}) < \infty$, by the same proposition. Similarly it cannot be that $U^* - \mu^* = -\infty$. Else, if $U^* > -\infty$, $B_{11}(\pi^*) \geqslant \underline{L}$ and $B_{12}(\pi^*) \in (\underline{L},U^*)$ and hence $B_{22}(B_1(\pi^*)) > -\infty$. If $U^* = -\infty$, then $B_1(\pi^*) = (-\infty,-\infty)$, but then $B_2(B_1(\pi^*)) \geqslant \overline{U} > -\infty$. This proves that the fixed point is interior. In turn then π^* and $B_1(\pi^*)$ are best responses to each other, and therefore they together give an equilibrium of the stochastic game.[14]

The case of a compact domain

The assumption that the state space of the stochastic process is the entire real line may be unsatisfactory for many readers. In many interesting economic examples there are natural restrictions on the range of values that the stochastic process can assume. Here we prove that the above analysis extends (in fact, it becomes simpler) to the case in which $E = [\underline{E}, \overline{E}]$, with $-\infty < \underline{E} < \overline{E} < +\infty$. In this formulation we have, of course, to specify the behaviour of the process at the two boundary points. We shall consider both of the standard conditions:

(R) $(X_t)_{t \geqslant 0}$ is reflected at $\underline{E}$ and $\overline{E}$

(A) $(X_t)_{t \geqslant 0}$ is absorbed at $\underline{E}$ and $\overline{E}$

See Karatzas and Shreve (1988) for a definition of reflected and absorbed processes. We also have to specify what actions are feasible to ensure that players' jumps do not take the game environment outside the state space. We do this in the most obvious manner: we say that $X(\tau -) + \xi_1(\tau) + \xi_2(\tau) < \underline{E}$ (respectively $X(\tau -) + \xi_1(\tau) + \xi_2(\tau) > \overline{E}$) implies that $X(\tau) = \underline{E}$ (respectively $X(\tau) = \overline{E}$.

Lemmas 4.1–4.4 and Corollaries 4.2 and 4.3 hold with no modifications so Theorem 4.1 holds. The best response is therefore well defined, and the best response to an (s, S) rule is an (s, S) rule. Similarly, the analysis of the monotonicity property of the best response is unchanged. In Proposition 5.1 if $L > \underline{E}$ the argument remains completely unchanged. On the other hand, when $L = \underline{E} = \hat{L}$, the conclusion is automatic. Finally, Proposition 5.2 holds without any modifications. We may conclude therefore:

Theorem 5.5. *Let the state space be* $[\underline{E}, \bar{E}]$, *and assume either* (R) *or* (A). *Then the stochastic game has an* (s, S) *equilibrium given by the two pairs* (L^*, θ^*), (U^*, μ^*), *with* $\underline{E} \leqslant L^* \leqslant L^* + \theta^* \leqslant \bar{E}, \leqslant U^* \leqslant U^* - \mu^* \leqslant \bar{E}$.

Proof. Obviously Π is a complete lattice. Now Tarski's fixed point applies immediately.

Later we will investigate other MPE.

THE FIRST-BEST SOLUTION

In this section we investigate the first-best or cartel problem for our stochastic game. The problem may be defined as follows: pick a set of (measurable) stopping times τ_i and an associated set of jumps to solve

$$\text{Max } E_x \left\{ \delta \int_0^\infty [\alpha r(x) + (1 - \alpha) r(-x)] e^{-\delta s} ds - c_0 \sum_{\tau_i} e^{-\delta \tau_i} \right\} \tag{21}$$

where $\alpha \in (0, 1)$. To simplify the exposition we shall concentrate attention on the symmetric case, ie to the case $\alpha = 1/2$. Without loss of generality consider the flow payoff to be $r(x) + r(-x)$ and denote this $\rho(x)$. Thus far all we have assumed about the flow payoffs r is that it is a monotone function. This does not of course have any implications for the sum ρ. We now make the further assumption:

(A5) r is a concave function.

Remark. It is immediate from (A6) that ρ is symmetric about $x = 0$, is concave, and attains a unique maximum at $x = 0$.

A natural conjecture for the optimal policy is that is will also be a two-sided (s, S) rule. Note that existing results that establish the optimality of (s, S) rules in single-agent decision problems cannot be directly applied here. For instance, Caplin and Sheshinski (1988) only allows for decreasing stochastic processes, whereas Harrison, Selke and Taylor (1983) impose an exogenous one-sided barrier in the problem.

Even though the optimality of a two-sided (s, S) rule seems likely in the first-best problem, what is far from clear is the relationship that this rule will have to the similar rule that was obtained for the Markov-perfect equilibrium. For instance, note that the first-best internalises all of the changes in returns due to a discrete change in the state. Part of any such change is the "cannibalisation" effect, that the returns of one player increase if and only if the returns of the other player decrease. This might suggest that the cartel is less likely to use discrete actions. On the other hand, in the first-best solution one can better control all subsequent jumps and hence it may be profitable to use a discrete action when this is not the case for a single player who fears quick retaliation. We prove the following two results:

Theorem 6.1. *The first-best solution is given by a two-sided* (s, S) *rule* $(L_c, L_c + \theta, U_c, U_c - \mu)$, *where* $-\infty \leqslant L_c < L_c + \theta = 0 = U_c - \mu < U_c \leqslant \infty$ *and is the form act iff* $x \leqslant L_c$ *or* $x \geqslant U_c$ *and then move the state to* 0. *Further,* $|L_c| = U_c$.

Note that since the game is symmetric there is always a symmetric equilibrium in the non-cooperative framework. It is clear that this symmetric equilibrium, denoted as $(L_n, L_n + \theta, U_n, U_n - \mu)$, is the relevant point of comparison with the first-best solution. In such an equilibrium $|L_n| = U_n$ and so one point of comparison will involve the respective band sizes $2U_c$ versus $2U_n$. Unfortunately it is not true that symmetry of the equilibrium implies that the jump destinations are zero, ie that $L_n + \theta = U_n - \mu = 0$. But suppose there is in fact such a *strongly symmetric equilibrium*, ie an equilibrium in which such destinations are zero. Then we have:

Theorem 6.2. *Suppose that* L_n *defines a strongly symmetric* MPE *and* L_c *is the first-best solution. Then* $L_c < L_n$ *(and so* $U_c > U_n$*).*

Proof of Theorem 6.1. The proof will follow by establishing that the value function for the first-best problem, denoted W, has the following property: W is continuous, is symmetric and attains a unique maximum at $x = 0$. That W is continuous can be established by arguments identical to those employed in the proof of Theorem 4.1. Symmetry follows straightforwardly from the symmetry of r. Finally in order to show that W is in fact strictly increasing on an interval $[L_n, 0]$ and strictly decreasing on $[0, |L_n|]$, we appeal to an argument identical to that in Lemma 4.2.

Proof of Theorem 6.2. Exactly as in an earlier section, let us define the operator $\Gamma_\rho W$ as

$$\Gamma_\rho W(x;a,b) = E_x\left\{\delta\int^{T_a \wedge T_b} e^{-\delta s}\rho(x_s)ds \right.$$

$$\left. + e^{-\delta T_a}W(a)\, I(T_a \leqslant T_b) + e^{-\delta T_b}W(b)\, I(T_a > T_b)\right\}$$

Suppose per absurdum that in fact $L_c > L_n$. Then, for $x \in (L_c, 0)$,

$$W(x) = \Gamma_\rho W(x; L_c + \varepsilon, \varepsilon) \text{ and } W(x - \varepsilon) = \Gamma_\rho W(x - \varepsilon; L_c, 0)$$

But from the spatial homogeneity of Brownian motion it follows that the distribution of T_c, with starting state x, is identical to the distribution of T_0 with starting state $x - \varepsilon$ (respectively, the distribution of $T_{L_c+\varepsilon}$ with starting state x is the same as that of T_{L_c} with initial state $x - \varepsilon$). So,

$$W(x) - W(x - \varepsilon) = \alpha_\rho(x) - \alpha_\rho(x - \varepsilon) + E_x e^{-\delta T_\varepsilon}[W(\varepsilon) - W(0)]$$
$$+ E_x e^{-\delta T_{L+\varepsilon}}[W(L_c + \varepsilon) - W(L_c)]$$

where

$$\alpha_\rho(x) = E_x \delta \int^{T_\varepsilon \wedge T_{L+\varepsilon}} e^{-\delta s} \rho(x_s) ds$$

and

$$\alpha_\rho(x - \varepsilon) = E_x \delta \int^{T_0 \wedge T_L} e^{-\delta s} \rho(x_s) ds$$

Dividing by ε and taking limits, it then follows from the differentiability of W and $W'(L_c) = W'(0) = 0$ that

$$W'(x) = \alpha'_\rho(x) \equiv E_x \delta \int e^{-\delta s} \rho'(x_s) ds$$

Similarly,

$$V'(x) = \alpha'_r(x) + E_x e^{-\delta T_L} V'(L_c) > \alpha'_r(x) \equiv E_x \delta \int e^{-\delta s} r'(x_s) ds$$

Since $\rho'(x) = r'(x) - r'(-x)$, it then follows that $\alpha'_r(x) > \alpha'_\rho(x)$. The above arguments have hence established that for all $x \in (L_c, 0)$, $V'(x) > W'(x)$. But then,

$$c_0 \geqslant V(0) - V(L_c) > W(0) - W(L_c) = c_0$$

which is a contradiction. Hence the theorem is proved.

In the light of Theorem 6.2 one may wonder if the state variable will be more volatile in the equilibrium of the game or in the first-best solution. The answer turns out to depend on the measure of volatility that one uses. If one chooses the variance of the invariant measure, that is larger for the first best than for the equilibrium. On the other hand, the expected number of jumps in the process is higher for paths produced at equilibrium. This difference is due, naturally, to the fact that in the equilibrium the state is displaced more frequently, but by a smaller amount.

A CHARACTERISATION OF MARKOV-PERFECT EQUILIBRIA

In this section we characterise the set of MPE of the stochastic game. In particular we show that any MPE must involve strategies which are simple generalisations of the two-sided (s, S)-rule equilibria studied thus far. Furthermore, the dynamic implications of MPE are straightforwardly analysed and not significantly different from that of two-sided (s, S) equilibria.

Consider any Markov strategy $\langle O, g \rangle$, where $O \subseteq \mathbb{R}$ is the set of states at which the player moves and for any $m \in O$, $g(m)$ is the jump destination. We will restrict attention to Markov strategies in which O is a closed set and will refer them as *closed Markov strategies.*[15]

Definition. A closed Markov strategy is said to be *interval-based* if the set M can be written as the disjoint union of (at most) a countable number of closed intervals, ie

$$O = \bigcup_n I_n, \quad \text{where } I_n \equiv [a_n, b_n], I_n \cap I_m = \phi, \quad \text{whenever } m \neq n$$

To ease notation, let us refer to player 1 as the L player and denote his closed Markov strategy by $\langle O_L, g_L \rangle$ (respectively player 2 will be referred to as the U player with closed Markov strategies written as $\langle O_U, g_U \rangle$). Interval-based strategies in the two cases are denoted $\langle O^I_L, g_L \rangle$ and $\langle O^I_U, g_U \rangle$, where $O^I_L = \cup_n I_{nL}$ and $I_{nL} = [\underline{L}_n, \overline{L}_n]$ (likewise $O^I_U = \cup_n I_{nU}$ with $I_{nU} = [\underline{U}_n, \overline{U}_n]$).

Theorem 7.1. *If there is an MPE in closed Markov strategies which do not overlap, ie, $O_U \cap O_L = \phi$, then it must be in interval-based strategies. Furthermore, the equilibrium pair $\langle \cup_n I_{nL}, g_L; \cup_n I_{nU}, g_U \rangle$ must satisfy the following property: the jump intervals alternate; ie, between I_{nL} and I_{n+1L} there must lie I_{mU} (and $I_{nL} \cap I_{mU} = \phi, I_{n+1L} \cap I_{mU} = \phi$). (A converse statement is true about any pair of adjacent intervals for the U-player). Hence, the number of intervals must be the same for both players, in equilibrium.*

Proof. The proof follows from the following lemma.

Lemma 7.2. *Suppose that $\langle O_U, g_U \rangle$ is a closed Markov strategy. Then any best response to this strategy must be interval-based.*

Proof. Since O_U is closed, its complement O^c_U is a countable union of open intervals, $O^c_U = \cup_n (a_n, b_n)$, $a_n < b_n$. We now show that in any closed Markov strategy that constitutes a best response, say $\langle O^I_L, g_L \rangle$, if $x_i \in O_L$, $i = 1,2$ and $a_n < x_1 < x_2 < b_n$, then it must be the case that $x \in O_L$ whenever $x \in (x_1, x_2)$.

Suppose to the contrary that for some $x' \in (x_1, x_2)$ we have $x' \notin O_L$. Define

$$z_1 \equiv \sup\{x < x' : x \in O_L\}$$
$$z_2 \equiv \inf\{x > x' : x \in O_L\}$$

Since O_L is closed, $z_i \in O_L$, $i = 1,2$. Furthermore, by hypothesis each of these sets is non-empty and hence z_i, $i = 1,2$, are well defined. Note that whenever $x \in (z_1, z_2)$, $x \notin O_L$. For any such x, we have

$$\begin{aligned} V(x) &= E_x\left\{\delta\int_0^T e^{-\delta s} r(x)\,ds + e^{-\delta T_1} V(z_1) + e^{-\delta T_2} V(z_2)\right\} \\ &= \alpha(x) + Ee^{-\delta T}\left[\max_z V(z) - c_0\right] \end{aligned} \tag{22}$$

where

$$T_i \equiv \inf\{t : X(t) = z_i \mid X(0) = x\}$$

Let $\theta_1(x) \equiv x - z_1$ and $\theta_2(x) \equiv z_2 - x$. Pick x so that $x_2 + \theta_2(x) < b_n$. Since $x_2 < b_n$ and $\lim_{x \uparrow z_2} \theta_2(x) = 0$, we can always find such an x (appropriately close to z_2).

Now consider the following strategy from initial state x_2: do not move for $x \in (x_2 - \theta_1(x), x_2 + \theta_2(x))$. At $x_2 - \theta_1(x)$ or $x_2 + \theta_2(x))$, move to $g_L(z_1)$ and thereafter follow the strategy $\langle O_L, g_L \rangle$. The two stopping times

$$S_1 \equiv \inf\{t : X(t) = x_2 - \theta_1(x) \mid X(0) = x_2\}$$

and

$$S_2 \equiv \inf\{t : X(t) = x_2 + \theta_2(x) \mid X(0) = x_2\}$$

have the same distribution as the stopping times (T_1, T_2): this follows from the spatial homogeneity of Brownian motion. Hence,

$$V(x_2) \geqslant \alpha(x_2) + Ee^{-\delta T}\left[\max_z V(z) - c_0\right] \tag{23}$$

So

$$V(x_2) > \alpha(x) + Ee^{-\delta T}\left[\max_z V(z) - c_0\right] \tag{24}$$

Since the proposed strategy is not necessarily optimal, inequality (23) follows. Since the immediate reward r is strictly increasing, (24) follows.

The right-hand side of (24) is equal to $V(x) > \max_z V(z) - c_0$. This last inequality is a consequence of the fact that, by hypothesis, it is inoptimal to move at x. Hence, it must be inoptimal to move at x_2, and that contradicts the maintained hypothesis. So it must be the case that if $x_1 \in O_L$, $x_2 \in O_L$ and $x_i \in (a_n, b_n)$, then $x \in O_L$, for all $x \in (x_1, x_2)$. This completes the proof of Lemma 7.2.

Since the sets O_U and O_L do not overlap, the intervals I_{nL} and I_{nU} are disjoint. The theorem is proved.

Remark 1. That the jump sets O_L and O_U not overlap is possibly a restriction that can be imposed without loss of generality. This is because the L-player always has the option to react to the U-player's move "immediately afterward". So if $x \in O_L \cap O_U$ (and hence the state ends up at $g_L(x) + g_U(x) - x$), an alternative (and payoff equivalent) strategy for the L-player is not to move at x and instead to move after the state goes to $G_U(x)$, and then move the state to $g_L(x) + g_U(x) - x$.

Remark 2. In the two-sided (s, S) equilibria, $I_L = (-\infty, L)$, $I_{nL} = \phi$, for all $n \geqslant 2$ (likewise for the U player). Furthermore, $g_L(x) = L + \theta$, $g_U(x) = U - \mu$, for $x \in I_L$ and $x \in I_U$ respectively.

Remark 3. Suppose that $g_L(x) = \xi_L$ and $g_U(x) = \xi_U$, ie, the jump destinations are uniquely defined. Then, the state eventually gets trapped in the two intervals containing ξ_L and ξ_U. Furthermore, the state transits between these two intervals depending on whether I_{Ln} or I_{Um} is hit. More generally, given any two interval-based strategies $\langle O_L^I, g_L; O_U^I, g_U \rangle$, the stochastic process generated by the strategies in a semi-Markov process whose communicating states are the intervals that intersect with $g_L(O_L^I) \cup g_U(O_U^I)$. Every other subset is transient.

CONCLUSIONS AND EXTENSIONS

In this chapter we analysed a class of continuous time stochastic games in which players can make discontinuous changes in the variables that affect their payoffs. We placed three principal restrictions on the games studied. The first was that only one-dimensional state formulations were analysed. Although many economic models do in fact confine attention to this case, it is an important restriction for our study in that it appears difficult to dispense with. Note that with a multi-dimensional state pace even the decision-theoretic problem is not fully understood. The challenge is twofold: first to find a sharp generalisation of the concept of an (s, S) rule which generates a similar response and then to show that the monotonicity arguments employed in this chapter can be appropriately extended.

A second restriction that was critical for the general analysis was the pure fixed costs assumption. In some special cases, as with linear payoffs, we have analysed a cost structure with a fixed cost plus a term proportional to the amount of change. The final restriction that facilitated the analysis is that we allowed players to only make discontinuous changes. A natural generalisation of the framework of this chapter would be to allow players to make both complementary discrete and continuous changes. We have some preliminary results with this formulation in the linear payoffs case and, again, hope to develop a more general theory.

Although we analysed a model in which the exogenous stochastic process is a Brownian motion, we feel that this specification is possibly inessential. The examples reported in the second section seem to suggest that two-sided (s, S) rule equilibria are true more generally than in this specific case. We have not however explored this in any detail.

Given the above restrictions, we showed that the stochastic game has a Markov-perfect equilibrium in two-sided (s, S) rules. For some parameter configurations, one (or both) player(s) may be inactive in equilibrium. We established that result by way of proving the optimality of one-sided (s, S) rules in a hitherto unsolved stochastic control problem and by establishing a strong monotonicity lemma on the best-response map. Although comparative statics are difficult to interpret in the presence of multiple equilibria, we showed by way of some examples that the higher the costs, the wider are the state intervals on which neither player chooses to act (and by extension, the more likely that players are inactive). We also showed that the first-best solution is also a two-sided (s, S) rule but in the symmetric problem has a wider band than that of the strongly symmetric equilibrium. Finally, we introduced a framework for continuous time games which is flexible enough to accommodate instantaneous reactions by a player to his opponent's moves. Since standard formulations are often unable to do this, our framework may be of more general interest.

Our initial interest in this problem arose from the theory of product innovations in a duopoly. In this context the state variable x can be interpreted as a measure of the difference in quality of the products of the duopolists. The two-sided (s, S) rules then have the following interpretation: whenever the difference in quality is too high, the firm with lower quality should introduce a better quality product. Hence it is perhaps worth commenting briefly on the relation of this chapter to the literature on dynamic innovations/industry evolution (Futia, 1980, Reinganum, 1985, Harris, 1988, and Ericson and Pakes, 1995). The principal difference is that we study discrete action effected at a fixed cost while the focus of all of the other papers is on continuous actions (which are hence adequately described by the associated variable costs). This difference in modelling may explain the consequent difference in arguments (impulse control versus continuous controls, monotone fixed points versus convex fixed points) and more importantly, the difference in results (periods of inactivity followed by sharp changes in the environment as opposed to less perceptible changes). A few other directed remarks may also be useful.

The model structure is closest to Harris (1988) who analyses the evolution of market share in a duopoly. Given the focuses on market share, Harris understandably studies the compact domain case above (with reflection at the boundary), when the *two players can only make continuous changes*. The Harris model belongs to a class of stochastic differential game whose theory has been well developed in several directions (see, eg, Friedman, 1976); in particular, existence of equilibria can be established by following standard fixed point arguments. In any model of dynamic competition there are clearly two possible equilibrium configurations; an increase in the initial asymmetry between the firms and alternatively, a constant interchange of market advantage so that no single firm is a permanent leader. Harris demonstrates that both types of equilibria can exist in his model and suggests the behaviour of joint flow profits, as a function of the market share, as explanation.

It is clear that we also get both behavioural possibilities. However, the intuition for the results seems quite different and is best understood by thinking momentarily of a model of multiple patent races (as in Reinganum, 1985). In such a model, the holder of the current patent works less than the other firm at getting a new patent since the incremental payoffs are smaller for such an incumbent firm; the increase in rent from a patent comes at the expense of lost profits on the current one. In our model too, a firm that "falls behind" in quality of product can then effect a discrete quality upgrade (which is similar to winning a new patent) and may indeed have a greater incentive to so act. However, this intuition, though instructive, is incomplete because we focus on innovations which are not drastic and patentable. The increase in returns from an innovation is therefore not exogenous but rather is determined by how soon the other firm may react to it. It is this "aggressiveness" of the market, and the magnitude of the fixed costs, which determines the advisability and size of change; in particular, a firm with the better quality product may actually be the one that innovates.

APPENDIX 1

In this section we provide a complete and detailed description of the game analysed in the chapter. We begin with a definition of time appropriate for our framework.

Our *time index set* T will have a special form. A time τ will be a pair $(\tau_1, \tau_2) \in \mathbb{R}_+ \times \mathbb{N}$. We give the index set the lexicographic order $\tau < \tau'$ iff $\tau_1 < \tau'_1$ or $\tau_1 = \tau'_1$ and $\tau_2 < \tau'_2$.

Each of the two players can decide whether to move at any time τ. We shall denote the choice by a variable that takes values in {0, 1}; 0 will correspond to "do not move" while 1 will stand for "move". If a player decides to move, then he can choose any displacement $\xi \in \mathbb{R}$ of the state variable. So the action of player i at time τ is a pair (m^i_τ, ξ^i_τ), where $m^i_\tau \in \{0, 1\}$, and if $m^I_\tau = 0$ then $\xi^i_\tau = \{\emptyset\}$; if $m^i_\tau = 1$ then $\xi^i_\tau \in \mathbb{R}$.

Note that we explicitly allow for a move to take place with zero displacement of the state variable.

At any time τ a *history* includes a description of three components: the evolution of the state variable, and the pair of moves and displacement of the state variable by the two players. Formally, the set of histories at time τ, H_τ, is given by the elements h_τ of the form

$$h_\tau = \{Z(\tau'_1) : \tau'_1 \leqslant \tau_1;\ m^i_{\tau'}, \xi^i_{\tau'} : i = 1, 2, \tau' < \tau\}$$

We shall not be interested in all such histories. In particular we restrict our attention to the set of histories for which the outcome, ie, the path of the state variable, is well defined.

Formally we define the set of *finite histories* at time τ, H^f_τ, as those histories for which $m^i_\tau \neq 0$ only for a finite set of τ', for both i = 1, 2. H^f is the set of all H^f_τ, for any τ, ie, $H^f = \cup_\tau H^f_\tau$.

A strategy for player i is a family of maps σ^i_τ, which describes for each τ and $h \in H^f_\tau$ the move and displacement of the state variable decided by the player:

$$\sigma^i_\tau(h) = (m^i_\tau, \xi^i_\tau) = (\sigma^{i,1}_\tau(h), \sigma^{i,2}_\tau(h))$$

Note that strategies are only defined on the set of finite histories H^f_τ.

To each set of histories H^f_τ, $\tau \leqslant +\infty$, we give a measurable space structure as follows. For any integer, m, an m-cylinder $C = C(A,(\tau^1, \cdots, \tau^m))$, where A is a Borel subset of $\mathbb{R}^m$, $\tau^i \in [0, \tau]$ is defined by

$$C \equiv \{h \in H_\tau : (Z(\tau^1), \cdots, Z(\tau^m)) \in A\}$$

B_τ is defined as the smallest σ-field containing all the cylinders, of all finite dimensions. Note that the family $\{B_\tau\}_{\tau \geqslant 0}$ is increasing (that is, $\tau_1 < \tau_2$ implies that $B_{\tau_1} \subset B_{\tau_2}$). The index set T is given the topology induced by the lexicographic order. (A subbase for such topology is the set of intervals $[0, \tau]$ and $(\tau, +\infty)^N$.) B(T) and $B(T_\tau)$ denote now the Borel σ-algebra on T and on T_τ respectively. (They are the same as the Borel σ-algebra generated by the product topology on $\mathbb{R}_+ \times N$.) We can at this point define a notion of measurability of the strategies.

Definition. A map σ: $T \times H \to \mathbb{R}$ is said to be progressively measurable relative to $\{B_\tau, \tau \in T\}$ *if and only if* the mapping $(\tau, h) \to \sigma(\tau, h)$ is $B(T_\tau) \times B(H_\tau)$ measurable for each $\tau > 0$.

This measurability condition on strategies requires the action taken at time τ to depend only on the history up to that time. It is therefore natural to assume the strategies to be progressively measurable.

It is clear that for any history in H^f_τ the state at time τ_1 is well defined. Indeed it can be written as (see Bensoussan and Lions, 1982 (p. 632) for details)

$$X(\tau) = Z(\tau) + \sum_{s<\tau} \xi^1_s + \sum_{s<\tau} \xi^2_s$$

(where the summations are taken over the finite set of times at which the displacements are non-zero for either player).

We next turn to the following question: given a pair of strategies (σ^1, σ^2) what is the (distribution of) outcomes associated with them? Our approach will be the following: for every sample path z of the underlying Brownian motion the strategies will induce either (a) a finite number of displacements over $[0, \tau)$ for all τ or (b) an infinite number of displacements by some τ. In the first case we present an algorithm to identify which finite history is the appropriate one. In the second case, we say the game terminates (and an account of the positive cost of each displacement, the payoffs are $-\infty$).

Fix $\tau_0 \geqslant 0$ and a finite history $h_{\tau_0} \in H^f_{\tau_0}$. For any $\tau' > \tau_0$ let the history $h_{\tau'}$ be defined by the path z, and no further moves by any of the players, ie

$$h_\tau = (z_{\tau'}, m^1, \xi^1, m^2, \xi^2)$$

where for i = 1, 2, one has $m^{i\prime}(\tau) = m^i(\tau)$ for $\tau \leqslant \tau_0$, and 0 otherwise; and similarly for ξ^1, i = 1, 2.

Then let

$$t_i = \inf\{s > \tau_1 : \sigma^{i,1}_\tau(h) = 1\}$$

If $t_i > \tau_1$ for i = 1, 2, then the original history h_{τ_0} has a continuation up to time $t_0 = \min\{t_1, t_2\}$. If this process can be iterated for a sequence of times tending to infinity, then there is a history in H^f associated with this pair of strategies. In any other case, the game terminates.

For completeness in the second case we set the terminal state equal to X_0, any arbitrary state, and the payoff to both players equal to $-\infty$.

Remark 1. In an earlier section we have shown that a pair of Markov strategies, under some restrictions, will always generate finite histories.

Remark 2. Since any history with an infinite number of moves terminates the game with payoffs $-\infty$, it is possible that there are trivial equilibria in which each player unilaterally terminates the game. Our interest is, of course, in non-trivial equilibria. Furthermore, simple restrictions can be placed on the strategies which rule out such trivial equilibria (see Dutta and Rustichini, 1991).

The general problem of determining conditions on strategy pairs (σ^1, σ^2) which give a well-defined outcome is an interesting question, but would take us far from the main topic of this chapter (see Stinchcombe, 1992).

APPENDIX 2

For the purpose of this example we limit ourselves to a Brownian motion with drift equal to zero, ie in our previous notation, m = 0.

Proof of Lemma 3.1. Fix a strategy for player 2 given by the pair (U, μ). We construct a function V and a pair L and θ which satisfy

$$\frac{\sigma^2}{2} V''(x) - \delta V(x) + \delta(x) = 0 \qquad x \in (L, U) \tag{A2.1}$$

$$V'(L) = 0 \tag{A2.2}$$

$$V'(L + \theta) = 0 \tag{A2.3}$$

$$V(U) = V(U - \mu) \quad \text{if} \quad L \leqslant U - \mu \tag{A2.4}$$

$$V(U) = V(L) \quad \text{if} \quad L \geqslant U - \mu \tag{A2.5}$$

$$V(x) = V(L + \theta) - c_0 \quad \text{for every} \quad x \leqslant L \tag{A2.6}$$

Pick a candidate strategy of player 1, say (L, θ). Equation (A2.1) is the differential equation which must be satisfied by the lifetime returns to any such strategy in the region in which no player moves (see, eg Bensoussan and Lions, 1982). The boundary conditions (A2.4) and (A2.5) are similarly satisfied by such returns. Now (A2.3) is the first-order necessary condition which determines the optimal θ by determining the landing position L + θ. Since the jumping position L is also chosen optimally we have the additional condition (A2.2). Equation (A2.6) gives the general form of the value function in the region where player 1 moves. Let $\lambda_\pm$ be the two roots of $(\sigma^2/2)\, x^2 - \delta = 0$:

$$\lambda_\pm = \pm \frac{\sqrt{2\delta}}{\sigma} \tag{A2.7}$$

For simplicity of exposition assume that $2\delta = \sigma^2$ so that $\lambda_+ = 1$. Further suppose that δ = 1. The solutions of equation (A2.1) have the form

$$g(x) = a_1 e^{-x} + a_2 e^{x} + x \tag{A2.8}$$

where a_1 and a_2 are parameters to be determined. The four unknown parameters (a_1, a_2, L, θ) will not be determined by the equations (A2.2) to (A2.5).

So the value function will in fact be described by

$$V(x) = \begin{cases} g(L + \theta) - c_0 & \text{if} \quad x \leqslant L \\ g(x) & \text{if} \quad L \leqslant x \leqslant U \\ g(U) & \text{if} \quad x \geqslant U \end{cases}$$

Equations (A2.2) and (A2.3) are solved for a_1 and a_2 to give

$$a_1 = \frac{e^L}{1 + e^{-\theta}}, \ a_2 = \frac{e^{-L}}{1 + e^{\theta}} \tag{A2.9}$$

In equation (A2.6) set x = L and substitute (A2.9) above to get

$$-\left[\frac{1 - e^{\theta}}{1 + e^{\theta}} - \frac{1 - e^{-\theta}}{1 + e^{-\theta}}\right] + c_0 - \theta = 0 \tag{A2.10}$$

Equation (A2.10) determines a unique optimal θ, denoted $\hat{\theta}$, which depends only on c_0 and not on (U, μ). Now we proceed to determine L.

We consider first the case $L \leqslant U - \mu$. Equations (A2.4) and (A2.9) now give

$$\left[\frac{1 - e^{-\mu}}{1 + e^{\theta}} e^{-(L-U)} + \frac{e^{\mu} - 1}{1 + e^{-\theta}} e^{L-U}\right] - \mu = 0$$

which is a simple quadratic equation in $e^{L-U} \equiv y$, with two roots $y_{\pm}(\mu)$:

$$y_{\pm}(\mu) = \mu\left\{1 \pm \left[1 - 4\frac{(1 - e^{-\mu})(e^{\mu} - 1)}{(1 + e^{\theta})(1 + e^{-\theta})}\right]^{1/2}\right\} \frac{1 - e^{-\theta}}{2(e^{\mu} - 1)} \tag{A2.11}$$

Since $L \leqslant U$, the root $y_+(\mu)$ is ruled out, and we find

$$L = U + \ln y - (\mu), \quad \text{for } L \leqslant U - \mu \tag{A2.12}$$

where ln denotes the natural logarithm. Note that $\ln y - (\mu) < 0$ for $\mu > 0$.

We now consider the case $L \geqslant U - \mu$. Equation (A2.5) gives, setting x = L – U, the equation

$$\frac{e^x - 1}{1 + e^{-\theta}} - \frac{e^{-x} - 1}{1 + e^{\theta}} - x = 0 \tag{A2.13}$$

which determines a unique negative x as a function of θ, with $x = 0$ if $\theta = 0$.

One can now check directly that as $L \to U +$, θ in equation (A2.3) tends to 0, and as $\theta \to +\infty$, $y - (\mu)$ in equation (A2.11) tends to $+\infty$. With $\hat{\theta}$ fixed from (A2.9), this determines a best $\hat{L}$. So the optimal jump is determined from (A2.9) and the jumping point from either (A2.12) or (A2.13). So Lemma 3.1 is proved.

Proof of Lemma 3.2. In the best-response map, constructed above, $(\hat{L}, \hat{\theta})$ is monotonic (componentwise); that is, if $U \leqslant U'$, $\mu' \leqslant \mu$ then $L(U; \mu) \leqslant L(U', \mu')$ and $\theta(U, \mu) \leqslant \theta(U', \mu')$. The statement is trivial for $\hat{\theta}$, and for L as a function of U, for both the case $L \geqslant U - \mu$ and the case $L \leqslant U - \mu$. Also, L is a non-constant function of μ only where $L \geqslant U - \mu$. We therefore need to prove our statement only in this case; equivalently, we need to prove that the function $x(\mu)$ is decreasing in μ. This follows immediately from the fact that the root y– is decreasing in μ (because $\mu/(e^{\mu} - 1)$ is decreasing, and $((e^{\mu} - 1)(1 - e^{-}\mu))/\mu^2$ is increasing in μ).

Remark. It may be interesting to point out that the optimal θ is increasing in c_0 (the fixed cost of every jump). The intuition is, of course, that once it becomes optimal to jump, then a higher fixed cost makes a larger size of the jump more convenient (because it makes jumps less frequent).

From Theorem 5.4, it follows that there is an equilibrium in this case. The band for the symmetric equilibrium can then be computed from (A2.11).

APPENDIX 3

Recall that $B_1(U, U - \mu)$ is the best response of player 1 to a strategy pair (U, μ) for U and μ finite, and is given by a pair (L, θ). In order to use a fixed point argument on the extended lattice, which includes the values $+\infty$ and $-\infty$ we now extend this function to infinite values of U and μ. So we define

$$B_1(\infty, U - \mu) = \lim_{U_\alpha \uparrow \infty} B_1(U_\alpha, U - \mu), -\infty < U - \mu < \infty \tag{A3.1}$$

$$B_1(\infty, \infty) = \lim_{U_\alpha - \mu_\alpha \uparrow \infty} B_1(\infty, U_\alpha - \mu_\alpha) \tag{A3.2}$$

$$B_1(U, -\infty) = \lim_{U_\alpha - \mu_\alpha \downarrow -\infty} B_1(U, U_\alpha - \mu_\alpha), -\infty < U < \infty \tag{A3.3}$$

$$B_1(-\infty, -\infty) = \lim_{U_\alpha \downarrow -\infty} B_1(U_\alpha, -\infty) \tag{A3.4}$$

Given the monotonicity result of Proposition 5.2, all of the above limits are well defined. Let us now establish a couple of useful properties of these limits. The proposition that follows establishes that the constructed best response of a boundary point is interior. For $B_1(U, \mu) = (L, L + \theta)$, we write $B_{11} = L$, $B_{12} = L + \theta$.

Proposition A.3. (i) *Suppose* $U - \mu < \infty$. *Then, there is* $\bar{L} < \infty$, *independent of* $U - \mu$, *such that* $B_{11}(\infty, U - \mu) \leqslant \bar{L}$;

(ii) $B_{11}(\infty, \infty) \leqslant \bar{L}$;

(iii) Fix $U > -\infty$. *Then, there is* $L(U) > -\infty$ *such that* $B_{11}(U, -\infty) \geqslant L(U)$.

Proof. (i) Let $U_\alpha \uparrow \infty$, with $U - \mu$ fixed and finite. Let L_α, θ_α denote the best response, and suppose, per absurdum, that L_α (and hence $L_\alpha + \theta_\alpha) \uparrow \infty$. Fix $\varepsilon < c_0$ and $\psi < 1$ such that $\psi c_0 + \varepsilon < c_0$. Pick $\hat{x} < \infty$ such that $r(\hat{x}) > \bar{r} - \varepsilon$, where $\bar{r} = \lim_{x\uparrow\infty} r(x)$ (which by A4(i) is finite). Further, let L_α, θ_α be such that $\hat{x} < L_\alpha + \theta_\alpha$ and $E_{\hat{x}} e^{-\delta\hat{T}} < \psi$, where $\hat{T} = \inf\{f : X(t) = \hat{x};\ X(0) = x_0 \in (\hat{x}, L_\alpha + \theta_\alpha)\}$, under the policy: "no jump till the first time $\hat{x}$ is hit, then jump to $L_\alpha + \theta_\alpha$ and follow the optimal policy thereafter".

The following sequence of inequalities is easily established:

$$V(x_0) \geqslant E_{\hat{x}}\left\{\delta\int_0^{\hat{T}} e^{-\delta s} r(x_s)ds + e^{-\delta\hat{T}}[V(L_\alpha + \theta_\alpha) - c_0]\right\}$$

$$> (1 - Ee^{-\delta\hat{T}})(\bar{r} - \varepsilon) + Ee^{-\delta\hat{T}}V(L_\alpha + \theta_\alpha) - \psi c_0$$

$$> V(L_\alpha + \theta_\alpha) - \varepsilon - \psi c_0$$

$$> V(L_\alpha + \theta_\alpha) - c_0 \qquad \textbf{(A3.5)}$$

But (A3.5) implies that not jumping is optimal at X_0, ie, that $L_\alpha \leqslant x_0$. In turn, the argument was completely independent of $U - \mu$. So, from (A3.1)–(A3.4), (i) and (ii) follow.

Lemma 4.4 established that for fixed $(U, U - \mu)$, $L > -\infty$. In fact, the same argument established that this jumping point lower bound could be chosen independently of $U - \mu$. That establishes (iii).

Finally, we report here a statement of Tarski's fixed point theorem (Tarski, 1995). We say X is a lattice if it is an ordered set with the property that $\inf\{x, y\}$ and $\sup\{x, y\}$ exist for every x, y, in X. We say that X is complete if sup(Y) and inf(Y) exist for every non-empty subset Y of X. A mapping $f : X \to X$ is said to be monotone increasing if $x \geqslant y$ implies that $f(x) \geqslant f(y)$. Then:

Tarski's Fixed Point Theorem. *Every monotone increasing mapping on a complete lattice has a smallest and largest fixed point.*

1 One could attempt to make this distinction in discrete time by introducing gestation lags to the realisation of some actions and calling such actions "discontinuous". The shortcomings of this approach are that (i) the gestation lags are necessarily exogenous and arbitrary, (ii) the sharp distinction between the two types of actions are lost, and (iii) the analysis is necessarily considerably more inelegant than it is in a continuous time formulation of the problem.

2 Note that all the references in the previous paragraph study single-agent or decision-theoretic problems only.

3 See also Milgrom and Roberts (1990) for applications of fixed point theorems for monotone maps in supermodular games.

4 Symmetry does not imply that the jump destination in equilibrium is 0. If that is the case we call the equilibrium strongly symmetric and the statement in the text refers to such an equilibrium.

5 Specifically, this characterisation refers to any MPE in which the state at which a player moves forms a closed set and the pair of such sets is non-overlapping.

6 Most of the discussion will in fact be for the non-compact domain case which is analytically the somewhat cleaner case. In many applications it may be more appropriate to have a compact domain. For this reason we develop the compact domain case fully as well.

7 The "total displacement up to time t" is well defined if the number of such displacements is finite. Whenever the number is infinite we say that the game "terminates" and assign an arbitrary termination date. This issue is analysed in detail in Appendix 1.

8 Assumption A3(i) is used in a crucial way only in the proof of Proposition A3. Of course, here and in (A3)(ii) we make use of the fact that, since the payoff function is monotonic and absolutely continuous, its derivative exists almost everywhere.

9 In some examples, for instance, the linear returns case (Example 3.3), we can allow the more general specification of fixed and proportional costs: (A4) $c(\xi) = c_0 + c_1 |\xi|, c_1 > 0$.

10 For example, the strategy in which $\theta = 0$ and L is finite, the outcome is a reflected Brownian motion with barrier at L; hence the path of the state variable is defined almost surely but also almost surely the payoff of player 1 is negative infinity.

11 The proof of Bensoussan and Lions consists of two main steps. In the first, they formulate a quasi-variational inequality (QVI) which is a general version of the system of equations A2.1–A2.6 that we formulate in Appendix 2 for the special case of linear payoffs. This QVI is defined in (1.7) of Chap. 4, p. 344. The connection of this QVI with the optimal control problem is explained in Section 1, Chap. 6, p. 615. The QVI has a regular unique solution as proved in the existence result of Theorem 1.2, Chap. 4, p. 345.

The second step proves that the value function is equal to the solution of the QVI and that an optimal Markovian policy exists. This is done in Theorem 1.1, Chap. 6, p. 619. The main difference between the problem they analyse and ours is in the boundary conditions. They assume a zero boundary condition on the value whereas in our case this is replaced with the condition $V(U) = V(U - \mu)$. Therefore in the first step, one needs to replace the space of fractions satisfying the zero boundary condition with the space of functions satisfying $V(U) = V(U - \mu)$. The arguments for the second step remain unchanged.

12 Strictly speaking we have shown that the best response is an (s, S) policy whenever $\mu > 0$. As explained in the remark following Lemma 2.3.1, the lifetime payoff of the first player is identically equal to minus infinity for all strategies when $\mu = 0$, ie any strategy is trivially a best response.

13 In fact, any solution of the equation is two times continuously differentiable, since the equation is linear with constant coefficients, and the function r(x) is continuous by assumption.

14 When $\mu = 0$ the best response for the first player can be defined as the limit of his best responses for $\mu > 0$; it is easy to see that the sequence of θs the best response must be bounded away from 0; hence the fixed point is not given by a pair of strategies with $\mu = \theta = 0$.

15 Since the set of jumping states, in any best response, is given by $M = \{x \in R: V(x) = \max_z V(z) - c_0\}$, this set is closed if the value function, V, is continuous. So the restriction is without loss of generality whenever the value function can be shown to be continuous by independent arguments.

BIBLIOGRAPHY

Bensoussan, A and J.L. Lions, 1982, *Impulse Control and Quasi-Variational Inequalities*, Gauthier-Villars, Paris.

Caplin, A. and E. Sheshinski, 1988, "The Optimality of (s, S) Rules", mimeo, Princeton University.

Constantanides, G. and S. Richard, 1978, "Existence of Optimal Simple Policies for Discounted-Cost Inventory and Cash Management in Continuous Time", *Operational Research* 26, pp. 620–36.

Dixit, A., 1989a, "Entry and Exit Decisions Under Uncertainty", *Journal of Political Economics* 97, pp. 620–38.

Dixit, A., 1989b, "A Simplified Exposition of the Theory of Optimal Control of Brownian Motion Control", mimeo, Princeton University.

Dutta, P.K. and A. Rustichini, 1991, "(s, S) equilibria in Stochastic Games with an Application to Product Innovation", University of Rochester discussion paper.

Ericson R. and A. Pakes, 1995, "Markov-Perfect Industry Dynamics: A Framework for Empirical Work", *Review of Economic Studies* 62, pp. 53–82.

Fudenberg, D. and J. Tirole, 1992, *A Course in Game Theory* (Cambridge: MIT Press).

Friedman, A., 1976, *Stochastic Differential Equation and Applications*, Volume 2 (New York: Academic Press).

Futia, C., 1980, "Schumpeterian Competition", *Quarterly Journal of Economics* 94, pp. 675–95.

Harris, C., 1988, "Dynamic Competition for Market Share: An Undiscounted Model", mimeo, Nuffield College, Oxford University.

Harrison, J.M., T.M. Selke and A.J. Taylor, 1983, "Impulse Control of a Brownian Motion", *Mathematics of Operations Research* 8, pp. 454–66.

Jovanovic, B. and R. Rob, 1990, "Long Waves and Short Waves: Growth Through Intensive and Extensive Search", *Econometrica* 58, pp. 1391–410.

Karatzas, I. and S. Shreve, 1988, *Brownian Motion and Stochastic Calculus* (Berlin: Springer–Verlag).

Milgrom, P. and J. Roberts, 1990, "Rationalizability, Learning and Equilibrium in Games with Strategic Complementarities", *Econometrica* 58, pp. 1255–78.

Parthasarathy, T., 1973, "Discounted, Positive and Non-cooperative Stochastic Games", *International Journal of Game Theory* 4, pp. 37–46.

Reinganum, J., 1985, "Innovation and Industry Evolution", *Quarterly Journal of Economics* 100, pp. 81–99.

Stinchcombe, M., 1992, "Maximal Strategy Sets for Continuous-Time Game Theory", *Journal of Economic Theory* 56, pp. 235–65.

Tarski, A., 1995, "A Lattice Theoretical Fixed Point Theorem and its Applications", *Pacific Journal of Mathematics* 5, pp. 285–309.

Index